Making Movement Modern

Making Movement Modern

Science, Politics, and the Body in Motion

WHITNEY E. LAEMMLI

The University of Chicago Press
Chicago and London

PUBLICATION OF THIS BOOK HAS BEEN AIDED
BY A GRANT FROM THE BEVINGTON FUND.

The University of Chicago Press, Chicago 60637
The University of Chicago Press, Ltd., London
© 2026 by The University of Chicago
All rights reserved. No part of this book may be used or reproduced in any manner whatsoever without written permission, except in the case of brief quotations in critical articles and reviews. For more information, contact the University of Chicago Press, 1427 E. 60th St., Chicago, IL 60637.
Published 2026
Printed in the United States of America

35 34 33 32 31 30 29 28 27 26   1 2 3 4 5

ISBN-13: 978-0-226-84578-4 (cloth)
ISBN-13: 978-0-226-84580-7 (paper)
ISBN-13: 978-0-226-84579-1 (ebook)
DOI: https://doi.org/10.7208/chicago/9780226845791.001.0001

Library of Congress Cataloging-in-Publication Data

Names: Laemmli, Whitney E. author
Title: Making movement modern : science, politics, and the body in motion / Whitney E. Laemmli.
Description: Chicago : The University of Chicago Press, 2025. | Includes bibliographical references and index.
Identifiers: LCCN 2025035006 | ISBN 9780226845784 (cloth) | ISBN 9780226845807 (paperback) | ISBN 9780226845791 (ebook)
Subjects: LCSH: Labanotation. | Dance notation—History—20th century. | Movement notation—History—20th century. | Human body—Social aspects.
Classification: LCC GV1587.L34 2025
LC record available at https://lccn.loc.gov/2025035006

♾ This paper meets the requirements of ANSI/NISO Z39.48-1992 (Permanence of Paper).

Authorized Representative for EU General Product Safety Regulation (GPSR) queries: **Easy Access System Europe**—Mustamäe tee 50, 10621 Tallinn, Estonia, gpsr.requests@easproject.com
Any other queries: https://press.uchicago.edu/press/contact.html

To MHH, who amazes me.

Contents

Introduction 1
1 Alien Gesticulations 18
2 The Lilt in Labour 62
3 The Dance Notation Bureau 97
4 Corporate Bodies 129
5 Moving On 164
6 From *Volk* to Folk 199
EPILOGUE Movement in the Digital Age 225

Acknowledgments 245
Notes 249
Works Cited 303
Index 325

INTRODUCTION

Late one afternoon at the height of World War II, a United States postal censor came across a document that gave him pause. Slicing open an envelope originating in England and destined for New York City, he discovered a cache of pages covered in what seemed like an incomprehensible hodgepodge of lines, tick marks, and polygons. Was it a coded Axis communiqué? Instructions for saboteurs? Unable to make heads or tails of the markings, he flagged the suspicious pages for additional review. Further investigation revealed that the censor's instinct was correct, if only in part. The document was indeed a German cipher, though not one intended for Nazi spies. It was a dance.[1]

It would have been hard, however, to blame the censor for his concern, as the black-and-white markings on the page bore little resemblance to the lively moving bodies they referenced. Leaps and turns had been transmogrified with the help of a system consisting of a multicolumn vertical staff and a set of symbols that evoked a marriage of mathematical notation and art deco metalwork. First published as "Kinetographie" in Germany in 1928, the technique was by then known as Labanotation (in honor of its creator, the choreographer Rudolf Laban) and was hailed by many in the dance world as a revolutionary mechanism for preserving a heretofore ephemeral art. With Labanotation's help, they promised, dance could at last achieve the kind of permanence associated with music, no longer fading away even as it was performed.

Had the censor known this from the start, he might well have heaved a sigh of relief and moved on to the next letter in his pile. The document contained no secret messages designed to undermine American national security, just technical discussions about the finer points of the system's orthography. Its destination was the Dance Notation Bureau, the technocratically titled organization dedicated to promoting Labanotation in the United States. In a larger sense, however, the censor was perhaps right to be cautious. Though most closely associated with the seemingly narrow domain of dance, Labanotation was a profoundly malleable tool with applications that had already extended far beyond the world of art, including use by the National Socialist state.

Rudolf Laban, in fact, had only recently departed a position overseeing dance in Joseph Goebbel's Propaganda Ministry, a post his work with Kinetographie had helped him to obtain. In particular, the Nazis were attracted to Laban's claims—allegedly rooted in the physiological and physical sciences—that movement was a direct conduit to the mind and that notation could thus function as a tool for reshaping human emotion and behavior. During his three-year tenure in the Nazi state apparatus (1934–37), Laban attempted to put that theory into practice. Notation in hand, Laban dedicated himself to identifying and eliminating degenerate "alien" movements from the German population and organized mass movement spectacles designed to knit the nation together in a new kind of embodied community.

A few short years later Kinetographie, by then repackaged as "Industrial Notation," would also be deployed by Germany's enemies, introduced into wartime British factories as a technology for managing and motivating labor. In the succeeding eight decades, moreover, it and other Laban-inspired recording techniques, including Labanotation, Effort/Shape Analysis, Aptitude Assessment, and Laban Movement Analysis, were taken up by an even wider range of institutions and individuals across the United States and Europe. They excited psychologists and psychiatrists as well as copyright lawyers and management consultants and piqued the interest of anthropologists, military officials, and computer scientists. These groups were not necessarily interested in dance, but they became captivated by the promise of recording anything from the way a young woman walked down a street to how a factory worker picked up a box.

Neither Labanotation nor its later derivatives were easy systems to use. It could take years for notators to perfect their craft and hours to notate even a few minutes of movement. Nevertheless, these varied groups pressed on, fascinated by the possibilities that recording, rationalizing, and remaking movement might present. They devoted time, energy, and substantial financial capital to learning and deploying complicated, unwieldy systems that promised to make the analysis of everyday human movement—long associated with the fleeting, the feminine, and the insignificant—into a new kind of science.

This book explores why. It is, therefore, not primarily a text about Kinetographie itself, nor about any of the specific Laban-inspired techniques that emerged in its wake. Instead, it uses the widespread interest in these systems as a guide to understanding how and why movement came to be a central object of scientific, political, and popular concern over the course of the twentieth century. In doing so, it opens up a window into new facets of twentieth-century culture, uncovering a world that conceived of human movement as the seat of an enormous power that could, among other things, reveal the innermost self, stabilize shaky economic systems, and reshape human political communities. As the following pages will reveal, movement notation was not only aimed at preserving a vanishing past but at the literal choreographing of modern life.

Observing Movement

We are almost always moving. Some of these movements are obvious: a determined stride down a hallway, a swan dive or belly flop that pierces the water of a cold swimming pool, Chubby Checker doing the twist. Others are more subtle: the shifting of weight in a chair, the flick of fingers across a keyboard, the sudden collective head-turn when a door opens in a quiet room. The forms vary, but the experience of movement is one we communally share. Aristotle held that the capacity for movement was a defining feature of animal life.[2]

As a result, movement has long been of at least passing interest to people in many societies. Evidence of attempts to both "read" and prescribe movement can be found around the globe and throughout human history—in codes for orators and actors; in prescriptive writing on manners; in religious, dance, and military manuals; in novels, poetry, and paintings; and in material objects and tools. Historians drawing on these sources have revealed, for example, the social roles ancient Greeks associated with particular ways of walking, sitting, and standing and the changing theological significance of the gesture in early medieval Europe.[3] The observation and regulation of movement, in other words, has frequently played an important role not just in the experience of daily life, but in the historical operations of race, gender, sexuality, and disability; of social class; and of political, economic, and religious power.[4]

The shape and intensity of that interest, however, has varied over time. One of the best-known modern academic analyses of movement is Marcel Mauss's 1935 essay "Techniques of the Body."[5] Here, the French sociologist and anthropologist sketched in vivid detail the multiplicity of physical methods by which people have carried out the routine activities of living. Positing that there were no universal techniques of, say, running or sitting or carrying an infant, Mauss sought to highlight movement's cultural and historical contingency. The walk of a girl raised in a convent (with fists tightly closed), he explained, was not that of a Māori woman (characterized by "a loose-jointed swinging of the hips") or an American nurse.[6] Mauss further proposed that these differences were produced and sustained through a mixture of social, biological, and psychological mechanisms, the product of an accretion of conscious instruction and innumerable subconscious moments of imitation, recognition, and memory.

But while it was clear to Mauss both that these differences existed *and* that they contained crucial cultural data, he observed that his fascination with movement was not necessarily widely shared. The study of movement was, in his words, an "ill-demarcated domain" and an "uncleared land," its scientific study at a "moment when, the science of certain facts not being yet reduced into concepts, the facts not even being organically grouped together, these masses of facts receive that posting of ignorance: 'Miscellaneous.'" This, Mauss wrote, "is where we have to penetrate."[7]

Mauss's (rather imperialist) call to arms was undoubtedly sincere, but it belied the extent to which change was already afoot. In the decades surrounding the turn of the twentieth century, movement had garnered renewed interest from a variety of constituencies, many of whom were—precisely as Mauss suggested—undertaking concerted efforts to move its study from the realm of miscellany and anecdote into the more regimented strictures of science. Though in broad strokes their interest in movement might have evoked comparable projects from the past, such efforts were increasingly characterized by a preoccupation with precision and permanence and often deployed new technologies in service of those ends. In the late nineteenth century, for example, Étienne-Jules Marey's chronophotography broke the movements of humans and animals into a series of sequential images, revealing transitory states previously invisible to the naked eye. Similarly canonical are the early twentieth-century cyclegraphs produced by efficiency experts Frank and Lillian Gilbreth, who recorded motion as part of their efforts to optimize labor, affixing small lights to workers' extremities as they operated drills, stamped requisition forms, and sorted screws, allowing the path of their movements to be traced over the course of a long photographic exposure.[8]

Indeed, a number of important studies have exposed how the regulation of movement became a key aspect of late nineteenth and early twentieth-century industrial practice. In factories across Europe and the United States, workers were subjected to intensive new regimes of discipline shaped variously by physiologists, floor supervisors, and material tools. Their bodies were seen as machines whose function could be profitably optimized for the purposes of political and economic stability. Such techniques, though implemented somewhat unevenly, affected the lives of countless individuals.[9]

As this book will demonstrate, however, scientific and technological management of human movement was not confined to the world of industry. By following the uptake of Laban-inspired tools, we can see how this interest in regulating and recording movement shaped not just industrial production but also white-collar offices, hospitals and city streets, anthropology and the arts. Though factory workers certainly do appear in the following pages, they are joined by physicians and physical therapists, Holocaust survivors, management consultants, ballet dancers, lawyers, IBM engineers, and the folklorist Alan Lomax. The story, moreover, does not stop in the first few decades of the twentieth century but continues into the postwar period and concludes in the near-present, showing how movement remained of interest long after the fixation on factory workers faded. The book aims, therefore, to create a more synthetic account of movement's study in the twentieth century, attending to the ways that interest in movement—and the technologies for studying it—crossed disciplines and geographies and considering a wider variety of actors than have often been examined.

This new lens also alters the story about what recorded and controlled movement was intended to achieve. While existing histories of industrial motion study rightly emphasize capital's interest in efficiency, mechanization, and discipline, the history that emerges from following the techniques outlined here is somewhat more complicated. Across the varied contexts of their use, the proponents of Laban-inspired techniques believed the widespread management of movement would produce not (just) increased profits or productivity, but something more metaphysically meaningful: a sense of joy, belonging, recognition, or healing that was deeply rooted in the physical body. They were frustrated by a modern world that seemed plagued by excessive rationalization, regimentation, and alienation and saw the cultivation of movement—with all its attendant associations with the romantic, the irrational, the feminine, and the divine—as a way of counterbalancing these tendencies. That they did so by turning a subjective and ephemeral experience into a regimented set of data, ripe for external control, is one of this book's central ironies.

Many users of these tools, in fact, turned to them because closely monitored movement seemed to occupy a productive middle ground between top-down control and individual freedom, promising two things simultaneously. First, they hoped that movement could provide a new source of expression and meaning in a seemingly disenchanted world. Second, they trusted that, with notation, this creative potential would never go too far: movement would be continually observed, broken down into its constituent parts, and put in the service of modern states, institutions, and bureaucracies. The book explores the results of this sometimes-contradictory endeavor, tracking how efforts to analyze, standardize, and control expressive motion became incorporated into the functioning of some of the central institutions of twentieth-century life.

Laban, Notation, and Information

Though largely absent from existing literature in the history of science and technology, both Rudolf Laban himself and the recording techniques he created and inspired garner immediate recognition in the dance world. Born in 1879, he was a key figure in modern dance in the first few decades of the twentieth century and is widely credited as an originator of a genre often known as German or Central European Expressionism, or *Ausdruckstanz*. Laban's own pupils included the luminaries Kurt Jooss and Mary Wigman, and his name still proudly adorns numerous institutions, including London's prestigious Trinity Laban Conservatoire of Music and Dance and New York's Laban/Bartenieff Institute of Movement Studies. At Laban's death in 1958, his obituary in the London *Times* described his career as one of "far-reaching influence," citing both his role as an aesthetic innovator and his "success in devising a dance notation" that became "the most widely used of all the notations that have been attempted to set down in score the steps, movements and patterns of the choreographer."[10]

Given this stature, Laban's life and work have received relatively scant scholarly attention even within dance history and dance studies. While he appears as a significant figure in numerous accounts of interwar modernism, full-length works that bring a critical eye to Laban or the notation systems he helped to develop are rare.[11] One significant exception is Lilian Karina and Marion Kant's *Hitler's Dancers: German Modern Dance and the Third Reich*, which situates Laban's later career within the larger story of modern dance's collaboration with National Socialism.[12] Shorter pieces by Carole Kew, Patricia Vertinsky, and Arabella Stanger have also illuminated particular aspects of Laban's thought and practice, including his mystical influences and his engagement with British physical education reform.[13] Still, the bulk of the monographs on Laban's life and work have been written by friends, protégés, or practitioners of the system.[14] While they are valuable as repositories of colorful stories, recollections, and technical explications, their goals are not the goals of this book.

INTRODUCTION 7

What follows is neither a biography of Laban—the man himself largely disappears relatively early in the story—nor a comprehensive treatment of Kinetographie, Labanotation, Effort/Shape, or Laban Movement Analysis as systems. The book does not, for example, center questions about the extent to which these later tools were faithful to Laban's original ideas or grapple with all the philosophical questions surrounding their efficacy as recording devices. Ample resources exist elsewhere for those who are interested in learning to practice Laban-based forms of notation or who want to further explore the intricacies of their evolution or contemporary use.[15] It is also worth noting that a number of other techniques for writing down movement emerged during the period covered here. Some of these tools—like Benesh and Eshkol-Wachman notation—are still in use and relatively well known, while others—like the "motography" developed by an Italian army lieutenant in the 1930s—faded quickly.[16] This book focuses on Laban-inspired techniques because of their longevity and unique cross-disciplinary impact, but the proliferation of similar projects suggests that the impulse to both value and manage movement extended beyond the bounds of this account. Accordingly, while the book will cast light on the complicated history of Laban-linked recording tools, the story might also be told in other ways, and evaluating whether these technologies "worked" as their creators intended is not my aim. Instead, this is a book about ambitions: a historical treatment of why movement recording and analysis sparked interest across so many domains in the twentieth-century United States and Europe, and one particularly rooted in the histories of science and technology.

For movement notation was very much a technology, a human-created system designed to impose a new kind of order on human bodies—and through them, on the world. Like other paper tools of its kind, it proved to be a quiet but powerful force: making visible, disciplining, drawing things together.[17] In some ways it was an artisanal technique, its practitioners occasionally half-jokingly referred to as "monks" not just for their mastery of an abstruse symbology but for the manual dexterity and aesthetic judgment required to produce scores that were themselves often works of art. Neither the notation's visual appeal nor the skilled labor required to create it should, however, distract from the fact that it was always a tool for transforming messy, moving bodies into neat, static information.[18]

In this way, the work of movement notation bears clear affinities to other large twentieth-century projects of standardization, quantification, and recording, from surveys to actuarial tables.[19] Its history should thus be placed alongside the profusion of other "big data" projects that have fascinated both historians and media studies scholars in recent years and upended our sense of the uniqueness of the twenty-first century's insatiable hunger for information.[20] Over the course of nearly a hundred years, its users tirelessly turned expressive human movement, once thought inherently elusive, into "objective" data ripe for replication, analysis, and storage. The large-scale possibilities inherent in this process became more obvious as the twentieth century wore on—perhaps particularly in the 1970s and 1980s, when Labanotation's conceptualization of the body as a discrete set of parts proved appealing to computer scientists and engineers eager to capture movement in digital form. Still, the fundamental intellectual and practical orientation toward "datafication" was always present.

Movement notation is also a project that is ongoing, as new kinds of information about human bodies in general and human movement in particular are being gathered at an increasingly rapid rate. FitBits gobble up steps in the tens of millions, novel surveillance tools rely on gait analysis, and at least one popular automated hiring software application has made decisions in part based on movement behavior.[21] Though there is currently little awareness of the history of these practices, a number of contemporary systems rely on techniques and technologies directly derived from the work of Laban and his followers. This book, therefore, speaks both to history and to our current moment, illuminating what was and continues to be at stake in efforts to understand, capture, and control human movement on a large scale.

The appeal of these recording techniques similarly cannot be understood without an attention to how they were built on a particular vision of the relationship between the body and the mind. As the following pages will reveal, Laban's motivation for creating a notational system was always tied to his fascination with physiology and his attendant belief that movement could directly mold the psyche. As he explained, his goal was not to produce a tool for recording dances that already existed (what he called *Tanzschrift*) but rather to use notation to change how dance—and ultimately, movement more generally—was produced, carefully selecting some movements and discarding others, ensuring that all were performed correctly (*Schrifttanz*). The aim, both for Laban himself and for many of his acolytes, was to activate new biological processes in the individual, reshaping their emotional and mental states to serve larger societal ends.

This story, therefore, must also be situated alongside histories of science that track the emergence and persistence of particular conceptualizations of mind-body relations. There is indeed a small but growing body of literature that has begun to think about movement through precisely this lens, tracing how scientific ideas about kinesthesia and proprioception have changed over time and how these concepts have reverberated in the wider world.²² Studies of what has been called "physiological aesthetics" have been particularly revealing, illuminating how early twentieth-century artists in the United States, Germany, Russia, and the Soviet Union attempted to use scientific principles to mold the sensory and emotional experiences of their audiences.²³ Other recent scholarship has demonstrated how an epistemology of nonrational "kinaesthetic knowing" became central to the development of design, art, and architecture education in early twentieth-century Germany.²⁴ As the art historian Robin Veder has summarized, it has become increasingly clear that much "modern art was about body movement," and practitioners drew on a variety of techniques—from breathing and posture exercises to Rolfing and the Alexander technique—to both produce their work and shape its reception.²⁵

One of the assets of this new scholarship is that it has begun to contest the idea that what might be called the "industrial" mode of twentieth-century bodily experience—in which human bodies were increasingly analyzed, broken down, and subjected to regimes of mechanical discipline—was entirely hegemonic. The twentieth century may have been the era of Taylorism, the modern medical gaze, and the commodification of cell lines, blood, and tissues, but it was also a moment when myriad other ways of thinking about the body flourished.²⁶ It was the age of Delsartian theater, which taught actors about the direct relationship between gestures and emotional states, and of handwriting instruction that emphasized the importance of rhythm and expression rather than the mimicking of individual strokes. Speech therapists examined the salutary effects of fluent movement on stuttering, and rhythmic gymnastics was a regular aspect of elementary education.²⁷ It was the era of *Freikörperkultur* and of widespread American and European fascination with yoga and other movement techniques of the so-called mystical "East."²⁸ Even Mauss's own 1934 "Techniques of the Body" essay culminates in a discussion of Taoist breathing practices and yoga, urging their study not just on sociological grounds but because of their potentially very real power to reshape human experience.²⁹

Indeed, the cultural historian Hillel Schwartz argued in a provocative 1992 essay that an entirely new kinesthetic—centered around rhythm, wholeness, fluidity, and the belief in a direct connection between the inner self and its outward expression—emerged in the decades surrounding the turn of the twentieth century. He called this new mode of prescribing and experiencing motion "torque" and traced its links to developments in late nineteenth-century neurophysiology and psychology.³⁰ Schwartz does not, of course, suggest that these alternative ways of moving were truly "free" but rather that they operated under different governing principles and were—at least sometimes—mobilized to serve different ends.

With this new scholarship, the complex contours of the twentieth-century movement landscape are coming into clearer focus. Most existing studies, however, have focused largely on one area of practice, following the application of new ideas about movement in dance, the fine arts, or sports. They have also tended to cluster their attention in the first few decades of the twentieth century, where these discourses were most visible. What is particularly useful about the study of Laban-based notation tools, therefore, is that their winding trajectory crosses disciplines, borders, and time periods, revealing previously hidden ties binding the study and control of movement in dance to later efforts in business, physiology, medicine, the social sciences, and engineering. Historians, moreover, will not be the only ones surprised to learn about these links. While some users of Labanotation, Effort/Shape, Industrial Notation, or Laban Movement Analysis were fully conscious of these techniques' histories, others would have been shocked to discover that the tool they relied on to select their director of cement plant operations or treat a patient with bipolar disorder emerged from a volatile blend of avant-garde art, nineteenth-century physiology, and resurgent German nationalism.

That said, the book's aim is not to demonstrate that these tools' users were all inescapably contaminated or constrained by their origins, or that those origins were anything but complex. Instead, it attends to historical particularity while simultaneously investigating what conceptual frameworks these actors shared and what problems they believed movement—if properly recorded and managed—might solve. Why, I ask, did these varied groups decide to pay attention to movement at all? What techniques, tools, and tactics did they mobilize in their efforts? And what were the consequences—individually, socially, politically, and otherwise?

In raising these questions, I take inspiration from scholars who have similarly traced how sound became a key political and technological problem in the nineteenth and twentieth centuries, "an object to be contemplated, reconstructed, and manipulated, something that can be fragmented, industrialized, and bought and sold."[31] Through the study of everything from the phonograph to new listening techniques to the design of opera halls and office buildings, they have revealed not just sound's profound historical contingency, but also how its transformation was a fundamental part of the human experience of modernity.[32] "Sound studies" is now a burgeoning interdisciplinary field, and though the cross-disciplinary scholarly analysis of movement is not yet nearly as robust, I hope this book might serve as one model of how "movement studies" might develop. Leading scholars in dance studies have already pushed the field to expand its traditional focus on performance to include the analysis of all forms of "socially structured movement," from sports to work to everyday gesture.[33] Other disciplines, however, have been slower to attend to movement's migration across domains of action; taking on this challenge may well result in transformative collaborations and new insights.

At the same time, the history recounted here might prompt scholars across the humanities and social sciences to reflect critically on the roots of our own interest in recovering, recording, and analyzing bodily experience.[34] In some ways we have much in common with the movement notators: a sense that the body has been unfairly neglected in part due to its association with women, racialized others, and manual labor, and an attendant desire to claim space in the historical record for an immensely important realm of endeavor. In the past few decades, work that has trained attention on the body has paid incredible dividends. Scholars of tacit and embodied knowledge have, for example, upended old understandings of the Scientific Revolution and demonstrated how crucial unwritten know-how has been to the successful operation of technical systems.[35] The story that follows, however, should remind us to be wary of inadvertently casting bodily experience as inherently more authentic, unmediated, or impervious to modern forms of rationalization.[36] While dance specifically and movement more generally have often been associated with resistance, joy, or solidarity, this book reveals how these very associations have also been mobilized in the service of far more morally complicated goals.[37] As the historian Roger Smith noted in a study of kinesthesia, "the reliance on touch and movement as authority for knowledge of reality has borne the weight of belief, desire and hope about individual people being a meaningful part of the world and of being agents."[38] This text explores the ongoing allure of that dream as well as some of its unanticipated consequences.

Organization

Chapter 1 ("Alien Gesticulations") begins by describing Kinetographie's genesis, asking how and why Laban became interested in translating the four-dimensional fleshiness of the moving body into something written, replicable, and universally valid. It follows Laban from his childhood as the son of the military governor of Bosnia-Herzegovina to the publication of Kinetographie in 1928, uncovering his development of a unique theory of movement and outlining how his thinking was shaped by influences as diverse as German Romanticism, nineteenth-century physiology, architectural theory, and contemporary physics. The chapter then moves to an analysis of Kinetographie in practice, focusing first on Laban's use of notation in coordinating movement choirs in the 1920s before turning to his time in the Nazi Ministry of Propaganda. Previously, scholars have taken at face value Laban's contention that his notation represented a neutral method for preserving important works of art. In this chapter, however, I argue that Kinetographie was fundamentally a technology for preserving and solidifying a racialized national identity, intended to facilitate the creation of a new kind of embodied community for an anxious, industrializing society.

Chapter 2 ("The Lilt in Labour") recounts Laban's emigration from prewar Germany and the transformation of Kinetographie from a tool of the National Socialist state to its role in British industry. In 1937 Laban lost his post in the Nazi government and left Germany, eventually settling into a position at the progressive Dartington Hall school in Devon, England. After arriving, however, Laban began using Kinetographie to record not dancers' leaps and undulations but the movements of industrial workers. In a collaboration with one of the UK's first industrial consultants, F. C. Lawrence, Laban introduced Kinetographie into canning factories, textile mills, tire plants, and assembly lines. While there were clear parallels between this work and the time-motion studies of scientific management consultants like Frederick Winslow Taylor, Laban and Lawrence were quick to distance themselves from these prior efforts. Instead, they claimed that the careful control of workers' movements would make them both more efficient and more fulfilled, transporting the same deep, mystical joy experienced in ritualized dance to the industrial workplace. Laban painted a picture of factories without conflict, populated not with aching, angry, alienated workers but with joyful, engaged, and spiritually sated citizens. This transformation, of course, was imbued with its own politics. As the chapter will show, Laban, Lawrence, and many of their clients hoped their efforts could ensure the industrial corporation's survival as disaffected workers increasingly challenged owners and managers.

Chapter 3 ("The Dance Notation Bureau") follows Labanotation across the Atlantic by focusing on the activities of the Dance Notation Bureau (DNB), the New York City organization founded in 1940 by several of Laban's female disciples. Drawing on the DNB's extensive institutional archives, this chapter chronicles how Labanotation was altered, expanded upon, and put to use in the midcentury United States. The bureau's stated aim was simply to popularize Laban's system, but this overarching goal translated into a variety of discrete tasks, from highlighting Labanotation's "scientific objectivity" to working to produce international notation standards to drawing attention to the need for cultural preservation in the age of the atomic bomb. Its leaders

INTRODUCTION

recorded the work of some of the twentieth century's most distinguished choreographers (including George Balanchine and Doris Humphrey), taught hundreds of students a year, and successfully lobbied the US Copyright Office to make Labanotation the standard medium of legal protection for dance.

Chapter 3 thus expands on some of the thematic questions surrounding the recording of movement introduced in chapter 1. It considers, for example, how Labanotation's users not only attempted to preserve artistic works for posterity, but also to coordinate movement across time and space, assert ownership, and imagine a very particular kind of global future. Additionally, it explores the consequences of notation's entry into American dance, demonstrating how ideas about the body, artistic creativity, ownership, and dance itself were altered alongside Labanotation's widespread use. Ironically, in their fervent effort to preserve the evanescent movements of the body for posterity—and in their advocacy for notation over filmed performance—the women of the Dance Notation Bureau fostered a vision of dance in which the human body was almost entirely absent. Relatedly, the chapter follows some of the earliest attempts to "mechanize" notation with typewriters and computers, revealing how Labanotation helped scientists and engineers begin to conceive of the body in "digital" terms.

As the twentieth century progressed, movement notation was used increasingly as an interpretive aid. In the 1950s, '60s, and '70s, a number of anthropologists, psychologists, and business consultants began to use new Laban-based techniques including "Action Profiling," "Movement Pattern Analysis," and "Laban Movement Analysis" as research tools. Chapter 4 ("Corporate Bodies") explores one key site for the application of these techniques: postwar management consulting. After World War II, "body language" emerged as a subject of popular interest, and chief among its boosters was Warren Lamb, a onetime student of Rudolf Laban. By the mid-1960s, Lamb was an internationally sought-after management consultant whose claim to fame was "Aptitude Assessment," a system that assessed employees' potential by observing their unconscious ways of moving. For Lamb, a movement profile was like a fingerprint: impossible to fake, it signaled a specific set of strengths and weaknesses and a unique personality profile. In his long career, Lamb was strikingly successful in selling his wares, advising large corporations like IBM, Mars, Monsanto, Hewlett Packard, General Motors, and British Petroleum.

Historians do not usually associate twentieth-century corporate boardrooms with the kind of bodily management that characterized artisanal production or industrial work, suggesting that midcentury office workers were subjected not to the tyrannies of Taylorist overseers but to the judgments of middle managers and the new cadre of psychologists populating human resources departments. This understanding is not incorrect, but this chapter will demonstrate that it is incomplete. The surprising popularity of Warren Lamb's system testifies to the fact that the management of the body remained central to the production of expertise in white-collar office settings, albeit in altered form. This chapter explores the historical reasons underlying the persistent importance of the physical body, including anxieties about the changing demographics of the office and a larger shift in the imagined relationships between the corporation, the employee, and work itself.

Chapter 5 ("Moving On") represents in part a mirror image of chapter 4, as it considers not the "ideal" bodies of the midcentury corporation, but the pathologized bodies of those experiencing mental illness and disability. In the decades following World War II, a diverse group of psychologists, psychiatrists, and "movement therapists" began to use Laban-derived systems to evaluate and treat their patients, including, ironically, Holocaust survivors and their children. They believed that such techniques, particularly Effort/Shape analysis, represented a revolutionary method for accessing and altering the inner lives of the often-uncommunicative individuals they confronted. Focusing largely on the careers of two of these individuals, Judith Kestenberg and Irmgard Bartenieff, the chapter describes a field very much in the making. Practitioners of movement therapy disagreed about many things, from the settings in which it should be performed to the mechanism by which it worked. Their educational backgrounds and the demographics of their chosen treatment populations—war veterans, psychiatric patients at Bronx public hospitals, victims of trauma, and everyday families—also varied widely. They were knit together, however, by an anxiety about the body's ability to incubate trauma, fear, and aggression and the consequences of that persistence for society at large. Chapter 5, therefore, illuminates how therapists used the moving body to negotiate the uneasy Cold War territory between expressive subjectivity and top-down control, individual autonomy and group cohesion.

The sixth chapter ("From *Volk* to Folk") describes how movement analysis attracted new interest within midcentury anthropology and folklore. As early as the 1950s, Labanotation had captured the attention of prominent anthropologists interested in bodily expression, among them Margaret Mead and Ray Birdwhistell. It was in 1965, however, that American folklorist Alan Lomax made Laban-based techniques central to a mammoth research project known as "Choreometrics," a decades-long effort to view, code, catalog, and preserve the totality of the world's dance traditions. Along with his collaborators, movement therapist Irmgard Bartenieff and her student Forrestine

Paulay, Lomax set out to study dance as a means of understanding the adaptive functioning of human culture and the history of human migration.

The chapter provides the first scholarly history of Choreometrics, illuminating how Lomax came to understand human bodily movement as carrying important and otherwise inaccessible information about social structures, work practices, and human history. Lomax also, however, conceived of his work as part of an activist project. He disdained the ways in which globalization had begun to destroy minority cultures around the world and reserved particular vitriol for the obliteration of "traditional" ways of moving. By planning to make both his findings and his data available for broad public consumption, Lomax intended to reverse this slide while "recalibrating" Americans' "perceptual apparatus" to fit a new multicultural world. Though separated by politics, an ocean, and more than half a century, Laban's and Lomax's beliefs about the power of the moving body were surprisingly similar. As they both recognized, contests about how human beings could and should move were also arguments about who belonged to the body politic—and who did not.

Finally, the epilogue considers the fate of Laban-based techniques from the mid-1970s to the present, focusing on their leap from paper to the screen. During this period, a growing community of engineers and computer scientists discovered Labanotation, excited by its apparent ability to translate movement into a discrete set of data points even a computer could understand. Experiments—first at the University of Pennsylvania's Department of Computer and Information Science and later at a variety of universities, private companies, and governments across the globe—progressed in fits and starts, but by the early decades of the twenty-first century Laban-based systems were being used for simulating human motion in animation, programming the movements of humanoid robots, and identifying emotional states through automated movement pattern recognition programs. To many of these systems' developers, Labanotation and Laban Movement Analysis were simply useful tools for solving a technical problem. Their decision to deploy them, however, had consequences, embedding in these new technologies assumptions about movement and its meaning first advanced nearly a century ago.

A few notators expressed anxieties about the use of Laban-based tools for these new commercial, military, and national security ends, but others saw these projects as a natural culmination of decades of effort. In many ways, the latter group was right. Movement notation was always an intensely political technology, one whose users consistently worked to make movement not just preservable or visible, but *useful*. They sought to convince the world that movement mattered and that, harnessed correctly, it could transform their societies. The contemporary push to master movement is very much a continuation of those efforts, and the history recounted here should help us bring a more critical eye to both its possibilities and its perils.

Finally, a note on sources. While whenever possible the book offers glimpses into the experiences of the innumerable individuals who had their movements analyzed or prescribed, available sources limit the ability to do so in a systematic way. Laban, for example, left behind pages upon pages of notes theorizing about movement's effects on the body and mind, but only one or two documents that speak, even obliquely, to the firsthand experiences of movement choir participants or industrial workers. Alan Lomax worked for years to create a complete record of the world's dance traditions, but we know almost nothing about the dancers whose movements he preserved. There is, of course, an irony here: even as notators argued fervently that movement was an essential part of the human experience, the actual humans whose movements they sought to catalog were often left out of their records. In some cases, this was by design, as movement therapists, for instance, appropriately sought to preserve the privacy of their patients. In others, the lack of attention

FIGURE I.1. Ann Hutchinson Guest with Labanotation teaching materials in Epping Forest, 1957. Courtesy of the Ann Hutchinson Guest Archive, copyright owned by the Language of Dance® Trust, https://www.lodc.org.

paid to individuals reveals less consciously chosen values or goals. These tensions are explored directly in several chapters and speak to many of the book's larger themes.

Ann Hutchinson Guest, the founding director of the Dance Notation Bureau and the person to whom that suspicious wartime missive was addressed, always argued that Labanotation's impact on the world could stretch far beyond the studio and the stage. Standing primly beside a blackboard, wielding a pointer and dressed in her signature calf-length skirt custom printed with a Labanotation score, she would instruct anyone who would listen about the possibilities inherent in this revolutionary tool. As a "universal" technique for recording movement, she would explain, Labanotation would produce entirely new fields of research, at last making movement a true object of scientific interest. She believed, moreover, that the resulting dividends would be more than purely intellectual, remarking prophetically that "Labanotation is not an isolated, abstract science. It is a practical science, a means to an end."[39] The following pages will uncover precisely what those ends were.

1

Alien Gesticulations

Symbols crawled across the sheets' centers and crowded their torn edges. Sometimes in pen and sometimes in pencil, the lines, whorls, blots, dots, and slashes formed the foundation of a project still unfinished. The first version didn't work. Neither did the second or the fifty-second. As the scraps of paper accumulated into drifts, patterns sometimes seemed on the verge of emerging but disappeared on a second glance; even the most determined would-be reader would have had trouble telling which way was up. Or, for that matter, that what they were looking at was the image of a body in motion.

But Rudolf Laban remained, as he almost always was, confident in paper's capacity to capture human movement. By the mid-1920s, the choreographer and movement theorist had spent decades searching for a system that would represent the movements of the human body—and dance in particular—on the page. Using musical notation as a model, Laban dreamed of a method of recording that could at long last describe dance "objectively" for posterity. No more would the art form be subject to fickle human memories, consigned to wither and die along with its creators. "What do we know of the art of dance in the past?" he would later write. "A few pictures and statues give us an inkling of the beauty of the movements. A few notes written in old forms of dance notation which we can barely decipher, inform us about some court dance-steps of the last two centuries. But an effective, serviceable notation, able to render the many faces of dance, has yet to be created and made universally applicable. . . . The dances of a Pavlova have already been buried with her. Must we also lose the works of our present dance-generation?"[1] Dependent on the body of the dancer alone, Laban feared, dance had neither a past nor a future. It was stuck in place, temporally and geographically, a ship forever languishing in the doldrums.

For Laban, the only solution was for dance to be disembodied, its ephemeral movements inscribed on the flat surface of paper, preserved permanently and in all their complexity. As the mounting pile of pages bore witness, this was not an easy task. A complete system would need to indicate the precise position and disposition of each leg, arm, foot, and hand at every moment, describe the dance's relationship to the musical score, and communicate how the body as a whole moved through space over time. Questions seemed unending: How to signify a jump or a bend? The precise speed of a step? A gesture's emotional nuances? How could the tool's legibility across languages and cultures be ensured? Laban was nonetheless sure that he was the person to overcome these challenges.

By 1928, albeit with the help of a number of collaborators, he believed he had. In May of that year, Laban presented his new system, dubbed Kinetographie, to a plenary session of the National Dancers' Congress in Essen, Germany. Armed with a farrago of shapes on an eleven-columned staff, Laban proposed a wholescale rethinking of dance preservation.

The reaction was swift and overwhelming. Already celebrated for his prominent role in the emergence of German Expressionist dance, Laban was hailed, once more, as a savior who would transport the art form as a whole into the modern age. As one newspaper reported, "For the uninitiated the triangles, oblongs, lines, dots and strokes are like a book with seven seals. To read movement from them seems to be impossible until instruction therein convinces him that the opposite is the case." Once appropriately enlightened, however, the invention's importance became "quite clear. The transmission of rhythmic gymnastic and dance exercises in the most simple diagrammatic form is no longer limited by space. Teachers on the other side of the world can get to know of new things in this way, things they have not seen with their own eyes."[2] The critic Hans Brandenburg noted that Kinetographie dwarfed even Laban's already matchless resume, writing that his earlier triumphs were "crowned by the dance script. This brilliant creation of Laban's is his egg of Columbus . . . one almost forgets the centuries-long path, the simplicity which makes the complicated seem easy when the solution is suddenly found."[3] *Singchor und Tanz*, the journal for German's professional dancers and chorists, called Kinetographie a "cultural deed of the highest order,"[4] and the well-known music publisher Universal Edition founded a new journal, *Schrifttanz* (Written dance), to discuss and disseminate the system.[5]

For months thereafter, Laban's face was plastered across special issues of the German-language dance press. International newspapers took notice as well, reporting excitedly on Laban's achievement, eager to read movement "as one reads a page from a book or a bar from a sheet of music."[6] By 1931, demand was such that Universal Edition published an English translation of Kinetographie, an event which the *Los Angeles Times* hailed as a key moment in dance's struggle to attain its "rightful place alongside the other arts."[7] New Yorker Irme Otte-Betz personally pledged to provide a translation of every issue of *Schrifttanz* for interested American readers.[8]

Interest in Kinetographie was not, however, confined to the dance community. Once movement was on paper, its traces moved not just across space, but across disciplines. As one commentator—reporting on surprising attention from fields as varied as mechanical engineering, film, industry, advertising, and medicine—noted, "To speak about the importance of the dance script at this point would be to carry coals to Newcastle. Today it matters more to show how Rudolf von Laban's life's work has already quietly spread in broad circles." It was, he continued, "the best proof of the great importance of Laban's invention that in a time when the economy is in a terrible state, when everyone can think only of his own immediate monetary interests, all circles of our cultural life which have anything at all to do with movement are turning towards Laban's creation with such obvious interest."[9] Even at a moment of massive economic, political, and cultural turmoil—or perhaps because of it—this set of scribbles captured the public imagination. Indeed, in its inaugural issue *Schrifttanz*'s editor, Alfred Schlee, provided an ambitious prediction of Kinetographie's potential civilizational impact, calling the system's birth "a step in the development of mankind . . . comparable to the impact of the alphabet on the art of words."[10] Something more than the precise preservation of leaps and turns seemed to be at stake.

This chapter is an account both of Kinetographie's genesis and of what came after. Though the technical details are vital, it is less the story of how Laban solved the intellectual problem of notation than of why he chose this problem to begin with and—even more significantly—why the world around him responded so eagerly to his intervention. The answer begins in the world of dance, but as Schlee and others prophetically suggested, it ultimately extends much further. What follows, therefore, does the same, opening with an account of the artistic scene in first decades of the twentieth century and its influence on Kinetographie's creation. The chapter then continues by exploring the moment's scientific context, revealing how Laban's reading of physiology and physics further convinced him of Kinetographie's necessity as well as of human movement's power to remold both individuals and larger communities. Finally, it follows Kinetographie in practice, uncovering how it functioned as a political tool, first during the Weimar period and later in the direct service of the Nazi state.

Making Notation

German dance in the first few decades of the twentieth century was in a time of transition. Though Laban may have admired the Russian ballerina Anna Pavlova—celebrated for the emotional intensity of her signature solo, "The Dying Swan"—it was clear to him that the age of ballet had passed. Much of the Central European dance world agreed.[11] Ballet was undoubtedly still practiced in the 1910s and '20s, but the "present dance-generation" had embarked on a wholesale rebellion against its cultural dominance, convinced that the restrictions of its highly formalized technique were inappropriate for the modern era. Instead of the courtly postures of the past, dancers appeared on stage barefoot and semi-nude, struck grotesque attitudes, took up themes of violence and death, imitated machines and raged against them.[12] The artistic scene was diverse in terms of genres, performers, and spaces, but it was bound together by a new sense of political engagement and a belief in the physical body as a uniquely powerful medium for natural and unmediated expression.

Dance was also increasingly construed as an activity that could lead to self-transformation, not just for professional performers but for the growing number of laypeople who took up the practice. In conjunction with projects of life reform and the broader wave of fascination with the body—*Freikörperkultur*, gestural theater, Dalcroze eurythmics, bodybuilding—sweeping Europe, enthusiastic amateurs could be found stretching and jumping in open fields and open schools, hoping to achieve everything from communion with nature to ecstasy to political liberation.[13] As dance scholar Kate Elswit has noted, "the most consistent characteristic of the specifically German physical culture movements that peaked between 1890 and 1936 was their situation of bodies as simultaneously both authentic—sites to reckon with truth—and manipulable—sites of work."[14] Dance also became an integral part of mass culture, with cabarets packed to capacity and avant-garde performers featured on the collectible cards inserted into cigarette packages.[15]

One of the artists whose picture would have been regularly plucked from those tobacco-scented confines was Rudolf Laban. His path to dance was winding. Born Rudolf Jean-Baptiste Attila Laban in December of 1879, Laban began life in Pressburg (now Bratislava). His mother was French, his father a career military officer from the Hungarian upper classes, and he spent his early years at his family's palatial home on the banks of the Danube. In 1878, however, Austria-Hungary occupied Bosnia and Herzegovina under the terms of the Congress of Berlin, and Laban's father was appointed military governor of the new territories. Laban's childhood was

FIGURE 1.1. Dancer Gert Ruth Lösser in Berlin. From Rudolf von Laban, *Gymnastik und Tanz* (Gerhard Stalling Verlag, 1926), 119.

thus spent only partly in the empire's center and partly alongside his father at military outposts in Sarajevo and Mostar. Laban senior was ultimately ennobled as a result of this military service; from the age of eighteen, young Rudolf added a "von" to his already sizable appellation.

Few sources on his early years exist, save Laban's own memoir, *Ein Leben für den Tanz* (A life for dance), published in Dresden in 1935. Within, Laban spun theatrical tales of a childhood full of enchantment and adventure: breathtaking sea journeys, glittering fetes, Sufi dervishes, days spent on a purebred steed evading capture by "sinister-looking Turks," and evenings bathed in candlelight, spellbound by his grandmother's stories of Napoleon's savagery and a mystical "blue flower."[16]

His father's military service also loomed large. Laban recounted, for example, the story of an afternoon in the late 1880s when the elder Laban was pursuing of a group of local rebels whom he suspected of hiding a cache of munitions in a "narrow, deep mountain-cleft."[17] After several of his burlier men tried to gain access to the crevasse with no success, little Rudy was summoned and promptly inverted. Moments later, young Laban found himself upside down in a dim mountain cavern, suspended only by a rope knotted around his ankles. From above, his father yelled at Rudolf to train his attention on the cave's walls, where the illicit arms might be stacked. Laban recalled, however, that he saw no rifles or revolvers. Instead, gazing at the cave's mineral formations, he felt transported to another world, surrounded by "glistening shapes and shadows" presenting a "movement-display" like none he had ever seen. Stunned, he became sure that "the stone was alive and possessed a will of its own in its tenacious, slow crystallisation, and its fight against destruction by sun or water." Upon being returned to solid ground, he was rendered nearly mute, with no report forthcoming. Annoyed, the expedition's leader decided that "the best thing to do would be to widen the opening and send a level-headed adult."[18] The story may or may not apocryphal: by 1935, Laban was a shrewd mythmaker, and the book itself was a clear act of public self-fashioning. It encapsulates, nevertheless, the distinctive intertwining of the romantic and modern, the artistic and militaristic, that undergirded both Laban's life and Kinetographie's creation.

Despite these early failures in martial spelunking, Laban was expected to follow in his father's footsteps. In 1899, he was sent to military school at the empire's Wiener Neustadt academy. There he would have received instruction in history, law, and the sciences, along with the "building of fortifications, terrain drawing, the major languages spoken in the monarchy, the handling of complex weapons, field exercises, drill, tactics, strategy, horseback riding, fencing, swimming, dancing and athletics."[19] The makeup of the student body was relatively diverse; though hierarchies could be found, the academy was broad-minded in its admissions policies and formally expressed no preference for cadets based on their ethnic, social, or religious backgrounds. This open-mindedness was to some extent strategic. The national and social mixing that characterized the joint army functioned as a tangible realization of an otherwise fractured empire, as did policies mandating service outside of one's homeland.[20] Laban, for example, would have encountered a number of ethnic Austrians under his father's command during his time in Bosnia, while military-age Bosnians were required to serve two years in Austrian territories.[21]

The training did not take—at least not in the way the academy had envisioned. A childhood fascination with his father's battalions "flying and tearing about in wonderful designs" and the chance to organize a "sword-dance" for his fellow cadets were not enough to sustain Laban's interest in military life. Against his family's wishes he departed after only a year to pursue a path in the arts.[22] (Laban alleged that he was dismissed for shooting a superior officer's cap off his head with a revolver. This is unlikely.) For years thereafter, he lived the peripatetic life of a turn-of-the-century bohemian: studying ballet and architecture in Paris, theater and painting in Munich, and spending much of World War I experimenting with nudism, vegetarianism, free love, and Freemasonry at the Monte Verità artists' colony in Ascona, Switzerland.[23] Largely cut off from family funds after his withdrawal from the military, he dabbled in graphic design and accounting to make ends meet.

Ultimately, however, Laban developed a special interest in the moving body, drawing in part on the time he spent at the Institut Jaques-Dalcroze at Hellerau in 1912. He founded his own schools of rhythmic movement in Germany and Switzerland, where he offered courses in gymnastics, eurhythmics, pantomime, dance, and harmonic movement. He also began performing his own choreography across the continent, premiering as many as a dozen

FIGURE 1.2. Rudolf Laban (far right) and dancers in Ascona, Switzerland, 1914.

works each year, both alone and with a small company. He was a key figure in the organization of the first German National Dancers' Congresses in 1927 (in Magdeburg), 1928 (in Essen), and 1930 (in Munich). His students—including the famous Mary Wigman and Kurt Jooss—fanned out across Europe, further elevating his standing. By the end of the decade, Laban had secured his place as perhaps the most important figure in the emergence of *Ausdruckstanz*, a genre defined primarily by its commitment to emotional expressivity and its opposition to traditional balletic forms.[24] His devotion to performance, however, soon gave way to a fascination with movement itself—and, more specifically, with how it might be recorded.

Laban, in fact, reported an interest in notation as early as 1908. In his memoir, he recalled roaming the streets of Paris with pencil in hand, watching behavior and attempting to record it symbolically. The project progressed in fits and starts for years; he experimented both publicly and privately, occasionally publishing sketches for possible systems. The particular contours of Laban's efforts may have been unique, but his interest in capturing the elusive movements of the human body was far from unusual. Beginning in the mid-nineteenth century, physiologists and physicians had increasingly turned their attention to the creation of devices intended to translate dynamic bodily states into lines on a page: the sphygmograph recorded the peregrinations of the pulse, the myograph made visible the force of muscular contractions, and the kymograph documented blood pressure on a circular drum. Fleeting qualities that had previously only revealed themselves to the trained doctor's touch were now inscribed on paper for anyone to read. Foremost among the champions of this new "graphic method" was the French physiologist Étienne-Jules Marey, whose laboratory in Paris served as a central site for experimentation.

In later years, alongside others like the American Eadweard Muybridge, Marey was also key to the development of chronophotography, a technique which revealed the gross movements of humans and animals by taking multiple photographs in rapid succession, then either superimposing them on a single photographic frame or arranging them in sequence. The resulting images garnered both scientific and popular interest: while Marey used chronophotography to aid his investigations in anatomy and physiology, Muybridge famously helped the industrialist and California governor Leland Stanford settle a bet by demonstrating that all four of a horse's hooves leave the ground when it gallops.[25]

In addition to resolving a number of kinesthetic mysteries, these images had undeniable aesthetic appeal. Chronophotography, for example, served as a visual reference point for early twentieth-century modernists like Marcel

Duchamp, who were similarly eager to experiment with new modes of representing time, movement, and the body.[26] At the Bauhaus, Oskar Schlemmer repeatedly returned to the theme of the human figure in space, both in print—as he did on the cover of one of the first issues of *Schrifttanz*—and on stage. In fact, a number of members of Laban's own company appeared in Schlemmer's *Triadic Ballet* in the early 1920s, a work famous for dissolving its dancers into abstract assemblage of cones, spheres, and wires, engaging in a play with darkness that blurred the distinction between two and three dimensions.[27]

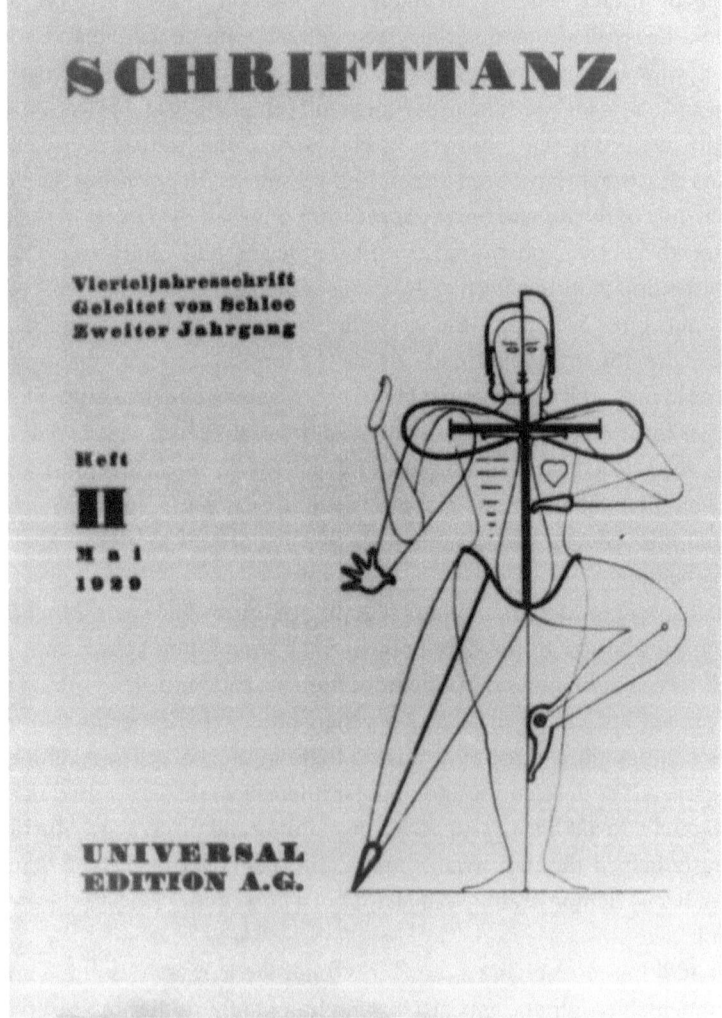

FIGURE 1.3. Cover of *Schrifttanz* 2, no. 2 (May 1929). Drawing by Oskar Schlemmer.

Laban was also not the first to conceptualize or execute a notation system specifically for dance. Various forms of notation had existed for at least several hundred years, and Laban himself was well-schooled in their history.[28] In a 1926 book, *Choreographie*, he grappled with this preceding work, paying particular attention to a technique known as Beauchamp-Feuillet notation. Developed for use in the French court at the end of the seventeenth century, Beauchamp-Feuillet notation was essentially cartographic in nature: the page upon which it was inscribed represented the dancing space as viewed from above. A single line indicated the dancers' path through space, while symbols on either side of the line signaled the position and movement of the feet and the vertical displacement of the body.[29] The resulting image was a kind of haute-baroque vision of the mail-order Arthur Murray diagrams that taught the foxtrot, tango, and waltz to twentieth-century Americans.

Initially, Beauchamp-Feuillet was a success: between 1700 and 1730, notators in France and England published collections of dances on a yearly basis, and at least 335 notated pieces survive into the present.[30] Laban himself pored over old Beauchamp-Feuillet texts at the Bibliothèque-Musée de l'Opéra during his days as a student in Paris.[31] By the 1780s, however, the system had largely fallen out of use, in part because it was ill-suited to capture gestures of the upper body, which had become increasingly central to European dance performance.[32] In *Choreographie*, Laban too ultimately rejected the method as insufficiently detailed and narrowly conceived, suited primarily for the relatively limited movement vocabulary of the French court style.[33] Laban was similarly skeptical of modern alternatives like film, at first largely for practical reasons. In addition to the cost of the stock, he noted that "the dancer has only to do a big jump and he has given the camera the slip; or a few rapid whirls and the stupid screen will register a shapeless cloud."[34] He proposed instead a new kind of universal notation system, one that would account not only for the wider range of movement used by the practitioners of *Ausdruckstanz*—a form without an established lexicon—but that would work equally well for all forms of movement, at all times, in all places.

In *Choreographie*'s remaining pages, Laban attempted to engineer such a system. Over the course of several chapters, he experimented with a variety of techniques and aesthetics, peppering the text with symbols evocative of Celtic runes and Babylonian cuneiform. The most complete system of movement writing in the book, however, was the "bodycross," or *Körperkreuz*, method. In it, the body was represented as a cross, the horizontal line dividing upper from lower body and the vertical line its right and left sides. Numbers, letters, or symbols were placed in each quadrant to indicate the direction in which the corresponding body parts would move, and additional signs indicated

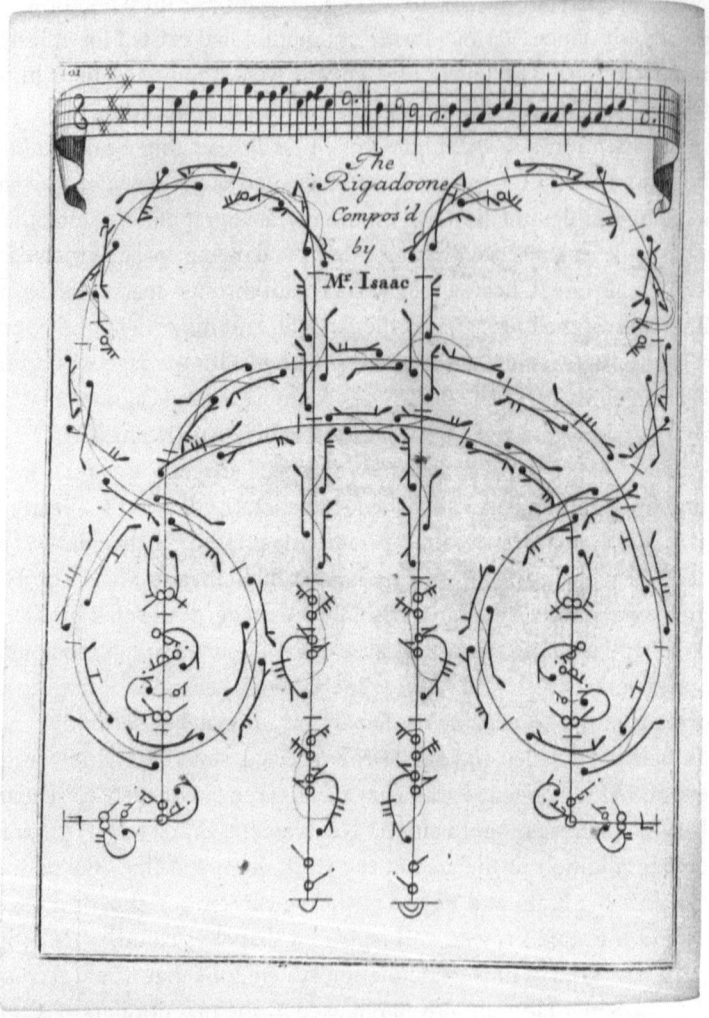

FIGURE 1.4. Beauchamp-Feuillet notation for a rigaudon, a French baroque dance. From John Weaver, Dancing Master, *Orchesography or the Art of Dancing . . . an Exact and Just Translation from the French of Monsieur Feuillet*, 2nd ed. (London, 1721).

movement qualities like speed, strength, and width. Crosses were placed one after another and read in sequence. In contrast to Beauchamp-Feuillet notation, the object at the center of the *Körperkreuz* was not the dancer's progress through space, but the body itself.

While the *Körperkreuz* method was relatively successful in documenting individual instances of movement, its users struggled to illustrate the transitions between movements and the passage of time. In 1927, however,

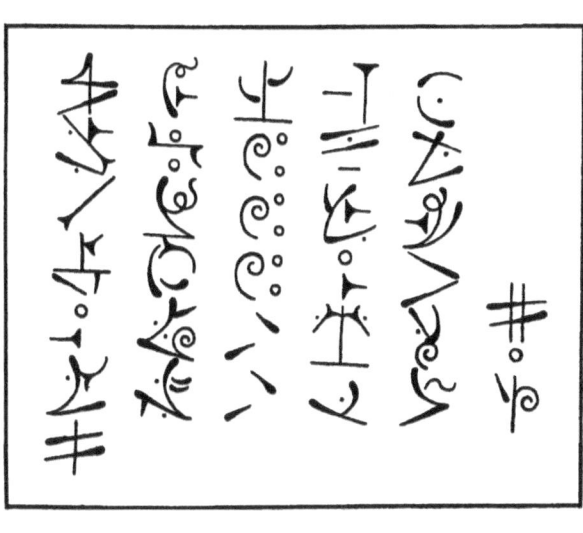

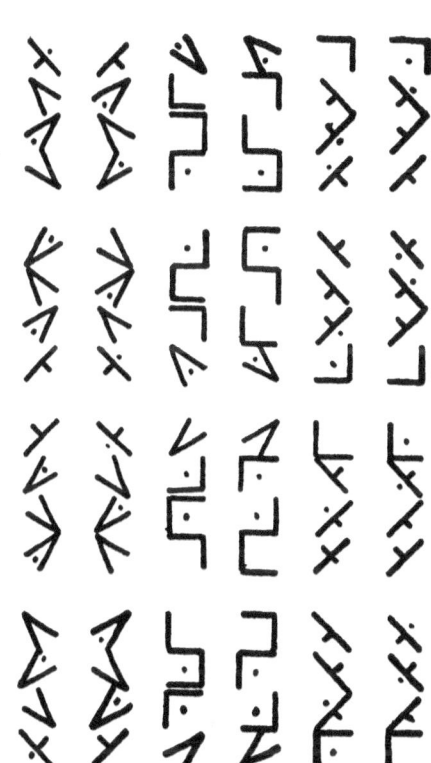

FIGURE 1.5A–B. Two experiments in notation. From Rudolf von Laban, *Choreographie* (Eugen Diederichs, 1926), 5, 72.

Dussia Bereska—a dancer, choreographer, and one of Laban's many romantic partners—proposed a solution. Instead of using a series of symbols, she suggested that movement be recorded continuously along a single score with duration indicated by a symbol's length.[35] From there, everything seemed to fall into place. Laban immediately began introducing the system to his students, publicly presented the work at the 1928 National Dancers' Congress, and published Kinetographie in the inaugural issue of *Schrifttanz* later that year.

As described in that issue (and elaborated in an accompanying manual), a blank Kinetographie score consisted of a set of parallel, vertical lines, with bar lines functioning to indicate time, much as they do in music notation. The center line served as the spine: actions taking place on the right side of the body were written on the right of the staff and actions of the left side of the body on the left. Symbols clustered around the center line indicated movements of the feet, and moving outward, invisible radiating columns provided space to indicate the movements of the legs, torso, arms, hands, and head.[36]

Each of the body parts could be represented as moving in any of one of three planes—forward and backward, side to side, and diagonally in either direction—and as being performed at a high, middle, or low elevation, relative to where the

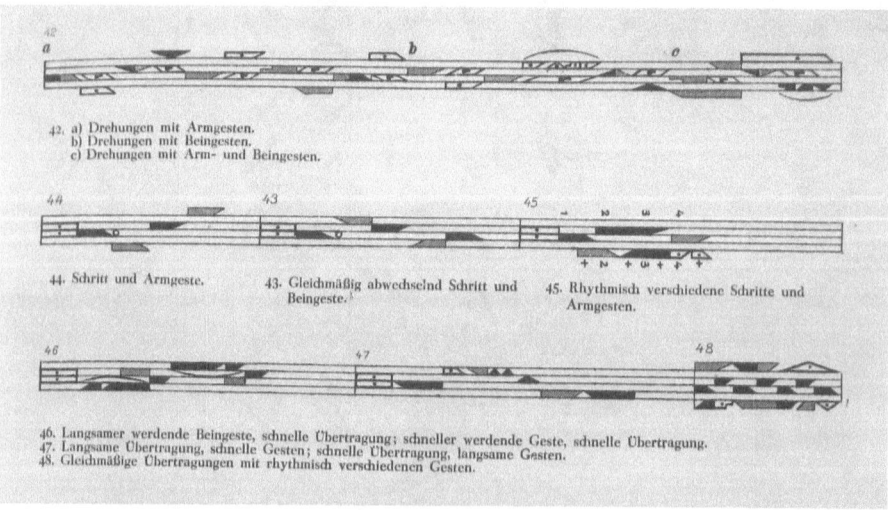

FIGURE 1.6A. Movement samples in an early version of Kinetographie. The middle line reads: "44. Step and arm gestures, 43. Evenly alternating step and leg gestures, 45. Rhythmically varying steps and arm gestures." Kinetographie was initially printed from left to right to facilitate its presentation alongside any accompanying music and because of the widespread availability of musical paper. Laban always suggested, however, that dancers rotate the page ninety degrees: the score would then be read from the bottom upward and its left and right sides would correspond naturally to the left and right sides of the human body. The vertical orientation ultimately became the standard orthography and is utilized in figure 1.6b. The staff itself was also eventually altered from five lines to three. From Rudolf von Laban, *Schrifttanz: Methodik, Orthographie, Erläuterungen* (Universal Edition, 1928), 3.

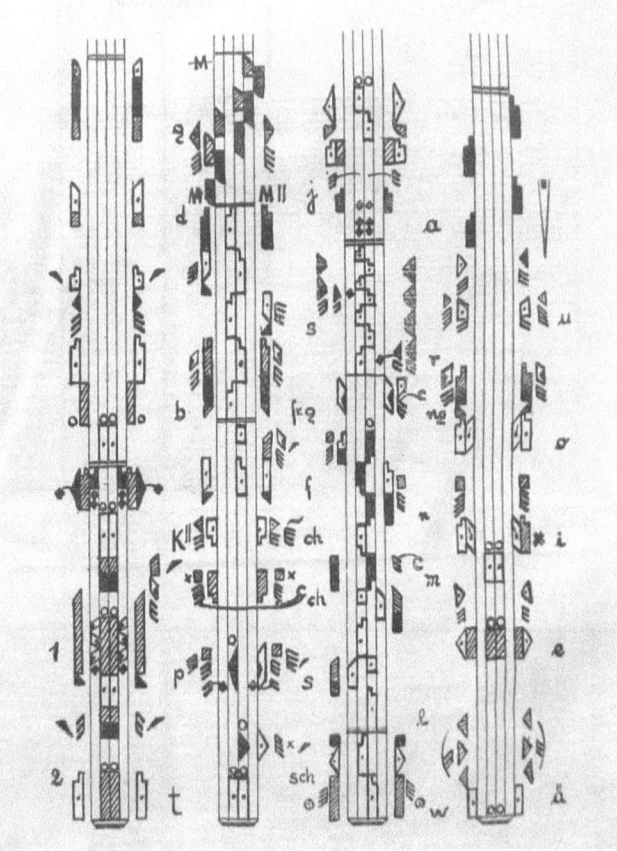

FIGURE 1.6B. Kinetographie accompanying Fritz Klingenbeck, "Schreiben und Lesen," *Schrifttanz* 2, no. 4 (1929): 76–78.

body part was normally found. The direction of movement was represented by a symbol's shape and its elevation by the symbol's shading. Leaps and swings became a series of blackened triangles, six-sided polygons, and trapezoids adorned with single dots. Additional symbols indicated movement qualities, such as how forcefully a given gesture should be performed.

Figure 1.8 provides an example of how the staff, the direction symbols, and the elevations could be combined to notate a simple dance step, here forward kicks on alternating legs. The first symbol appears in the column for the right foot and indicates forward movement at mid-level: a step. The second shows forward movement of the left leg at a high level (that is, raised from its normal position): a kick. The third represents a mid-level movement forward with the left leg: another step. The fourth shows another kick, this time with the right leg. This is notation at a very elementary level, but it suggests the central elements of Laban's system.

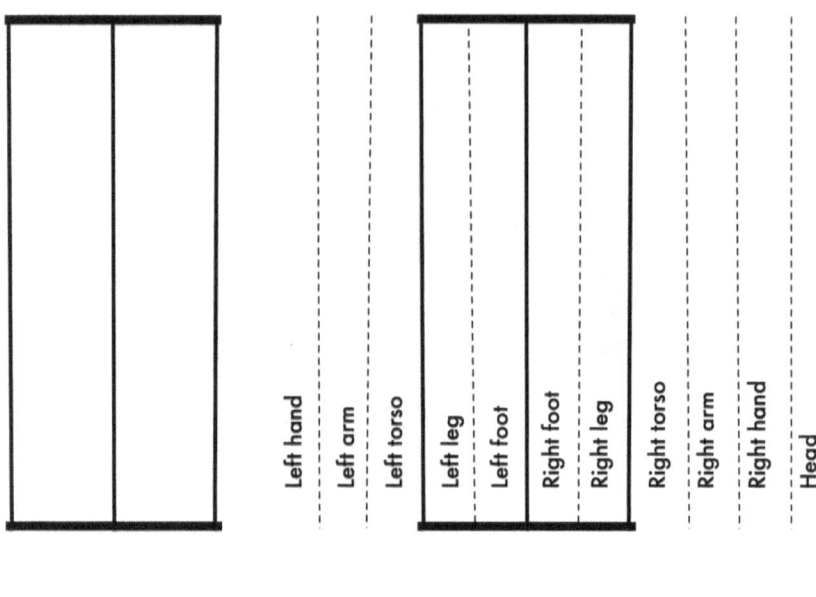
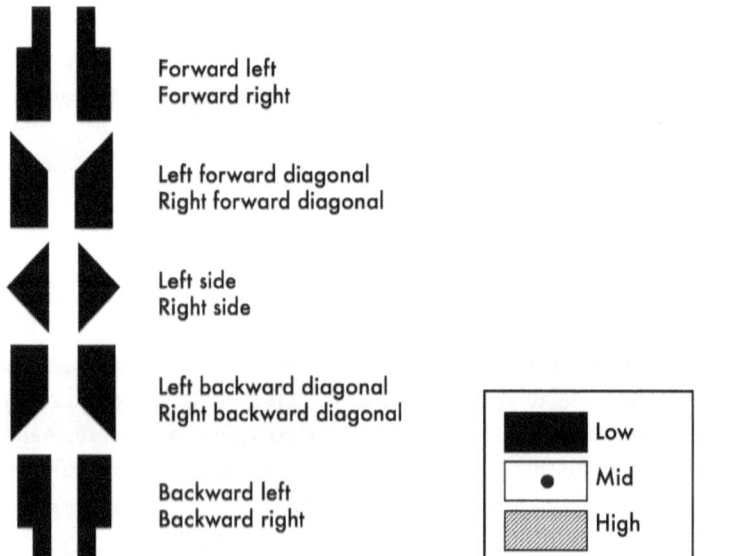

FIGURE 1.7A–D. Key to Kinetographie: (a) blank Kinetographie staff, (b) columns indicating location for notating each body part, (c) direction symbols, (d) elevation symbols. Prepared by author.

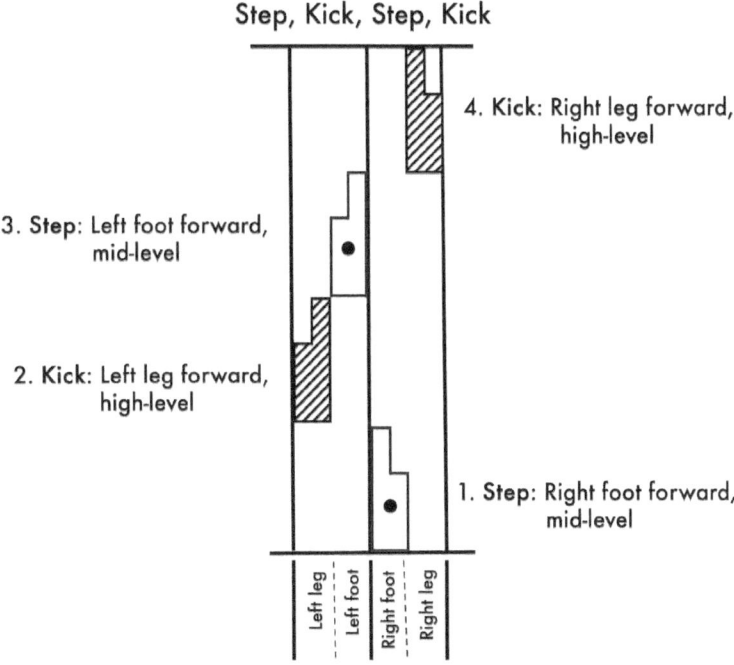

FIGURE 1.8. Simple Kinetographie notation for kicks on alternating legs. Prepared by author.

The other key component, though one that Laban himself did not remark upon, was the system's appearance. Laban's fascination with occult knowledge—his Freemasonry, Rosicrucianism, and enthusiastic Orientalism—had never deserted him, and earlier iterations of the script had seemed intended to conjure the inscriptions of ancient sorcerers.[37] The new version was unquestionably a product of the early twentieth century: its abstraction, rigid lines, hard edges, and distinct geometries invoked a new, self-consciously modern sensibility. Visually, Kinetographie was in simultaneous conversation with modernist art (including the Bauhaus, Constructivism, De Stijl, and the Wiener Werkstätte), with industrial development, and with new modes of graphic representation in the sciences.[38] Kinetographie's ostensible purpose was to record art, but the system also served as an aesthetic statement in its own right, one that positioned the script as uniquely forward-looking, rational, and scientific.

Laban and his supporters reinforced this argument constantly, touting Kinetographie's revolutionary character, in particular its promise to capture universal forms of human movement, rather than the vocabulary of any one dance style.[39] With Kinetographie in hand, Laban argued, the "science of living movement" could progress: dance could be broken down into its parts, accurately described in its entirety, and easily preserved and distributed. As a result, like music scholarship—which had recently begun to make inroads

FIGURE 1.9. Rudolf von Laban with Kinetographie, ca. 1930.

in the German university system with its own chairs, journals, and critical editions—dance too might ascend to a new rung in the intellectual hierarchy.[40]

This envisaged modernity was, it is important to note, thoroughly gendered. In his 1927 talk at the German Dancers' Congress in Magdeburg, for example, Laban linked some critics' reluctance to adopt a dance script—their "support [for] dance illiteracy"—with their fondness for "flowery descriptions of predominantly feminine bodies," continuing by claiming that notation, once embraced, would further highlight and solidify the influence of the great male choreographers. "As in all building," he ventured, "the design of a dance artwork will . . . remain predominately an activity of man."[41] Elsewhere, the critic Hans Brandenburg contended that the earlier, widely praised aesthetic reforms of female innovators like Isadora Duncan ultimately amounted to little more than girlish rebellion. Laban's work, on the other hand, fused the old and new, promoting novel forms of expression while simultaneously subjecting them to new methods of control. "A woman as a dancer or artist in general can instinctively do the right thing," Brandenburg wrote, "but only the male spirit can find laws, make them and compel general obedience to them. And Laban's work has arisen entirely from the male spirit. His every deed was dictation."[42]

Such boasts exploited false dichotomies between the modern, the intellectual, and the masculine, and the premodern, the instinctual, and the feminine. They also obscured the crucial role many women—Bereska in particular—played in Kinetographie's genesis.[43] Brandenburg's words did, however, accurately suggest the growing scope of Laban's ambitions. While existing histories of Kinetographie have focused on its role as an innovative recording tool, closer study reveals that it was much more. In Laban's most expansive plans, Kinetographie was not merely a method for recording dance. It was a technology for shaping human movement, an endeavor that, for Laban, had the potential to remake not just the individual body, but the entirety of the social, spiritual, and political world. Understanding how Laban arrived at such a conviction requires an excursion into the broader scientific milieu of which he was a part.

Movement's Laws

Laban's interest in dance as a world-shaping practice drove his involvement with utopian artist colonies in the 1910s, but he began to articulate it particularly explicitly and to connect it with scientific discourses in his 1920 book *Die Welt des Tänzers* (The world of the dancer). This publication was Laban's effort to articulate a coherent theory of dance, its history, and its role in the wider world. On these terms, it does not necessarily succeed. Composed in five "rounds," or *reigns*, the book is less a single, structured argument than a series of loosely connected meditations. Later critics—including some of Laban's own followers—would tar it as incoherent, but under closer examination clear themes emerge. Key among them was Laban's belief that dance was more than an art of performance, a proposition embedded in the text's very title. *The World of the Dancer* was not an analysis of dance as it might be experienced by an imagined audience, but an exploration of dance's effects on dancers themselves.[44]

Across the entire book, Laban made the case that dance was best understood as a particularly powerful form of self-cultivation, or as he put it, a "method for the rhythmic regulation of emotional things."[45] Many dancers, he remarked, knew this truth instinctively, having experienced firsthand how movement altered their moods, generated feelings of harmony and community, or provoked new forms of perception. Laban also rooted his basic contentions in personal, visceral knowledge, observing that "the physical-emotional-intellectual-educational power of dancing has been proved to me by a thousand practical experiences."[46] Moreover, he noted that—though largely ignored in the twentieth century—movement's hidden power had once been broadly acknowledged, discussed and employed by figures ranging from Plato and Pythagoras to medieval European mystics and Sufi clerics.[47]

Laban was determined to both draw on and surpass these predecessors. Not content with simply observing the existence of ties between movement and mind, he called for a new program of scientific study to uncover the exact contours of their relationship (*Tanzwissenschaft*). He attacked the "false premise" that the meaning associated with certain movements or gestures was a mere social construct, contending instead that a given movement's "symbolic significance" derived from its "formal structure."[48] There was, in other words, a natural law that linked the drooping shoulders of the defeated man with his sorrow and humiliation or the fist, propelled forward and upward, with joy, pride, or excitement. And though Laban argued that "reason" could never provide a final explanation for the "mystical, occult, intuitive" elements of dance's power, he confidently asserted that some form of objective analysis was a vital tool in the artist's arsenal.[49]

By the time of the book's publication, there was no shortage of scientific resources to support Laban in this claim. He found particularly useful work in physiology, which since the middle of the nineteenth century had been a discipline in ascendance. New discoveries had increased the field's stature, and new institutes, academic societies, and faculty chairs proliferated. Moreover, as the century wore on, its practitioners increasingly turned their attention to questions about sensation, perception, and subjective experience once thought to be the sole domain of philosophy. Led by figures like Hermann von Helmholtz, Gustav Fechner, and Wilhelm Wundt, physiologists began to examine not just how a muscle might twitch under electrical stimulation, but whether, when, and how that event would be processed by the mind. This new research program reached its apotheosis in Wundt, who—in subjecting human consciousness to regularized laboratory study—has been credited with founding experimental psychology as a discipline.[50]

These investigations, however, also moved beyond traditional laboratory walls, including via a boundary-blurring field of study the historian Robert Brain has called "physiological aesthetics." Flourishing from the 1880s to the 1910s, its practitioners—both scientists and artists—sought to use the new tools of the experimental physiologist's laboratory to examine the effects of particular aesthetic choices or techniques on the human sensorium. The polymathic Sorbonne librarian Charles Henry, for example, used graphic recording technologies to document both the media of art itself—colors, sound waves, odors—and human physiological reactions to it—breath, movement, heartbeat—in linear form. The symbolist poet Gustave Kahn sought to remake poetry, overturning accepted poetic forms and arguing that meter was best rooted in the particular physiological rhythms—the beating blood, the bated breath—of the individual poet. This search for a set of new aesthetic rules governed by the internal operations of the human body shaped European artistic modernism in profound ways, infusing practice and curricula across the continent.[51] It was also concomitant with a view of the artist as a kind of engineer of the human sensorium, a figure capable of controlling "the effects of their art on the viewing or listening public, altering bodily, mental, or emotional states and even transporting the beholder to a new or unexpected state of feeling or consciousness."[52]

Such discourses pervaded Europe, and undoubtedly shaped Laban's own artistic education in the century's first two decades. But even as many former proponents of physiological aesthetics began to distance themselves from the program in the aftermath of the cataclysm of World War I, Laban embraced it all the more tightly.[53] Though Laban made few direct references to particular works in the main text of *The World of the Dancer*, his interest in engaging with scientific discourse is clear and pervasive. Detailed discussions of the skeletal, muscular, and nervous systems are woven throughout, and the book's bibliography includes anatomical studies by the German physician and anatomist Johann C. F. Harless, the French anatomy professor Mathias-Marie Duval, and the embryologist and anatomist Julius Kollman, one of Laban's professors at the Munich Kunstakademie. In addition, the bibliography contains an entire section devoted to "Works on the Physiology of Facial Expression and Movement," including Wilhelm Wundt's *Principles of Physiological Psychology* (1874), Charles Darwin's *The Expression of the Emotions in Man and Animals* (1872), Guillaume-Benjamin-Amand Duchenne's *Physiology of Movements* (1867), and Carl Ludwig Schleich's *On the Switching Mechanism of Thoughts* (1916) and *Thought Power and Hysteria* (1920).

Though these texts varied widely in their arguments and particular objects of study, their authors worked from the common premise that human emotional and psychological life was profoundly bound up with the body's physical experience—and that these relationships could be studied. Wundt's physiological psychology seemed to be a special inspiration for Laban, who, echoing Wundt's call for an integration of the two fields, argued that physiology's laboratory methods might be profitably applied to the "so-called emotional or intellectual functions."⁵⁴ Laban held that studying the body's spatial orientation would quickly lead any perceptive observer to the "remarkable conclusion that each emotional state coincides with a quite definite body tension," producing changes of "mood and thought" according to underlying "gestural laws."⁵⁵

In *The World of the Dancer* and his 1926 *Gymnastik und Tanz* (Gymnastics and dance), Laban's discussions of these connections were largely empirical. In later years, however, Laban sought a more concrete mechanism for the movement-mind link and believed he had found it in research on the nervous system. His notes reflect a particular interest in the work of three individuals: the Scottish surgeon and artist Charles Bell (1774–1842), the Danish physician and physiologist Carl Lange (1834–1900), and the British physiologist John Newport Langley (1852–1925).⁵⁶

Though Bell, Lange, and Langley were separated by time and geography, Laban saw their work as part of a coherent corpus, one that illuminated important ties between nervous activity, physiological arousal, and emotional experience. Laban likely first encountered Bell through his 1806 book *Essays on the Anatomy of Expression in Painting*, which although more than a century old was still widely assigned to art students.⁵⁷ Within physiology, Bell was best known for delineating the distinct roles of the sensory and motor nerves. Prior to Bell's work, the fibers of the central nervous system were generally understood to "promiscuously" transmit, receive, and interpret both movement and sensation. Bell argued instead that though the motor and sensory nerves might appear to form a single fiber, they remained functionally distinct.⁵⁸ Sensory nerves carried information about the environment to the brain, while motor nerves transmitted signals between the brain and the muscles, organs, and tissues. He contended, moreover, that if the sensory nerves were the conduits for intellectual life, conveying information about the properties of the material world to the brain, then the motor nerves provided the medium for emotional expression, transmitting signals between the brain and the muscles, organs, and tissues. Bell also posited that this mind-body relationship was bidirectional: muscular movements not only expressed internal emotional states but could play a role in transforming them. Lange's famous 1885 work "On Emotions" advanced a related premise, arguing that physiological arousal was the root cause of all emotional experience. Alongside the American psychologist William James, he would eventually be credited with what is now known as the James-Lange theory of emotions.⁵⁹

Langley's work dealt primarily with the autonomic nervous system, which is responsible for controlling relatively "involuntary" bodily movements: the beating of the heart, the contraction and dilation of the pupils, and the expansion of the lungs. As Langley outlined, the sympathetic and parasympathetic systems represented its two primary divisions, and the two systems operated in concert. The sympathetic system worked to move the organism to a state of agitation: accelerating the heartbeat, slowing digestion, dilating the pupils, and increasing blood flow to the muscles while decreasing it elsewhere. Meanwhile the parasympathetic worked to move the organism toward a calmer state: slowing the heart, stimulating digestion, constricting the pupils, and expanding the blood vessels.[60]

Drawing on these texts, Laban developed his own account of movement's effects on the body, mind, and soul. Paraphrasing Bell, he contended that it was the "heart and the lungs as well as the whole breathing apparatus" that helped to produce the "human passions," and he praised Lange's empirical descriptions of the physical correlates of joy, anger, fear, sorrow, embarrassment, and tension.[61] He also echoed Langley's work, noting that that the activation of the sympathetic nervous system induced "biologically negative" results, preparing the organism for "emergency and defence." In contrast, he explained that stimulation of the parasympathetic system was "biologically positive," producing a feeling that was "essentially expansive, expressive—pleasant."[62]

Laban also went further, adding his own observations about the structural arrangement of the physical nerves within the body, noting that the layout of the parasympathetic nerves was "more straight-lined," whereas the sympathetic nerves formed a "more curved or almost circularly flat and disc-like" pattern. This configuration, he contended, provided additional proof of the relationship between bodily carriage and emotion: the "upright fearless carriage bringing the spine into a stretched position corresponds to the positively stimulating general effect of an active sympathicus," while the "collapsed contractual state" was both the cause and consequence of the "straight and electrified parasympathicus."[63] Laban saw this distinction—between sympathetic and parasympathetic, reactive and assertive, Apollonian and Dionysian—as reflecting a fundamental and perhaps biological dichotomy at the core of the human psyche. On this basis he concluded that performing flowing movements that extended out into space would promote "self-expression" and "give rise to such positive virtues as richness, domination, sensuality," while movements that collapsed inward were primarily "self-disciplinary" in nature, fostering "poverty, chastity, and obedience, which may be considered more negative virtues."[64]

Again and again, Laban argued for a deep reciprocity between mind and body, insisting that it was not only the body that expressed the mental state, but the mental state that might be transformed by the movements of the body. In doing so, he not only drew on the works of Wundt, Bell, Lange, and Langley, but capitalized on a broader strand of German holistic biology that sought to "reground the mind in the body and to reanimate the body with the mind."[65] "The influence of mental tensions on bodily behavior and states is generally acknowledged," Laban noted, but "the reverse is sometimes doubted." In fact, he continued, "the curing influence of body movements on mental flow is even *stronger* than that of mental flow on body movement."[66]

In a letter to the gymnastics innovator Rudolf Bode, Laban made this connection once more, writing that the "new art of dancing" produces "ethical values in contrast to the purely external aesthetic effect of the old stage dance. An all-around, strictly natural and thorough training of the organism is being striven for and achieved."[67] Bode agreed, citing the letter in his 1916 book *The Modern Dance*, and noting that "the transformation of feeling into muscular activity produces a clear sensation and strengthens thereby the power of decision, providing the precondition for the desire for great things, i.e. it can strengthen the foundation of the ethical personality."[68] Dance could be a method for shaping the self.

Laban suspected, moreover, that movement's effects could reverberate across space, transforming not just individuals but larger communities. He rooted this conviction too in insights from science and engineering. Like others of his era, Laban was fascinated by new discoveries in electromagnetism, radioactivity, and atomic structure—and new devices like the x-ray, wireless telegraph, and broadcast radio. These innovations seemed to be fundamentally remaking ideas about form and space, as both empty air and solid objects were reconceptualized as constituted by invisible forms of oscillating energy.[69]

Laban was further influenced by work in biology that postulated the existence of protoplasm, a pulsating, semi-fluid, nitrogenous material suspected to be the fundamental substrate of all life. Popularized by figures like T. H. Huxley, among the theory's most notable claims was that seemingly disparate organisms were shaped by the same fundamental force, made manifest in the protoplasm's unceasing vibratory activity. As Huxley summarized to a rapt audience in an 1869 lecture, "There is some one kind of matter which is common to all living beings. . . . [T]heir endless diversities are bound together by a physical, as well as an ideal, unity." As the medium of all cells, protoplasm bound "the animalcule and the whale," "the fungus and the fig tree," "the flower which a girl wears in her hair and the blood which courses through her youthful veins."[70] Others, including the German biologist and embryologist Ernst Haeckel, would take this argument further, arguing that there was not just little difference between the human and the plant or animal, but between organic and inorganic life writ large. Certain crystals, Haeckel contended in a 1917 work, had been shown to participate in this pulsating symphony, their molecules wriggling rhythmically under the microscope's gaze.[71] Haeckel additionally believed that the protoplasm held the key to evolution, arguing that hereditary data was stored in each organism via a set of characteristic waveforms governing various bodily functions.[72]

Citing Haeckel, Laban advanced the idea that all organisms—from amoebas to humans—were not solid, differentiable entities, but rather a "momentary harmony of all imaginable vibrations, a meeting point of waves of all kinds," constituted by "the smallest living organisms in their millions through which thousands of millions of tensions vibrate."[73] Movements and gestures shaped these microscopic vibrations, gradually refashioning the body at the basic cellular level.[74] These waveforms then propagated outward, producing excitations of space capable of affecting nearby individuals. Some of the effects were immediately evident while others burrowed into the body silently, shaping the human receivers without their conscious knowledge. If "odors, magnetic light vibrations, chemical waves, stifling heat and other emanations extend quite far beyond the limits of the body's skin," Laban argued, there was reason to believe movement did the same.[75]

As he summarized in *The World of the Dancer*, "Man has gestures which he can use to influence the actions of other men and animals around him," explaining that movement's effects were disseminated via the "strange recording apparatus which surrounds our bodies. . . . The steady pulse of the body spreads around the world like waves. It beams its own pulse or the waves of the strange pulse which strike it into space."[76] One should envision each individual, Laban suggested, as enclosed in a shell of nervously charged atmosphere, produced by the radiation of their unique nervous energy into the surrounding space. When these shells came into contact, the stronger would generally transform the weaker, though a gradual harmonization of force might also occur. "Under certain conditions," he elaborated, "a person's gestures can immediately affect the N. System of another person. There must be a positive (active) and negative (receptive) person. This stream of nervous energy can flow from a leader to a group as a form of magnetism, from one person to another as in love, fear, spiritual fascination, etc. This may be beneficial or destructive."[77] Indeed, Laban believed that such interactions seemed to be a primary "means of forming contact between two or groups of people." With enough force, moreover, Laban argued that "the irradiation of one person is intensive enough to affect the 'shells'" of many others. In such cases, the flow of energy would become "extremely powerful and free," and its "rhythmical surging" might "prove uncontrollable." Soon, the atmosphere would come "alive with the shapes and efforts irradiating out from the whole crowd now contained in one great shell."[78] Movement thus shaped both the individual and the group, amplifying its potential not just as an art form, but as a form of politics.[79]

This constellation of influences was made concrete in Laban's "icosahedron." One of a series of training devices he began constructing in the 1920s, it was composed of thirty white enameled tubes joined together at twelve vertices to produce a regular twenty-sided polyhedron just large enough to hold a full-grown man or woman. The structures were designed to help Laban's pupils visualize the spatial "kinesphere" that surrounded their bodies and to shape their movement in precise ways. Drawing on the principles Laban deduced from his own idiosyncratic blend of physiology, biology, physics, geometry, and crystallography—plus a dose of Neoplatonic mysticism—the students who clambered into the device's interior were instructed to direct their movements to particular points on the polyhedron in a particular order.[80]

Observers were not always sure how to react to these strange pieces of latticework. The Bauhaus artist László Moholy-Nagy, who featured the icosahedron in a 1927 work titled *Das Korsett*, seemed intrigued by the structure's promises to simultaneously mold body and mind. In a 1928 article for the *New York Times*, the American dance critic John Martin felt obliged to correct a "large mass of gratuitous misinterpretation" about the icosahedron, scolding readers that "as a matter of fact, it is neither the instrument of torture which it has been sometimes been suspected of being, nor anything else of a weird and unnatural character."[81] Even Laban's admirers were apt to call them "cages," and the image of the dancers' fleshy, flowing limbs against the polyhedron's metal fixity is undoubtedly an arresting one.[82] Were they prisons or portals? For some, it was difficult to tell.

Laban himself had no use for such anxieties. For him, distinctions between free expression and control, art and science, were at bottom illusory. The imposition of disciplinary structures did not necessarily imply the erasure of the self; in fact, only through constraint could a deeper, more authentic kind of fulfillment be achieved. It was, in fact, precisely this notion of dance's explosive power that motivated Laban's development of Kinetographie. If movement were such a potent force of nature, then understanding, recording, and ultimately controlling it was the modern era's most crucial task, and Kinetographie was the tool for doing so. Deducing dance's effects was just the first step in a longer process. Indeed, even as he composed *The*

FIGURE 1.10A-B. Students practicing "movement scales" in icosahedra. From Rudolf von Laban, *Choreographie* (Eugen Diederichs, 1926), 14, 32.

FIGURE 1.11. László Moholy-Nagy, *Das Korsett* (The corset), 1927. © 2025 Estate of László Moholy-Nagy / Artists Rights Society (ARS), New York.

World of the Dancer, Laban had his next project in mind. "What I have to say in my second book 'The Dancers Script,' about movement ideas and the perfection of choreography into a usable modern dance script cannot," he wrote, "be broadcast without the preceding clear indication of the essence of the dance and the dancer."[83] Stymied by technical challenges, his plans to publish did not come to fruition for another eight years. When Kinetographie did emerge, however, it was quickly put to work.

Notation in Action

Initially, Kinetographie's most enthusiastic reception came from the dancers and choreographers who constituted Laban's immediate community. Within just a few years of the system's publication, its partisans were already constructing hagiographic histories of its emergence. According to the critic Fritz Klingenbeck, for example, previous attempts to instruct dance students in notation were largely unsuccessful, resulting in "heated verbal exchanges and differences of opinion . . . more or less war. One student had observed a movement in this way, and the other had interpreted it and written it down otherwise." With Kinetographie, he claimed, "this was now all changed," difficulties melting away as the new script dispatched any lack of clarity.[84] In another characteristically sunny article, "The Movement Script is Easily Learned by Everyone," Laban's student Albrecht Knust praised Kinetographie's universal applicability, comparing it favorably to "several centuries" of previous attempts that had been either too rooted within specific movement vocabularies (most often ballet) or insufficiently comprehensive to succeed.[85] It is not entirely clear, however, to what extent Kinetographie shaped the daily practice of theatrical dance in its first few years of existence. It was certainly integrated into performances and curricula at the extensive network of schools run by Laban and his affiliates, but less evidence exists as to whether or how quickly Kinetographie changed how European choreographers, dancers, and directors not directly affiliated with Laban actually went about their work.

Laban, however, envisioned uses for Kinetographie far beyond the bounds of the theatrical proscenium. In 1929, officials in Vienna asked Laban to organize the city's Festzug der Gewerbe (Pageant of the trades), intended to serve as an event to unify the city's workers at a moment of increasing political fracture.[86] Laban, described as the event's "designer as well as director, screen-painter and dance-master," swiftly conceptualized a massive parade in which each trade group would process along a four-mile route, performing dance-like sequences derived from their working movements. The plan was inspired, at least in part, by the economist Karl Bücher's influential book *Arbeit und Rythmus*, first published in 1896 and in its sixth edition by 1924. Drawing on historical, anthropological, and physiological data, Bücher argued that nearly all "primitive" human communities had used rhythm to facilitate their productive labor. Whether swinging scythes or grinding grain, humans were at their most content and productive when their working movements occurred at a recognizable tempo, the hammer's blow ringing out a syncopated melody. Bücher contended, however, that this attention to rhythm had been largely lost in the modern period, with disastrous consequences.[87] Laban, similarly concerned with the enervating effects of modern civilization, sought to put Bücher's theory into practice.

The festival was set for early June, but Laban arrived in Vienna several months in advance. He set up his headquarters in the former imperial riding school but spent most of his time out in the city, observing and recording in Kinetographie the "traditional" movement elements of a dizzying panoply of crafts—from confectioners and furriers to carriage decorators, barbers, blacksmiths, meat smokers, newspapermen, gamekeepers, bakers, and gunsmiths. Once he had committed their movements to paper Laban returned to Vienna, using his notation as the basis for each group's festival choreography. As even the cavernous riding school could hold only a fraction of the planned participants, Laban then disseminated the movements back to his rehearsal directors via these written scores.

When June arrived, nearly ten thousand participants assembled on the *Ringstrasse*, and a million spectators lined the route. Excitement was high as the audience anticipated an entirely new kind of celebration; for the first time, wrote *Singchor und Tanz*, "dance formed the basis for a folk festival of huge proportions."[88] As the procession wound its way forward, the fruits of Laban's efforts came into view, including, among others, a "blacksmiths dance according to old blacksmith rhythms," a "stamping dance . . . performed according to the rhythm of the street pavers," and "a keg-binders dance, which Laban composed according to a historical coopers dance from the sixteenth century."[89] The festival also included work that fell outside the bounds of the traditional craft guilds: agricultural workers joined in with "harvest and grape gathering dances," as did groups of laborers who worked primarily with new industrial machinery. As *Singchor und Tanz* reported, "working belts, welding and riveting machines hissed past, their droning noise sometimes being the only accompaniment to the novel work dances." It was "apparent from the rhythms," the article continued, "that this was not just a superficially observed simulation of the various work methods, but a bringing to light of an almost forgotten treasure of craft culture through profound and painstaking study."[90] Elaborate costumes often completed the effect.

Not all reviewers were so impressed. Even the journal *Der Tanz* (generally stalwart in its support of Laban's work) noted that "the public who had paid dearly for grandstand seats were partly disappointed and not entirely without justification." They had been promised "a creative, new shape of the processional idea based on movement rhythm" but experienced instead a marginally coordinated parade of the marginally coordinated. This was not, in the author's view, Laban's fault, but a predictable result of the

FIGURE 1.12. Bakers at the Festzug der Gewerbe, 1929. Trinity Laban Archive Collections, London, UK.

constraints of working with people with "little physical skill." Laban's vision was thwarted by reality, from the blacksmith who became suddenly self-conscious when confronted with hundreds of expectant eyes to the guild heads who resisted direction and insisted on simply walking, "solemnly robed and perspiring, ahead of their products."[91] Kinetographie offered a fantasy of studied, coordinated movement on a mass scale, but its realization was far messier.

Laban himself remained upbeat, both about the fundamental premise underlying the event—the idea that movement could serve as a personal, social, and political balm—and its execution. After all, the point of the exercise was not the aesthetic experience of the audience but the embodied awakening of the participants. As he recalled several years later, he believed firmly that his efforts had:

> enlightened the guilds, from the master down to the apprentices about their own traditions and ... arouse[d] their enthusiasm for them. In most cases this was completely successful and, even after many years, I had the satisfaction of hearing from one or the other that I had given them more than just a festival and a momentary advertisement. Young men, who have since become masters have told me that because of the pageant they turned to their trades with far more understanding and love, and that for them the revival of old traditions especially helped to make their work pleasanter and lighter.[92]

Self-serving as this account may be, its significance lies in Laban's conviction that his new ability to record and share movement was not simply a neutral technique for preserving the past. It was a tool for stirring the spirit on a mass scale.

Vienna, in fact, was only one instance of Laban's involvement with large-scale movement events. In the late 1920s and early 1930s, demand for Laban's work with what he called "movement choirs" (*Bewegungschor*) was at its height. Often gathering on fields, beaches, or riverbanks, the choirs consisted of anywhere from five to five hundred individuals. Recruited from a local town, church, school, or political organization, the participants were taught a sequence of movements to be performed largely in concert: swaying, stretching to the sky, falling to the ground.

Here again notation was crucial. In his autobiography, Laban recounted his pleasure at how written dance allowed him to synchronize these movements from a distance: "I was often obliged to dash by plane from one place to the other to look after things. The procedure was usually like this: we would be commissioned to arrange a celebration or festival for a special occasion. In good time, we would form a choir of young people from all walks of life, who enjoyed movement, and give them body training. My choric works were written down in dance notation and rehearsed by the movement choir from

FIGURE 1.13. Movement choir. Original caption: "Martin Gleisner, Leiter der Bewegungschöre Rudolf von Laban in Jena and Thüringen." From Rudolf von Laban, *Gymnastik und Tanz* (Gerhard Stalling Verlag, 1926), 80.

this, very much as an orchestra would learn and rehearse a musical work from the score."[93] With Kinetographie, Laban's exact choreography could be performed across the nation regardless of his physical presence, a process of collective corporeal engagement he hoped might create a new form of transcendent community.

The violence of World War I, the dissolution of the Austro-Hungarian empire, and the ongoing fragmentation of the postwar period had left Laban concerned about the prospects for European civilization. His choreographic works often explored how movement could unite "diverging tendencies" into a "true togetherness," and he wrote admiringly about the folk dances of rural peoples. These rituals, he observed, produced a profound sense of collective vitality and ease, allowing for individual expression even as the group's common characteristics "gr[e]w stronger."[94] Though reports from choir members themselves are scarce, at least one young participant, thirteen-year-old Helga Hain, seemed to agree: "Among German children, who, as we did to begin with, start fighting every time they meet, this should be introduced . . . so that they can learn to understand each other. . . . This helps us to live in the community and later it is not so difficult to get by in the great community of life without which there is no peace, because in peace all nations and people must stick together." Hans-Joachim Kurras, also thirteen, had a similar—if somewhat idiosyncratic—reaction to his participation in a Laban-affiliated children's dance group, remarking that dancing in pairs had not only rid him of bodily tension and improved the color balance in his paintings, but had cured his instinctive dislike of girls and small children.[95]

Benedict Anderson famously ventured that print technology made it possible for people to imagine large, linked communities that had previously enjoyed no special form of togetherness. In the broadest sense, that vision is exactly what Laban hoped his system would accomplish—the creation of a sense of bodily intimacy, a nation that felt, as Anderson put it, a "complete confidence in their steady, anonymous, simultaneous activity."[96] That "confidence," again, would not be a matter of blind faith. Just as printing had (theoretically) stabilized the written word, Kinetographie would stabilize the moving body. In fact, though Laban began working with movement choirs before Kinetographie was finalized, he—and many of his supporters—felt that the form could only be practiced properly with notation in hand. As Laban's friend, the art critic Martin Gleisner put it, the "lay dance written down in dance script" was an absolute "precondition" for the movement choir's propagation. This claim was not just a pragmatic one about efficient instruction or widespread accessibility, particularly as most choir participants were never trained to read the script. It was an assertion about movement's power and the need to control it. Gleisner tellingly wrote that it would have been, in fact, "irresponsible to propagate movement choirs, and encourage people to become movement choir leaders," without the existence of notation, reiterating that "the circle around Rudolf Laban would not spread the idea of the movement choir so widely without the certainty of this possibility."[97]

A consensus was emerging that movement—because of its acute effects on the body, the mind, and the community—had to be managed properly. A gesture, an emphasis, a limb out of place could spoil the whole effect, leaving the participants in a choir not only unchanged, but worse off, disconnected and dissolute. Amateurs could not be "allowed to dabble irresponsibly."[98] Recording movement graphically was thus also a means of understanding and controlling both individuals and groups with a new degree of exactitude. As Laban explained, "the writing down of [a] composition is both a stimulus and a controlling means for [the dancer's] abilities and for *his temperament*, which a developed art of the dance cannot be without."[99] Kinetographie was an interventionist tool, designed to produce new kinds of selves and new kinds of communities, a technology for carefully harnessing the body's primitive, explosive power to remake the masses for a modern world.

Once again, however, debate emerged about whether Laban's grandiose plans were realistic, let alone politically desirable. In 1927, he used Kinetographie to coordinate the public performance of a large movement choir in Magdeburg. Though the system was not yet finalized, Laban was pleased with the

FIGURE 1.14. Image from a call for movement choir participants hosted by the Association for the Promotion of Laban Movement Theory, early 1930s. The aesthetics of Kinetographie are echoed in the depiction of the dancers' bodies. Trinity Laban Archive Collections, London, UK.

results: eighty men and women moving in synchronous spectacle in a piece he called *Titan*. One review of the event began in a seemingly positive vein, the author noting the striking degree to which "the individual dancer has disappeared. Only groups move together and against each other." He continued, however, remarking: "Laban is proud of the fact that this choral work was studied from his own written descriptions in dance notation, but somehow these choral celebrations have something a bit frosty about them.... The choral work itself was good, but the choral celebrations seemed like a good piano piece played by a laic figure who is pleased if he can at least get the technical difficulties over and done with properly. Nevertheless the large building of the Busch Circus resounded with loud applause."[100] Mixed reactions like this one—characterized both by a sense of notation's potential and by a nagging fear that it elevated control and regimentation over authenticity—would persist through Kinetographie's career.

The applause, though, would soon grow louder still. Though the moving body was central to a number of Weimar-era cultural and political projects, its management became even more crucial to the Nazi agenda taking shape across the 1920s and early 1930s. Within months of Hitler's ascent in 1933, the party set out to reinvent the entire cultural landscape, a wide-ranging project which included the founding of a dance division within the Reich Culture Chamber, itself a division of Reichsministerium für Volksaufklärung und Propaganda (Ministry of Public Enlightenment and Propaganda). In a letter to Propaganda Minister Joseph Goebbels supporting the creation of the new division, the critic Fritz Böhme wrote the following:

> If we really want to arrive again at healthy conditions in the entire field and for the first time to seriously ensure an authentic German form of expression, a unified leadership is essential. . . . So taken in hand, dance could function as a constructive and formative force. It could defend racial values and ward off the influx of alien movements and gestures, which are confusing to the German character and are undermining the German attitude. It could be easily built into the structure of propaganda. It could be effectively employed against instinct-uncertainty and weaknesses, which alien gesticulations create and which work themselves out in the will and serve to lame it.[101]

Among the foremost figures in German dance, Laban was an obvious candidate to lead the division, and with Böhme's support he was officially appointed to the post in 1934. Complete information about his work during this period is not available, as many of the relevant records were destroyed in air raids in 1944. It is known, however, that his initial primary duties included the centralization of dance education and performance through the direction of the German Dance Theater, the German Master Workshops in Dance, and the German Dance Festivals. For the first time, set syllabi and national examinations governed entry into the profession for both dancers and teachers; public performances were similarly regulated. In a 1935 letter to Otto von Keudell, Laban's immediate superior at the Ministry of Propaganda, he wrote that "racial characteristics stamp themselves in the movements, especially in the rhythm, in the posture of the body, and in the use of the body parts."[102] His new division would closely monitor their expression. As Böhme had hoped, movement became yet another realm in which Germanness might be defined and articulated.

Accounts by Laban's disciples have tended to portray him as a reluctant recruit to the Nazi cause, either forced into service under the implied threat of violence or slowly seduced by the prospect of state support for an art form long characterized by financial precarity. Recent scholarship, however, has made it clear that Laban was an enterprising and enthusiastic participant in the regime's activities. Early in 1933, he seized the political moment and proposed the creation of a new archive at the State Opera that would record and preserve in Kinetographie "traditional" German dances and work movements, formally codifying an imagined vision of German movement culture. The same year, while still the director of the State Theater in Berlin, Laban removed all non-Aryan pupils from his school, years before such expulsions became fully compulsory.[103]

Laban's writings also reveal a preexisting sympathy with Nazi racial ideology. In response to a questionnaire in *Singchor und Tanz* in 1928, he argued that variations in dance style could be traced to differences in racial character, noting that the European "seeks the meaning of awakening," while the "Negro approaches dance through touch alone." He added that it would be futile to attempt to assimilate "ethnographically alien or hostile influences" into Western dance culture because of the incongruent spiritual and intellectual attributes they embodied.[104] At the 1930 Dancers' Congress in Munich, he mused about the role movement choirs could play in the construction of a "new folk dance movement of the white race," contrasting the "purity" of the choirs' "primordial driving force" with "fashionable social dances which show an invasion of foreign racial movements."[105] Here Laban was likely referring to swing, which the Nazi Party would later work to ban due to its associations with both Jews and African Americans. Tellingly, Laban discussed human disease in much the same terms as the influx of "foreign" dance styles. Arguing that viruses and bacteria were lower forms of life, he contended that their deleterious effects came from their vibrational patterns' disruption of the normal state of human body, resulting in a "wild spreading of foreign tissues."

Disease, he wrote, is "always a sign that somewhere within the organism a process of *reversion* has started."[106]

Once officially in the state's employ, Laban also used Kinetographie to continue his work with large movement choirs, now renamed *Gemeinschaftstanz*, or community dances. In the 1920s, many movement choirs had no particular political affiliation, while others—including those led by Laban's onetime friend, the Jewish socialist and art critic Martin Gleisner—were organized by explicitly left-wing groups.[107] After 1933, however, all such choirs came under control of the state, and a dedicated ministerial subdivision—the Reich League for Community Dance—was established in 1935 at Laban's prompting. Its aim was clear. At a 1934 meeting of movement choir leaders, Laban gave two lectures, one devoted to compositional considerations in choral dance and the other to a history of "how each nation and each race has come to have the type of movement which is peculiar to it."[108] Böhme, also in attendance, gave his own historical account of the use of movement as a political tool in the "teaching of nationalism."[109]

For Böhme, dance was a particularly potent weapon because of its direct links to the "biological." Adopting Laban's language of "waves" and "emanations," he compared dance favorably with other art forms, noting in 1934 that "the individual has the defense of intellectual criticism and conscious discussion against the spoken word and the character portrayed in a play. This does not apply to the dance, it and its effects are delivered for better or worse. . . . [T]he rhythmic factor opens his soul without him noticing, and then into the opened soul there enters unhindered" the effects of the "measured gesture."[110] Movement acted immediately and unambiguously, and neither spectator nor dancer had the power to resist its surreptitious authority.

Many also saw the movement choir itself as a literal embodiment of a new kind of relationship between the individual and the state. Rather than striking out as soloists, participants in a choir moved collectively or in small groups, guided not by individual creative impulses but by a larger all-encompassing plan, itself disciplined by a written script. As Brandenburg put it, "true" German dancing was once—and would be again—"a communal art, an artistic prototype of that national community which today is trying to evolve." Laban, in his view, had not only created a "German movement script which can become as epoch-making for the dance as notation has become for music"; he had created a new aesthetic form to "express heroism, martyrdom and subordination under one leader." In combining the "age-old urge" to bodily expression with modern techniques in "the newest of the arts," Laban had found the key to forging a new kind of community.[111]

Nor were the new movement choirs confined to Germany's legal borders. In the summer of 1935, for example, Laban was engaged to organize the movement elements for the annual Wagner festival in Sopot (Zoppot) in the Free City of Danzig.[112] A semiautonomous city-state created by the Treaty of Versailles and carved out of East Prussia, the region existed in part to provide Poland with access to a port on the Baltic Sea. Despite Poland's special status, however, the majority of the population considered themselves German and resented their new status; nationalists within Germany referred to Danzig as the "open wound in the east." Beginning in 1930 the Nazis gained significant power in the region, and in the elections of 1933 they achieved a controlling majority. The Wagner festivals had been held since 1909 and had once attracted a relatively diverse audience, but the event, like much of public life in Danzig, acquired an explicitly nationalist character following the Nazi takeover.[113]

Laban had committed to the project early in 1935, but he became ill shortly thereafter and sent his former student Albrecht Knust in his place. Knust's goal was to recreate Laban's vision—three movement choirs of thirty participants each—based on the notation the two men had produced back in Berlin. The groups were composed of young people and segregated by gender, an increasingly common practice. The boys' group hailed from the local chapter of the Hitler Youth. In a series of letters, Knust reported the proceedings back to Laban, noting that there was only "very good news from Zoppot." It was, he wrote, "the first time that I am teaching a *new* group dance according to the kinetogram. And I am very excited about it. All the little considerations—how to create the specific move, how to lead from one group formation into the next—have already been worked out in advance." His local contact, moreover, was "an ideal coworker . . . equally interested in the lay choir and the notation." They worked together on the notation each morning in the open-air bath, and Knust assured Laban that "we need have no worry that Schwartzer will in every way rehearse exactly as we want it."[114] In Sopot, therefore, Kinetographie not only helped to construct the very idea of a uniquely German movement culture, it became part of a concerted political effort to extend that culture's geographic reach.

The most famous of Laban's contributions to the National Socialist cause was a composition for the 1936 Berlin Olympics. Titled *Vom Tauwind und der neuen Freude* (Of the warm wind and the new joy), the piece was a massive display which featured one thousand amateur dancers from forty different towns, set to lyrics from Nietzsche's *Thus Spoke Zarathustra*. It was scheduled to be performed at the dedication ceremonies for the new Dietrich Eckart Theater, a twenty-two-thousand-seat venue constructed by Goebbels as part of the Olympic complex and intended to host the international dance competition that would accompany the games. Plans submitted to Otto von Keudell, sketched the outlines of a performance that would trace the German people's rise from the "horrors of war and the miseries of the postwar period" to the establishment of a unified and victorious national community. Various choirs would emerge as the performance progressed, until "in fresh, natural joy in the festivity, the choir of all swings and finds its way to the German summer wheel [an ancient emblem often bearing a swastika], the symbol of our ancestors for our people's community, of eternal light, to which we all strive."[115] The lay dancers would be accompanied by choral speakers and singers, but movement was to remain at the performance's center. Its animating thesis was that "choric movement is the expressive language of our time," and the simultaneous movement of bodies from across the nation would prove the proposition.[116]

A pamphlet published about the piece, *Wir Tanzen*, took pains to highlight the critical role notation played in the production. In a section titled "Dancers Write Their Own Script," Knust mused that "if one learns that the devotional play 'Of the Warm Wind and the New Joy' was rehearsed in over twenty places at the same time, and that the thousand students only had six days to rehearse in public during the Community Dance Week," one asks oneself involuntarily how it was possible to bring unity into profoundly different rehearsals."[117] The answer—of course—was Kinetographie. Knust declared that this new technology was the only way in which the directors of community dance groups could convey their works "precisely and unequivocally," without concern for clarity or distortion.[118] Notation thus literally knit together the moving bodies of the emergent German *volk*. In doing so, it sought to both liberate and regulate the surging forces the dancing body produced. As another contributor to *Wir Tanzen* explained, "At present in Germany people from all classes are again dancing. They realize that the great events of life leave us silent and let only the body act." These powerful moments and movements, however, could not go unanalyzed, unrecorded, or unmanaged. As the writer continued: "If even the spontaneous gesture can make one freer, stronger, and happier, how much more will the dance awaken a new feeling of life when it uses instead of the spontaneous gesture a well thought-out movement?"[119] Dionysian release and Apollonian rationality would merge.

But despite the reams of Kinetographie scores, the weeks of dispersed rehearsals, and the forty thousand *Reichsmarks* budgeted for the production, *Of the Warm Wind* was never performed. Two days before the dedication ceremony, Goebbels attended the dress rehearsal and left unsatisfied. His diary entry for June 21, 1936, notes: "Dietrich Eckart Theater. Rehearsal of

FIGURE 1.15A-B. Rehearsal photographs of *Vom Tauwind und der neuen Freude* published in *Wir Tanzen*. Short notation excerpts were also included in the pamphlet. In *Wir Tanzen* (Reichsbund für Gemeinschaftstanz in der Reichstheaterkammer, 1936).

dance piece—free adaptation of Nietzsche, badly done and artificial work. I prevent a lot. That is all too intellectual. Goes around in our costume but is not really one of our own."[120] The performance was canceled. A public rationale for the decision was never offered, but the historian Marion Kant has speculated that, regardless of his own tastes, Goebbels suspected the piece would not be to Hitler's liking. Previous attempts to introduce the Führer to the work of other leading members of the German avant-garde had gone poorly; Hitler preferred the light, strong, athletic bodies that would appear in Leni Riefenstahl's *Olympia* to the relatively darker, experimental aesthetic that often characterized *Ausdruckstanz*.[121] Regardless of the reason, the cancellation represented the beginning of the end of Laban's career in Germany.[122]

That same month, Rudolf Cunz replaced Otto von Keudell as Laban's superior at the Reich Culture Chamber. From the start, Cunz proved relatively hostile. He echoed Goebbels's anxieties about Laban's "intellectualism" and, perhaps more damningly, began to point out irregularities in the finances of the institutions under Laban's control. In August of 1936, Laban made several requests regarding personnel, salaries, and organizational structure that were summarily denied. Before the situation came to a head, however, Laban fell ill with intestinal ulcers and left Berlin for a sanatorium. By the end of August, he had written to Cunz requesting a leave of absence from his duties until the end of October; several months later he wrote again, asking to be relieved of his management responsibilities and retained at a minimal salary as a consultant. Cunz was delighted; the problem seemed to have solved itself.

The fight, however, was not truly over. Laban had retained his seat on the board of the Reich League for Community Dance, and in 1937 a contest broke out over its control. Under pressure to demonstrate strong leadership, Cunz suggested the league be reorganized, maintaining its core purpose—the production of "the living expression" of the Aryan people—but stripping from its charter any references to specific theories, systems, or figures. Laban, in essence, was to be symbolically excised from the organization he had created. Laban responded with decisive, albeit foolish, action, arranging a weeklong training course for the league in Bad Homburg without seeking the approval of the ministry. Once Cunz was made aware of the plan, he too sprang into battle. League dancers were immediately told to boycott the training and, moreover, to lodge a formal complaint against Laban. Soon after, Cunz began to take steps to dismantle the league entirely.

Pugnacious as always, Laban refused to give up, mounting an angry defense of his work in the service of the state. In a letter to his assistant Marie-Luise Lieschke he fumed, ". . . if Dr. G is *really* such an enemy of our movement choirs and our piece (*The Warm Wind*), if he or others in a *really* idiotic way scent Oriental Masonic elements in the notation scheme or the harmony-doctrine, then . . . the error must be corrected. *Second*, the man must be informed about the methods of his helpers."[123] Significantly, Laban never turned against Nazi ideology itself. Instead, his complaints were always directed either at individuals or the ministry's bureaucratic structure. He may have railed against whatever "pig" was "plotting against" him, but he always maintained his dedication to the cause. Those scheming against Laban had to be thwarted, in his words, "Not because of me but for the sake of the future of German dance!"[124]

His defense failed. Laban had never been popular with many elements of the regime, including the Fighting League for German Culture, the Hitler Youth, and the German Workers' Front, and complaints against him rapidly mounted. He was accused of financial improprieties and homosexuality, and his previous membership in Masonic organizations came back to haunt him. Cunz renamed the Reich League for Community Dance the "German Dance Community" and placed it under his personal control. Removed from all official posts, Laban's membership in the Reich Theater Chamber was canceled; he was advised to return to his long-abandoned work in the graphic arts. Shortly thereafter he left Germany for Paris, and then Paris for England, where he would spend the remainder of his life.

Endings and Beginnings

Laban's eager cooperation with the Nazis has always sat uncomfortably alongside his venerated status as one of the founders of German modern dance, the man who not only helped reshape the form but gave it its own written language. He remains even today a towering figure, celebrated by legions of acolytes who often downplay or simply ignore this period in his career. Most existing biographical treatments of Laban similarly gloss over his role in the National Socialist state, painting his initial decision to collaborate as a regrettable but understandable capitulation to forces beyond his control and citing his eventual expulsion from Germany as evidence that his own views and practices remained fundamentally at odds with Nazi ideology. (In one widely circulating apocryphal story, Goebbels publicly rebukes Laban for his supposed commitment to individual forms of expression, barking that "in Germany there is room for only one movement, the Nazi movement!")[125] These claims have been bolstered by histories of German art that see 1933 as a decisive moment of rupture, characterized by the state's wholesale rejection of the "degenerate" avant-garde. But while this narrative holds true in some cases, scholars like Susan Manning and Marion Kant have demonstrated that it is largely inaccurate for dance. They note, for example, that many of the leading figures of Weimar *Ausdruckstanz* found homes within the Nazi state apparatus and that the movement itself survived, relatively unaltered, into the postwar period.[126]

Laban himself never publicly expressed regret for his work with the Nazis. In fact, he rarely spoke about it at all, save for occasional remarks that, as an artist, he saw himself as apolitical. In oral histories conducted decades later, Laban's friends and colleagues recalled that he seemed to maintain a naive faith that his work could be separated from the context of its development and use. Even as the regime's violence and antisemitism became clearer and clearer, Laban remained fundamentally unmoved. In 1934, he attended at least one private dinner with Hitler, Goebbels, Goering, and their wives, then turned up afterward, near midnight, on the doorstep of Felicia Sachs, a Jewish friend and former artistic collaborator. Sachs recounted that Laban was "overwhelmed" by what he had heard: gleeful stories of the confiscation of the property of Jewish Germans, of torture and murder. He was worried too about his own safety and status, particularly in the wake of the leadership purge of the Night of the Long Knives. Still, Sachs remembered that Laban could not seem to shake his enchantment with the "charm of the inner circle," noting the "lovely family feeling" and remarking on the Führer's floral arrangements. He saw his invitation into "this closest family" as a "great honor," a sign that his artistic agenda was truly appreciated.[127]

It is impossible to determine definitively the exact proportions of opportunism, conviction, and self-preservation that fueled Laban's work with the Nazis. We do know, however, that his ideas about movement had long been racialized and that, at the very least, he was quite willing to ignore the peril of even those close to him for his own gain. The regime's embrace of mass movement spectacles as political weapons also neatly aligned with Laban's aesthetic proclivities. In his 1935 autobiography, Laban wrote fondly of the lasting influence of the military troop movements he witnessed as a child, writing that they embodied "a united front which I assumed could only come about through steady comradeship and loyalty: loyalty of one to another, and of the individual to the whole, to the all-embracing unity, to the fatherland." "Only art," he continued, "matched up to this ideal."[128] And while this passage's particularly emphatic invocation of "the fatherland" might be read as a function of the moment of its publication, it speaks nonetheless to Laban's long-standing interest in how movement—appropriately managed—might be used to form new kinds of modern communities.

Still, when evaluating this history, the question of Laban's personal political sympathies is not the most interesting one. Instead, it is far more productive to focus on how Kinetographie itself was a natural fit for the National Socialist program. The historian Jeffrey Herf has detailed how the Nazis embraced an ideology of "reactionary modernism," embracing simultaneously a future-oriented program of technological development and a *volkish* authoritarian nationalism rooted in an imagined past.[129] This fusion is precisely the vision Kinetographie sought to create: a community rooted in expressive, romantic movement that was simultaneously constrained by a novel paper technology of centralized control. In one essay, the critic Rudolf Bach sought to explain the movement choir's appeal, observing that "the German of today is striving with an inner longing because of deep dissatisfaction with the mechanistics of the age," yearning for a communal, cultic "celebration of the creative, divine powers of existence." This "deep passionate demand," he continued, "is met by hardly another artistic form of expression so purely and strongly as it is by the choral dance."[130] Bach neglected to mention, however, that this cultic power would not be loosed indiscriminately but rather, as another critic put it, crafted "with the same expertise and precision with which the engineer builds a machine."[131]

At first, the subtlety of this synthesis may have been part of Kinetographie's appeal. Laban's dancers, clad in loose tunics, moving with graceful ease on mountains, beaches, and fields, *looked* free; the silent paper tool that shaped their gestures faded into the background. In fact, as Laban himself lamented, it was always a struggle to convince audiences to "see the impersonal dance work," distracted as they continued to be by the "dancer and his personal peculiarities."[132] Though the reality was more complex, Laban's choreography seemed like a repudiation of both the aristocratic formalities of ballet and the hollow regimentation of the Tiller Girls.[133] It promised, instead, an art of dance that was more individual, more human.

Under the Nazis, however, what had once been a productive tension between surface and structure became a liability. Goebbels and others were preoccupied with *Of the Warm Wind*'s formal qualities—its sometimes feminine gestural vocabulary, its mystical referents, its abstraction and stylistic links to other avant-garde movements—a focus that seems to have blinded them to the ideological potential of the technology that underlaid its creation.[134] In fact, however, Kinetographie was an impeccable articulation of the Nazi promise to provide a new model of the relationship between control and release, modernity and tradition. The body itself could retain its role as a signifier of the natural, the irrational, and the emotional, while being simultaneously harnessed for the creation of an efficient and modern totalitarian state. Exploiting the presumed relationships between physiological response, movement, and mind, Kinetographie would allow those who wielded it to shape both the bodies of their subjects and their innermost selves.

But if Nazi officials were incapable of seeing Kinetographie's promise—or simply sufficiently frustrated with Laban himself to throw the baby out with the bathwater—the technology quickly found eager adopters elsewhere. By 1937 the system had been defined and the synthesis achieved; the question that remained was what to what ends Kinetographie would be put in Laban's new life in England.

2

The Lilt in Labour

In 1936, United Artists released Charlie Chaplin's *Modern Times* to great acclaim. A satire of the conditions of the industrialized world and an international hit, the film begins with Chaplin's character the Little Tramp at work on an assembly line, wrenches in both hands, tightening a never-ending stream of nuts. The factory in which he works is a machine designed for the extraction of maximum efficiency: its lines run at increasingly inhuman speeds (the Tramp falls behind after a moment's pause to scratch an armpit or swat an insect), surveillance is ever present (a supervisor appears on a giant two-way video screen in the bathroom when he tries to sneak a cigarette), and attempts are made to mechanize even the workers' lunch breaks (resulting in a disastrously comedic sequence in which the Tramp is subjected to a "feeding machine"). In his autobiography, Chaplin wrote that he conceived the film after a conversation with a reporter about Detroit's "factory-belt system," a "harrowing story of big industry luring healthy young men off the farms who, after four or five years at the belt system, became nervous wrecks."[1] Like the Detroit auto workers, the Tramp too eventually suffers a breakdown, running amok through the factory, spritzing his fellow workers with oil, and taking his wrench to anything remotely resembling a nut. At one point, in the film's most iconic image, he is sucked into the factory's gears, a literal cog in the machine.

But as he runs, he also dances. Skipping jauntily with an oil can, he performs an *échappé battu*; pushing down a giant lever, he pauses briefly, grinning, in an *attitude derrière*. In contrast to the twitching "Forditis" he exhibits earlier in the film, Chaplin's vigorous, whole-bodied, dancerly movement functions as an embodied repudiation of the factory's systems of control and rigidity. It is also, quite literally, a form of industrial sabotage, disrupting the assembly line's orderly functioning.[2]

The association of dance with nonconformity, self-expression, and resistance has, as in *Modern Times*, often made it a useful foil for the regimentation of industrial production. There is also a real history of dance being linked to radical labor movements, particularly in 1930s New York.[3] In World War II Britain, however, dance—via a reformulated version of Kinetographie—was brought inside the factory not to disturb its functioning but to facilitate it. This set of practices, called "Industrial Rhythm," was built on the premise that properly directed working movements could simultaneously produce maximal efficiency and a deeper sense of transcendent fulfillment.

The fruit of a collaboration between Rudolf Laban and the British cost and works accountant F. C. Lawrence, Industrial Rhythm appeared to many—at least at first glance—to be yet another iteration of the kind of scientific management associated with Frederick Winslow Taylor, Frank and Lillian Gilbreth, and Charles Bedaux. The particulars of such systems varied, but all promised to increase output and reduce costs through a rationalization of the work process. Laban and Lawrence, however, believed they were doing something quite different. While they too sought new kinds of control over production, their larger stated aim was a complete reconfiguration of the relationship between workers and their work, an end to drudgery and a flowering of new life within factory walls. The project was, like many others of its era, an attempt to respond to increasing worries about the costs—individual, cultural, and political—of industrial capitalism. It was also, for the better part of a decade, a relative success. Among others, Laban and Lawrence's clients included the Manchester Dock Workers, J. Lyons Tea, Pilkington Tile, the Royal Air Force, and the factory from which emanated the British Army's mammoth supply of Mars Bars. In 1947 they published a book, *Effort*, that saw at least two printings. They spoke repeatedly at meetings of the Manchester Association of Engineers, and one of their top associates toured almost every chapter of the American Institute of Industrial Engineers, drawing crowds of more than a hundred listeners in New York, Columbus, Louisville, Fort Worth, Dayton, and San Francisco.

One might be surprised to find Laban, whose most recent resume entry was "Nazi dance minister," at the heart of such discussions. This chapter is in part the story of this seemingly unlikely transition, as notation went from a tool for fortifying German national identity to a mediator of British working life. Upon closer examination, however, threads emerge tying these disparate moments together. Just as Kinetographie was deployed in Germany to reconcile expression and regimentation in service of the Nazi state, it captured the British imagination because of its utopian promise to salve the wounds of the "age of industrial man" by turning factory work into a kind of premodern festive ritual. Movement, in this vision, would function as the bridge between the body and the soul, craft and industry, the past and the future. Ultimately, the factory would be sacralized and its workers—and the society in which they lived—redeemed.

This imagined transformation, of course, had its own politics. As the chapter will demonstrate, Laban, Lawrence, and many of their clients also hoped that their efforts could safeguard industry's power as it faced increasingly widespread, confrontational, and effective challenges from the forces of labor. Their solution to modernity's woes would not be found in the structural transformation of the economy but in the remaking of the individual human body.

From Nazi Germany to English Boarding School

Laban's initial invitation to England was largely the result of the efforts of Kurt Jooss. One of Laban's former students, Jooss danced with Laban from 1920 until 1924 before founding his own company, Die Neue Tanzbühne. In subsequent years, Jooss developed a significant reputation within the *Ausdruckstanz* movement and garnered particular praise for his 1932 ballet *The Green Table*, a dark satire on war and its aftermath. Unlike Laban, however, Jooss refused to fire the Jewish members of his staff—including his primary collaborator, the composer Fritz Cohen—after Hitler's ascent to power. Instead, he left the country in early 1933, crossing the border with his company in tow just eighteen hours before police arrived at his house. Now without a home base, the company, by then called Ballet Jooss, continued on their planned tour of Europe and the United States.

In attendance at one of the Ballet Jooss's London performances were Dorothy Payne Whitney and Leonard Elmhirst. Elmhirst was an English agricultural economist and Whitney an American heiress and philanthropist, known for—among other activities—the key financing she provided for the creation of *The New Republic* and the New School for Social Research. Brought together by a common interest in progressive causes, the two married in 1924 and almost immediately hatched a plan to create a new kind of entity dedicated to the renewal of rural England. In 1925 they settled on a location: a fourteenth-century estate in Devon featuring eight hundred acres of monumental stone structures, gardens, and sprawling fields. First built in the late fourteenth century for John Holland, Duke of Exeter and half brother of Richard II, the complex—known as Dartington Hall—had long since fallen into disrepair. When Elmhirst came to assess the property in 1925, he found the estate inhabited primarily by rats, along with a tenant farmer who lived in the main hall's courtyard and sowed his crops in the remnants of its roofless buildings.

It was here, however, that Elmhirst and Whitney planned to establish their institution, intended to revolutionize both education and rural life. In what they came to call the "Dartington Experiment," a progressive school, arts instruction and performance, productive farmland, and light industry would be knit together in a single entity. This union, they believed, would provoke a wholesale economic, cultural, and spiritual regeneration of the Hall's surrounding area—devastated by an agricultural depression that began in the 1870s—and eventually serve as a prototype to be emulated elsewhere, providing a new kind of remedy for modernity's upheavals.

Dartington Hall itself was modeled on a community in Santiniketan, Bengal, where Elmhirst had spent several formative years. Founded by the Nobel Prize-winning poet and social reformer Rabindranath Tagore, the Santiniketan complex included an ashram, school, library, and gardens as well as, eventually, the adjacent "Institute for Rural Reconstruction." For Tagore, the project was intended to serve as a practical mechanism for sparking new life in the rural area, but also as a demonstration that it was possible to embrace modern technologies of production without succumbing to the ills that had plagued industrializing Europe or replicating colonial structures of power.

Though Elmhirst had studied history and theology as a Cambridge student, intending to pursue career in the church, a stint with the YMCA in India during World War I led to a change in direction. Upon his return from service Elmhirst left England for Ithaca, New York, where he pursued a degree in agriculture at Cornell University and came into contact with Tagore during one of the poet's tours of the United States. Inspired by Tagore's ideas about education and development, Elmhirst eventually followed him back to India, serving as both Tagore's personal secretary and, between 1921 and 1924, the director of the Institute for Rural Reconstruction (later renamed Sriniketan). Whitney—who had inherited the equivalent of four hundred million dollars upon the death of her father, William Collins Whitney, in 1904 and met Elmhirst as the result of his fundraising activities for the Cornell Cosmopolitan Club—helped fund the venture. Tagore's influence thus shaped Elmhirst and Whitney deeply, and the three would remain in close contact until the end of Tagore's life.[4]

Indeed, as at Santiniketan, Whitney and Elmhirst's vision was for Dartington Hall to be not just an efficient site of production but a community dedicated to the sustenance of life in the broadest sense. As a prospectus noted in 1925, "The gap that has existed in education has created a gap in our national life between the artist and the factory, between the man of science and the humanist, and not least between the life of the town and the country."[5] Dartington would remedy that disjuncture by fully integrating agricultural and industrial work with a liberal education and a reverence for the arts. Children would be educated in classrooms, but also in in the Hall's chicken coops and timber yards, where their instructors would be drawn from the estate's existing employees. Those employees, in turn, would join in Dartington's intellectual and creative life, attending lectures and participating in theatrical performances and classes in the fine arts. Artists in residence would serve as both teachers and as ambassadors to the surrounding area, drawing in the sometimes-skeptical neighbors with a wealth of new cultural opportunities.[6] Within a decade, the Hall acquired a following among elite British progressives: Aldous Huxley sent a child there, as did Ernst Freud and Bertrand Russell, the American documentary filmmaker Robert Flaherty, and the scientist J. D. Bernal.

By the time of Jooss's London performance in July of 1933, Dartington already had an established arts program, but Whitney, ever eager to expand, was taken both by the company's immediate practical plight and by *The Green Table*'s message. Whitney and Elmhirst offered all fifty members of the company a new home; they set up shop in 1934 as Dartington's first official resident performance group.[7]

In 1937, Laban at last decamped Germany for Paris. Once there, he attended the occasional event—including the Second International Congress of Aesthetics and the Science of Art—but kept largely to himself. Often ill, he spent most of his days in a small, run-down apartment with his former collaborator and lover Dussia Bereska, who passed the majority of her days in an alcohol-induced stupor. Jooss paid Laban a visit in late 1937 and was dismayed to find his former mentor in such an abject state. He convinced Laban to join him at Dartington, obtaining both the consent of the Elmhirsts and an official temporary residency permit. Laban lived initially in the Jooss family home; later he was given his own small apartment and studio in Dartington's Barton House.

Laban took time to adjust to his new surroundings, particularly as his health at first remained poor and his English weak. He devoted hours each day to writing and rewriting his theories in what was sometimes known as his alchemist's studio, where papers piled up like snowdrifts and half-completed models of icosahedra dangled from the ceiling and perched on every available surface.[8] Thanks in part to Jooss, however, Laban's ideas had long been part of Dartington's daily life. Notation was integrated into the standard coursework for dance students, as were twice-weekly classes in "Choreutics" and "Eukinetics."[9] Though Laban did little teaching himself, students revered him from a distance, and by 1939, his English significantly improved, he delivered a series of sweeping lectures on the history of dance.

Moreover, though Laban's initial decision to make a home at Dartington was born more of expedience than of philosophical or political affinity, once there Laban found himself increasingly captivated by Dartington's mission. He compared Dartington to the life reform colony he had attempted to establish in Ascona, Switzerland, in the years just before World War I, recalling its interconnected roster of priorities: "Sometimes arts, sometimes health, sometimes the production of alimentation, clothing, and buildings, sometimes came waves of spiritual researches."[10] Laban was also taken by the Elmhirsts' efforts to create a true community sensibility, and to that end proposed to establish local movement choirs, designed to produce both fellow-feeling and "psycho-physiological harmony."[11]

For a former member of Hitler's government, Laban initially had shockingly little trouble with the immigration authorities, or with his fellow Dartington denizens. Indeed, Laban was eager to make reference to his work in Germany, citing it as a precedent and even directly referencing the booklet on movement choirs he produced for the 1936 Olympics.[12] Jooss's high praise for Laban undoubtedly helped smooth the transition, as did Elmhirst and Whitney's extensive network of political contacts.[13] In a number of press reports, Laban was incorrectly referred to as Czech, rather than German or Hungarian; in others, he was portrayed as yet another refugee, driven from Germany because of his commitment to free expression.[14] When England entered the war in 1939, however, his legal status became somewhat more precarious. Along with all other foreign "aliens," he was required to relocate away from the southern coast, eventually moving with Lisa Ullmann, one of Jooss's dancers, to a cottage in Wales. He continued receiving a small allowance from the Dartington trustees but was warned that the fund would soon run dry. In letters, C. C. Martin, the head of Dartington's Arts Department, urged Laban to seek other employment to ensure both his financial stability and his place in the country.[15]

Ultimately, it was Martin who ultimately made the connection that would structure the next chapter of Laban's—and, more importantly, Kinetographie's—life. An accountant named F. C. Lawrence had been a member of the Dartington Hall Board of Trustees since 1930, and, in 1941, Martin suggested that he and Laban meet. In a letter to Laban in August of that year, Martin described Lawrence as a "prominent business man in the North of England who is interested in a study of the movements which working men have to make to do their job." "Businessmen in this country," Martin continued, "are coming to the conclusion that there is a right and a wrong way of performing all such movements, and it has been suggested to me that the script might make a very useful basis for the recording of such movement and its study. It might then be possible when such a record has been compiled, to suggest better and more economical movements for performing the same function." Martin noted that Laban would hear from Lawrence soon and counseled him to keep an open mind. While factory management might at first appear a rather mundane fate for his groundbreaking artistic tool, Martin shared his own belief that the project had revolutionary potential: "If the script could be made available to industrialists at the present time it would be performing a national and even world-wide service of a kind which you never envisaged for it."[16]

Laban, as it turned out, did not need much convincing. He assured Martin by letter that "the idea to use my script for the notation of industrial operations interests me tremendously, especially as I am not unacquainted with the workers need of movement-regulation though rather more from the recreative point of view."[17] Confident that his work for the 1929 Vienna festival would provide an adequate foundation, Laban prepared for the new endeavor. By mid-September, he had already met once with Lawrence's assistant and further steps toward collaboration, including a meeting with Lawrence himself and a factory site visit, were on the calendar.

Motion Studies

At the time he reached out to Laban, F. C. Lawrence was a practitioner of a relatively new British occupation, that of the "cost and works accountant." This field took shape in the United Kingdom in the years surrounding World War I; its members defined themselves by their focus on recording, analyzing, and controlling the costs of production—not, as had been done in the past, across an enterprise as a whole, but at the level of a particular division, process, or product. As one such accountant explained in 1920, "each minute of each worker's time that is paid for, each pennyworth of material *used* (not purchased), and each penny of other expense incurred, must be so recorded that it will be possible to allocate them to the work in connection with which they are incurred."[18] The aim, in essence, was not merely to document, but to manage, maximizing profits by strategically shifting resources across an enterprise.[19]

Cost and works accountants were further distinguished from existing professional chartered accountants by their training. Most cost and works practitioners had little formal instruction in accounting but began their careers as engineers, rooting their authority in their hands-on knowledge of the nitty-gritty practicalities of production. Lawrence himself followed this trajectory, receiving a bachelor's degree in engineering and practicing as an electrical engineer before gradually transitioning into accounting and management. In the early 1930s he established a successful independent partnership, Paton Lawrence & Co., providing "consulting in costing and factory organization." By 1936 he was sufficiently well known to snag a slot as one of five headliners at the Business Efficiency Exhibition in Birmingham, a nine-day event featuring five hundred experts and drawing more than thirty thousand total attendees.[20]

As Lawrence's presence at the Birmingham exhibition suggests, the rise of cost accounting was also intertwined with a growing national interest in scientific management. While there has been some scholarly debate about the precise extent to which scientific management in general—and motion study in particular—penetrated industrial practice in the United Kingdom, the average factory owner or worker certainly would have been familiar with the concept and perhaps with its execution as well.[21] Whether via Taylor's stopwatch, the Gilbreths' elaborately charted "therbligs," films, and cyclegraphs, or Bedaux's B units, the object of such studies was to subject each and every movement a worker made to close inspection, analysis, and reform, all in the service of increasing production.[22] Some praised the ingenuity and "scientific spirit" of such interventions, but others noted that work managed in this way became so controlled, so parceled out, so divorced from the exercise of judgment that it could not help but be alienating. Moreover, though exponents of scientific management almost uniformly paid lip service to fatigue reduction and industrial welfare, critics contended that their true aim was profit at whatever cost. Use of the Bedaux system, which was particularly widespread in Britain, precipitated at least thirty-six strikes between 1929 and 1939, and the Bedaux central office eventually recommended that its consultants avoid using the term "Bedaux" in the field, so poisoned had its association with increased speed, reduced pay, and diminished autonomy become.[23]

Lawrence's consulting firm employed various forms of motion study, but Lawrence himself had long been unsatisfied with the practice. Some of his complaints were pragmatic—his own attempts at recording motions had resulted in the "utmost difficulty in making useful records, either by description or by sketches."[24] His other objections, however, were more foundational. "I don't care for it," he said later. "It doesn't get at the thing, but people go round with a stop watch and bit of paper, instead of looking at the person."[25] To Lawrence, scientific management seemed divorced from the realm of the human and both practically and morally troubling. His hope was that Laban's involvement might begin to resolve both these concerns, his notation providing a workable recording method and his artistic and philosophical commitments producing a less degrading mode of control.

Kinetographie itself, moreover, was fairly easy adapted to novel ends. Ultimately, Laban and Lawrence arrived at a system in which the original methods for indicating direction, level, and length of movement were retained, but the staff itself was compressed into just two parallel lines. On and around the first line were written the movements of the body, while the second line held symbols representing the materials, parts, or tools a worker would employ.[26] As Laban and Lawrence explained in a 1942 booklet aimed at potential clients, this new "Industrial Notation" was intended to capture the "full bodily activity of an operator" at all stages of the working process, showing not only the position of each working limb but the "exact duration" of both a task as a whole and its subcomponents.[27] The advantage of this system over films or verbal descriptions, they argued, was clear. Industrial notation was flattening and economical, recording only the most salient movement characteristics to produce a single "static image which [could] then be easily studied" and compared with others.[28] An evaluator could, for example, use the notation to devise and record an ideal working procedure or to determine "whether the time spent for the whole operation or for single parts of it is justified."[29] He could also observe and notate what workers were doing in practice, assessing them both in relation to the company's standard and one another.

But while Kinetographie was the scaffolding upon which Industrial Notation was built, it was not the system's only element. In addition to recording

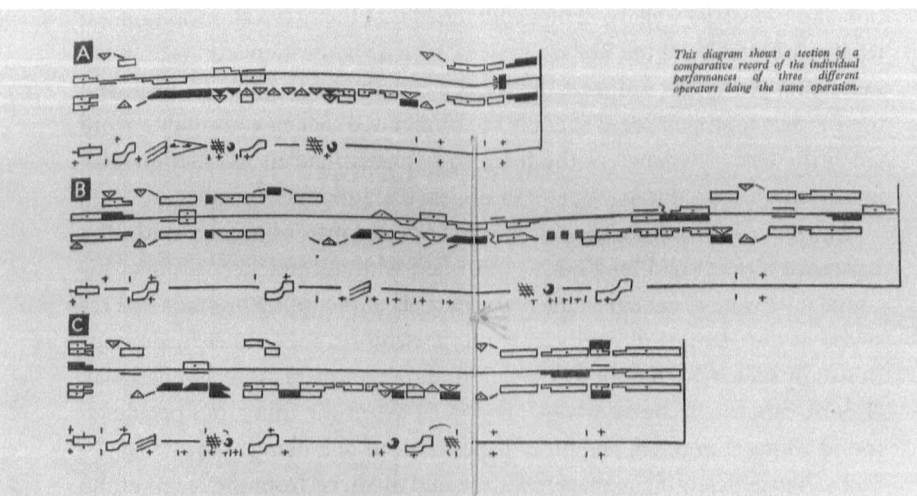

FIGURE 2.1. Comparative notation for three operators. From "Laban Lawrence Industrial Rhythm and Lilt in Labour," 1942, 8–9. Dartington Hall Trust Records, T/AD/3/D/6.

working movements themselves, Laban and Lawrence were interested in capturing *how* they were performed. To accomplish this task, they developed a second type of notation, the "Effort Graph," which conveyed information about four different movement qualities: "weight," "space," "flow," and "time." Each of these qualities, Laban posited, existed along a continuum. Weight could be light or strong; flow could be free or bound; the use of space could be direct or indirect; and time could be sustained or sudden.[30]

Early Effort Graphs rendered the status of these four qualities on four superimposed axes. Time and flow appeared as parallel, congruent horizontal lines, bisected perpendicularly by a third congruent line representing weight (or "mass"). Space was represented by two perpendicular half-lines joined at a point and placed in upper right-hand corner. Four empty boxes placed on the axes indicated which quality a given movement expressed: a box on the far left side of the time axis, for example, would signal to the reader that the movement was very "sustained," while a box on the mid-right side would indicate it was moderately "sudden." Later Effort Graphs, however, abandoned this attempt to capture gradations, allowing only binary choices. Dispensing with the movable boxes, these graphs included only the half of each line that best described the trait (that is, a movement could be either "sustained" or "sudden" but not in between), producing a set of symbols resembling half-finished hash marks.

Effort Graphs were used in several ways. First, they recorded the movement qualities believed to be necessary for a given job or task; these were known as "Job Effort Graphs." A drilling operation, for instance, might require the use of strength (weight) and speed (time), while the assembly of an intricate mechanism would necessitate a light touch (weight) and a controlled, bound use of movement (flow). Second, they were deployed to capture employees' "natural" movement proclivities: the way they preferred to move if operating without constraints. These were known as "Personal Effort Graphs." Finally, Effort Graphs were sometimes placed alongside Kinetographie scripts, providing more detailed information about the movement qualities prescribed for each subset of a given task.

The purpose of the Effort Graphs was, much like the simplified Kinetographie notation, essentially comparative. In assigning workers to jobs, Laban and Lawrence suggested that the overlap—or lack thereof—between the "Job Effort Graph" and the worker's "Personal Effort Graph" be considered. Ideally, workers would be matched with tasks that suited their individual movement preferences. In cases in which this was impossible, Laban and Lawrence would provide training exercises designed to "change or to develop the personal habits and inclinations of movement until they conform with the required action."[31]

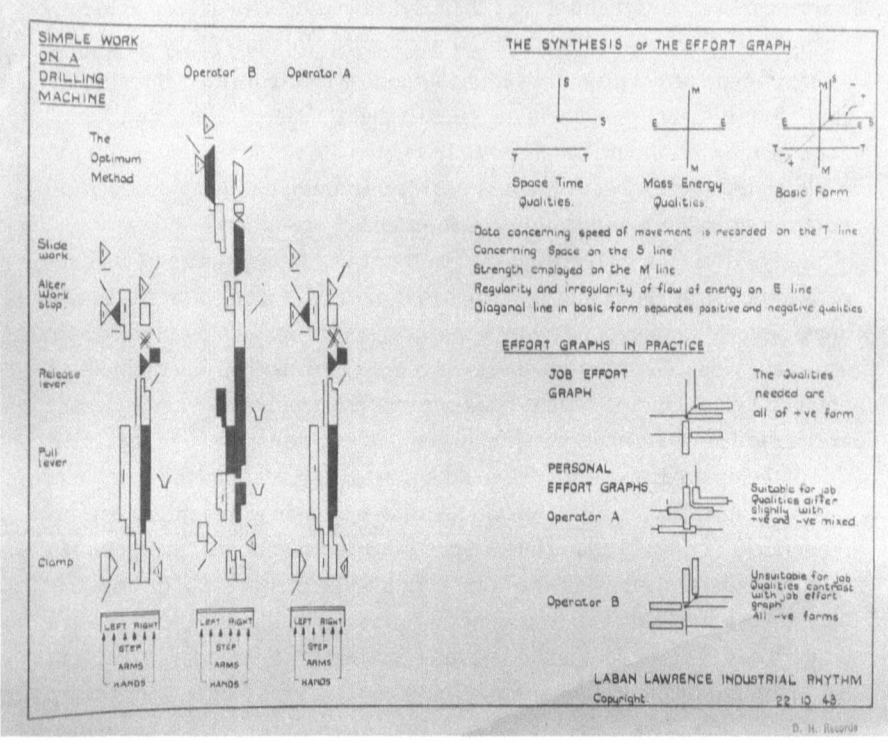

FIGURE 2.2. "Simple Work on a Drilling Machine." Dartington Hall Trust Records, T/AD/3/D/7.

Within these tools in hand, Laban and Lawrence—with the assistance of Lisa Ullmann and Lawrence's niece, a mathematician—devised a basic procedure for industrial consulting. During the team's initial visit to a factory, Laban, Ullmann, or eventually one of their assistants would spend several hours in observation: asking questions, watching workers' existing movements, and recording them in both forms of notation. Back in their offices or studios, they would devise new procedures, record them, and then return to the site multiple times over the course of several weeks, arms heavy with sheaves of paper. These notation scores, of course, could not be simply passed out to the factory's workers, none of whom were—at least initially—capable of reading the script. Instead, members of the Industrial Rhythm team would demonstrate the new movement sequences and qualities, which employees were expected to mimic. The team's second major task was teaching regular movement courses to small groups of employees. To get the best results, Laban and Lawrence recommended that the courses occur a few times a week and last between thirty and sixty minutes. In general, they consisted of sets of exercises designed to increase workers' awareness of their bodies and inculcate new ways of moving useful for the tasks to which they were assigned.

An ancillary purpose of the courses was to identify existing employees with a natural facility for movement observation; these workers would be offered additional training in the rudiments of Industrial Notation and put in charge of continuing the program after Laban and Lawrence had departed. The making of Effort Graphs was thought to be particularly easy to pick up, something foremen "of average intelligence" could be trained to do relatively quickly, allowing them to deploy "this means of control after a few days exercise."[32] In practice, the process was not quite so smooth. Though the Industrial Rhythm team did locate a number of individuals willing to undergo the training, log books indicate that it took far longer than a few days to achieve the most basic grasp of Industrial Notation. Even those capable of understanding the system's nuances often lacked sufficient time to practice their new skills.[33] One trainee, a Miss Bartlett at Dartington Hall, was also the factory's welfare officer and was already strained to her breaking point by managing everything from paperwork to the acquisition of clothing coupons to outbreaks of dysentery.[34] Though she devoted as much time as she could to notation study, it was never her primary priority and progress was slow.

Despite the challenges, Laban and Lawrence persisted with the training programs, as they needed internal experts who could ensure that the movement standards Industrial Rhythm established were upheld over time.[35] It would be their job to regularly observe and notate employees in action, reporting on charts and graphs whether their movements were performed both in the sequence indicated and with the prescribed movement qualities.[36] One aim of this continual monitoring was to identify workers who were struggling and provide opportunities for intervention. A worker in the Dartington sawmills, for example, had originally been evaluated as diligent and capable, but later Effort Graph assessments indicated that her work had begun to flag. In response, the Industrial Rhythm team suggested compensatory physical exercises and a smaller, more appropriate, work chair. They believed this intervention would quickly remedy the issues; as an internal report summarized, "As Mrs. Johnson was already assessed last year as a fairly skilled worker, her present decrease of efficiency stated in the graphs must have some removable cause."[37]

In other cases, however, Industrial Notation would be deployed more punitively: to decrease the wages of workers who either refused or were incapable of following the program. Though Laban and Lawrence never mandated that the companies they worked with adopt a particular wage structure, they suggested frequently that notation could be used to "rationally" adjust pay in line with the "effort" a particular worker was putting forth, particularly in situations in which total output was affected by external factors—as on an assembly line in which any individual's piece rate was limited by the flow of materials and the pace of their coworkers. Supervisors were sometimes asked, for example, to notate an employee's actions over the course of several consecutive days and record on a chart whether their movements were correctly performed. At the end of the pay period, workers would receive a substantial "bonus" if the notation showed they were in compliance with the Industrial Rhythm program, while those who failed to measure up might earn wages as much as 30 percent lower.[38]

In many ways, therefore, Industrial Rhythm did not look terribly different from existing forms of motion study or scientific management.[39] Laban and Lawrence too argued for individualized pay structures, enacted precise control over working methods, valued efficiency, and spoke of "defects" to "be remedied" and "perfect performances" to be "perpetuated."[40] Given the increasing resistance to such methods, both men acknowledged this association as a challenge: "We have felt some slight resistance to our work," Laban remarked, "because of working in the shadow of Bedaux."[41] The Industrial Rhythm team, however, consistently stressed that their primary objectives were different. Focusing only on an unending increase in output, as other forms of scientific management did, they maintained, would ultimately be self-defeating. Long hours and high speeds led inevitably to wastage, nervous exhaustion, "and a greater loss of man-hours, then a restriction at optimum-output." As an internal memo sketching out the planned work at one company noted, "To offer them 100% maximum output, or any increase of output at all" would be "like promising a patient an increase of fever instead of a balance of his well being."[42] Their goal instead was to change workers' experience of their own labor.

What this meant in practice was that the movements Laban prompted workers to use were designed not to be merely efficient, but pleasurable. With proper control, he theorized, hours of stamping metal or packing boxes could transform from drudgery into a kind of dance, enhancing life rather than sapping it. As Lawrence summarized: "There are rules of rhythm just as there are rules for everything else worth while in life, and these can be codified by Laban in a practical form so that the industrial controller of effort can apply them and thereby give beauty even to the most monotonous and despised work in a factory."[43]

One can gain insight into Laban's aesthetic vision through descriptions of the movement courses the Industrial Rhythm team offered. While the instructors varied their content in line with the specific work the participants did, there were clear commonalities across courses. Nearly all included exercises prompting workers to alternate moments of physical tension and relaxation as well as the contraction and expansion of their bodies. The coordination of breath and movement was another constant refrain, as were instructions to "swing" or "shake" the arms, legs, and feet. Though, other than an occasional drum, the classes were rarely accompanied by music, attention to rhythm was key to the experience. In the vast majority of cases, participants were asked to move in syncopated or "lilting" patterns rather than to regular, metronomic beats. Laban's "movement scales," patterned sequences in which participants reached toward predetermined points on their personal "kinesphere" (often envisioned as an icosahedron, but sometimes also an octahedron or cube), formed a final part of the activities.[44]

In short, both the visual and the kinesthetic experience of the class stood in marked contrast to popular understandings of industrial movement. Instead of the unvarying clanging of metal machine parts, there were the subtle cadences of modern dance. Instead of the constrained and repetitive gestures of assembly-line workers, there were full-bodied motions interrupted by moments of complete relaxation. This did not, however, mean that participants' movements were entirely free: there were undoubtedly still *wrong* ways of moving, whether at work or at rest. Instructors' notebooks, for example, commented on exercises that were "not well done" and logged dissatisfaction with individuals who could not shake their "stiffness" or moved in a "cramped and heavy" manner. "We live in a time of speed and machinery," Laban wrote, "and are bound to accommodate our motion to the various and differentiated rhythms in which our present life becomes more and more involved."[45] For him, though, this did not require adopting mechanical motion or its "frenzy of hyperactivity."[46] This was not the technological sublime of the ceaseless chugging of factory machinery, but something new and different.[47]

Ironically, given Laban's own racial prejudices, his vision for Industrial Rhythm had much in common with the new aesthetic forms developed by Black dancers and musicians in the interwar years. As Joel Dinerstein has chronicled, genres like jazz, tap, and swing were similarly designed as both an embrace and a critique of the new "Machine Age," taking on its speed and dynamism but imbuing it with looseness, flow, rhythmic play, and an individualistic, improvisatory character.[48] Industrial Rhythm sought to do the same, but with an important caveat: the movements performed would be determined by Laban, Lawrence, and their cadre of notators, rather than the individual mover, guaranteeing that that the ensuing performance was both productive and never *too* wild or unrestrained.[49] The somewhat paradoxical nature of this approach is perhaps best distilled in the words of Industrial Rhythm's slogan: "To Bring That Swing and Lilt in Labour, Which Makes Efficiency a Pleasure."[50]

At the very least, as the *Birmingham Mail* reported in August of 1942, Laban's work promised to put a new spin on an old tradition: "Mr. Laban, while preaching the same doctrine [of motion study], puts an altogether more romantic colouring on it, and demonstrates in a prosaic manual process an essay in lyricism. In this, his Czecho-Slovakian flair for song and dance gives him an élan which the more sober and staid Midland factory workers will almost certainly lack."[51] As Laban once more became Czech, motion study metamorphosed into a dance.

Wartime Experiment

Within about a year of Laban and Lawrence's first meeting, Industrial Rhythm was off to a surprisingly strong start: initial clients included Tyresoles, J. Lyons Tea, St. Olave's Curing and Preserving Company, Mars Confections, Portland Cement, and Dartington Hall itself, where the team spearheaded efforts to manage the estate's wartime production of both agricultural and manufactured goods. These contracts were facilitated in part by Lawrence's deep connections in the British business world but also by the unique labor situation facing wartime Britain. With millions of men removed from their industrial jobs and sent to the front, labor shortages in key industries quickly emerged. In response, the government passed the National Service Act of 1941, which provided for the near-universal conscription of women (and men of nonmilitary age) into war work.[52] Ultimately, more than 2.5 million workers were added to the labor force, 2.2 million of them women.

A massive increase in output, however, did not immediately follow. In late 1941, Minister of Labour Ernest Bevin complained that production still lagged by 30 to 40 percent, and industrial capacity remained a central point of anxiety throughout the early 1940s. There were manifold reasons for this shortfall, ranging from supply chain issues to machine shortages and the expected bureaucratic headaches of creating a war industry overnight. Moreover, as the *New Statesman* reported in March 1942, many employers simply did not know what to do with an influx of inexperienced women, showing "no willingness" to provide training but "complaining that women's labour is no use."[53] Low worker morale also represented a serious ongoing challenge.

Many conscripts, it turned out, were not as eager to participate in Britain's fight as the government had hoped.

An illustrative example of this tension can be found in the story of Peggy, a twenty-year-old factory worker who was conscripted in 1941.[54] Following her registration, Peggy was assigned work in the machine shop of a hastily constructed factory in a previously obscure country town. The hours were long—twelve-hour days, five to six days a week—and the work dull, with hour after hour of repetitive drilling. Peggy did everything in her power to avoid it, whether sneaking out early with friends, purposefully missing the morning bus after a late night on the town, or operating her machine in a "slapdash manner," resulting in numerous broken drills. Wartime regulations meant that any of these infractions could have resulted in severe penalties, including imprisonment, but one observer noted that "threats and warnings from the authorities have no effect on her whatever, except to provide food for the entertainment of her neighbours on the bench."[55]

Peggy was not alone. Though there were certainly a number of individuals—both men and women—zealously devoted to both the cause of Allied victory and to their daily work, others were less dedicated. One reporter, having spent several months in one of the factories, noted that "between three and five in the afternoon more slacking and idling goes on than one would have thought possible in a wartime factory. Sometimes one can look along the bench and see not more than one girl in four actually working. But the others are rarely doing anything that could be definitely picked on by a foreman, such as knitting or reading. One will be sitting with her hand on the handle of her machine, as if just about to pull it down, and yet somehow not doing it; another will be patting her hair; another staring for the moment out of the window; another just settling down after a visit to the cloakroom, and so on."[56] Many expected that the fight against fascism in itself would have motivated action, but "apathy" ran rampant, and many women reported feeling disconnected both from their work and from the war effort as a whole.

This was not terribly surprising: industrial work was often profoundly unpleasant. As one newspaper reported in 1943, "One can never get away from the essential fact that the operator is human, not a machine. . . . These people come from all walks of life, and not all of them are suited by temperament to the work. Some hate machinery, some hate working in perpetual artificial light, some hate the noise, the dirt, the smell, the heat; some hate the boredom of endless repetition."[57] Moreover, motivating unenthusiastic workers by traditional means—for example, with the threat of unemployment—was difficult in the wartime context, as new laws made it illegal to discharge, transfer, advertise for, or hire new employees without the consent of one's local National Service Officer.[58] Even government propaganda seemed of little use. As historian Arthur McIvor notes, "The wartime crisis, state intervention, the revival of trade union power and the return to full employment combined to drastically shift the balance of power in the workplace from capital to labour."[59] As a result, companies began to experiment with novel industrial welfare programs intended to push employees toward productivity with honey rather than vinegar. Such efforts further drew on newer currents in the burgeoning British science of work, which had begun to ostensibly refocus on questions of laborers' health and well-being.[60]

Industrial Rhythm seemed perfectly engineered for these circumstances, promising to unleash hidden torrents of spontaneous enthusiasm that would transform reluctant workers into contentedly industrious citizens. Businesses' openness to new tactics may have also derived from a belief that women would be more malleable subjects, less likely to revolt against intrusive changes to their work processes than their male predecessors.[61] Indeed, Laban and Lawrence openly acknowledged this fact, explaining to potential clients that while "older employees and skilled male workers" might require additional convincing, as they "usually estimate further training as superfluous," the new "women workers quickly and eagerly react to movement exercises."[62]

One setting in which Industrial Rhythm was tested was the factory floor at Mars Confections. In 1942 Laban and Lawrence were hired by the company and embarked on a visit to its main production plant in Slough, twenty miles west of London. The factory—like the rest of wartime industrial Britain—was in the midst of a speed-up, producing millions upon millions of the caffeinated and highly caloric chocolates to supplement the mess kits of British soldiers. The company sought to motivate workers with reminders of the significance of their service, insisting that for both "the fighting forces" and the "kiddies in the Shops" it was the extras like Mars Bars that made it "easier to stand the strain . . . to work, to fight and to play the game." But morale continued to suffer.[63] Years later, Lawrence described the scene in the wrapping division, where about sixty young women worked: "The girls were at the end of a conveyor stretching right back to the confectionary kitchen . . . facing an endless parade of little rectangular bits, marching down there about twelve abreast. It was a horrible sight to see these things coming at you." Each worker picked up and wrapped a new Mars Bar every 3.75 seconds, for a total division output of approximately 58,000 chocolates per hour, 518,000 per day, and 15 million per month.[64] Fatigue and demoralization set in quickly; Lawrence immediately observed the frequency with which "their fingers or their minds crash," resulting in mistakes, wastage, and the persistent problems of low attendance and high turnover.[65]

Indeed, turnover played a central role in convincing Mars to retain Laban and Lawrence's services after their initial visit. "It would appear to us," Laban and Lawrence noted in their proposal, "that the strain under which some of the operators must work will manifest itself in non-attendance and in an increase in Labour Turnover, which are both expensive and disruptive of continued high output." They also reminded Mars that—given the wartime labor situation—the issue was unlikely to resolve itself, writing that "at the present time recruits cannot be specially chosen from applicants for vacancies and you suffer from the employment of those whom normally you would not employ."[66] A week later, Mars formally engaged Laban and Lawrence for a full study and training program.

Following a day of observation by Laban, Lawrence, and Ullmann, the Industrial Rhythm group diagnosed Mars with an epidemic of "over-hastiness" and an excess of muscular tension. The solution they offered was simple. Over the course of about a week, they would teach the employees new ways of physically approaching their work, both on the line itself and in series of daily, forty-five-minute preparatory classes. For the wrapping line, for example, they sketched out an intricate set of hand and arm movements that required the women to alternate moments of strain with those of relaxation, producing a pleasing rhythm that repeated every few seconds.[67]

The plan was communicated both to Mars's management and to the workers themselves in a pamphlet titled "What we would like to tell your Workpeople about 'LILT IN LABOUR.'" Laban and Lawrence noted that the workers had likely already noticed the strange figures skulking around the premises with pen and paper and explained that their observations had "shown that the rhythm of work can be improved by showing you some exercises that are specially made up for you and your job by specialists who know, as only they can, just what is needed."[68] Some of the coming interventions would be familiar, they promised, including the placement of tools in easier-to-reach locations and the adjustment of the heights of chairs and tables. Others—including the notation and accompanying movement classes—might seem more unusual. They were nonetheless, as Laban and Lawrence were always at pains to communicate, designed for both the good of both the factory and the workers themselves.

The benefits for employees, moreover, would extend beyond the cure of aching backs or fatigued wrists. As the two men wrote to the women of Mars, Industrial Rhythm would help workers "find out for instance, as soon as one side of your body becomes tired, how to make a slight change in your

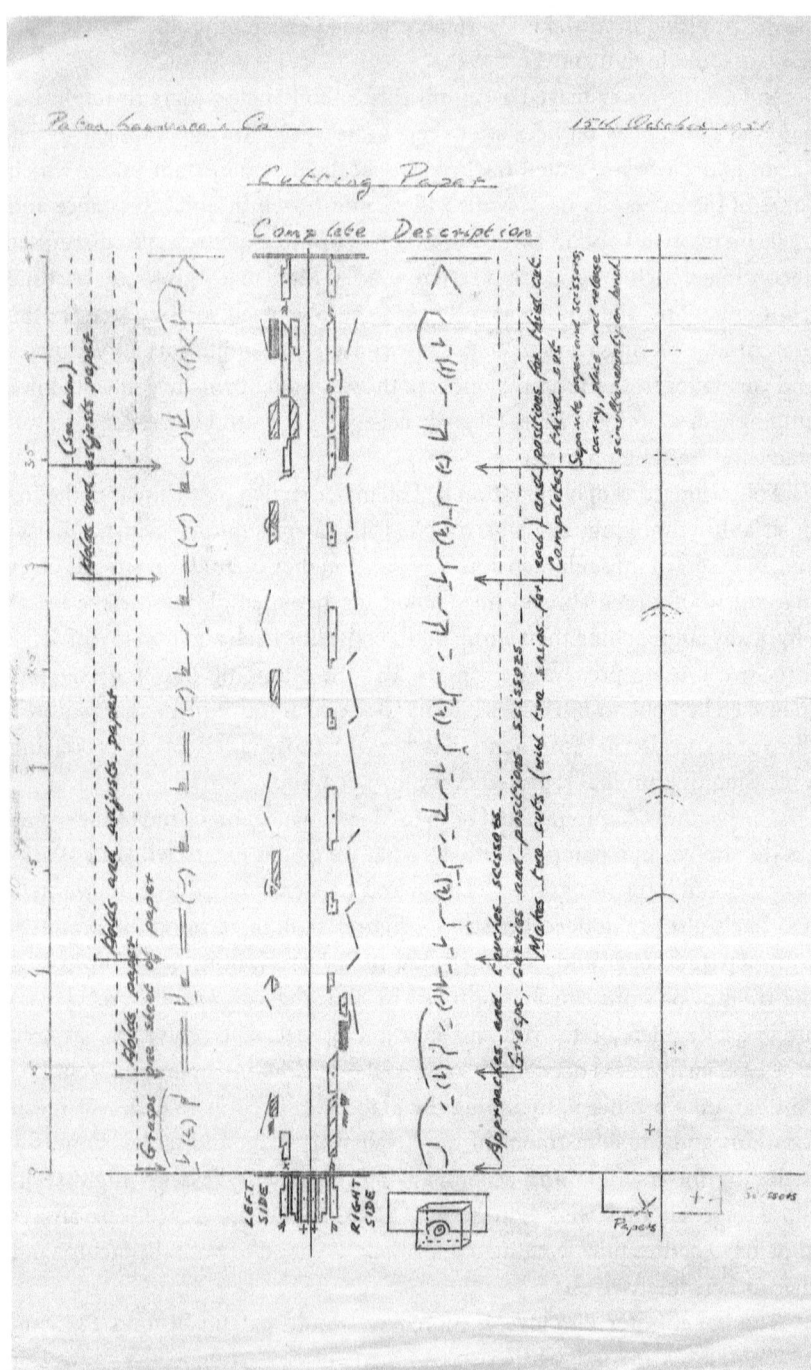

FIGURE 2.3. Personal Effort Assessment using Labanotation of someone "Cutting Paper. Complete Description," October 15, 1951. Archive Ref No L/E/62/66. From the Rudolf Laban Archive, University of Surrey, © University of Surrey. No reuse without permission.

movements to relieve it." The ultimate effects of their movement revolution, however, would be far more profound:

> You will find that your own personality will become stronger and freer, that you can keep up your interest in the work all the time and not be overcome with tiredness. . . . [I]t will also show you how to discover the kind of action and work which will give you the inner sense of ease and calmness from which you will get the greatest satisfaction in your work. And this satisfaction and not a driver for greater output is the real aim of the training in the "lilt in Labour."[69]

In fact, Laban and Lawrence posited that the returns from this method would be so great that it might eventually dwarf the motivating influence of other factors, up to and including wages. "The times," they wrote, "when it was thought that satisfaction was only to be had through money have passed."

Inquiries from prospective clients began arriving at a steady pace; eventually, Lawrence had to turn down at least one interested businessman on the grounds that the team was already working for his competitor. Though their methods varied slightly as the substance of work changed—tinning bacon, boxing tea, mixing aggregate, chopping wood—the underlying format remained largely the same: an initial evaluation, followed by the production of new notation scripts, the introduction of daily movement classes, and the evaluation of their effects.

Industrial Rhythm's aims also remained constant. Laban and Lawrence repeatedly emphasized how the introduction of dance-like rhythms into industry could eliminate the persistent plagues of turnover, fatigue, and inattention. At Glaxo, for instance, Laban introduced a new method for capping medical vials, contending that the new scheme would extract the drudgery from even this most dull and repetitive act. "Because the Capping operation properly performed has a natural rhythm that is agreeable to female workers," he noted, "it can in suitable conditions be performed continuously by them. Consequently, for a girl with correctly applied effort and automatized working to the correct rhythm, with supplies of caps and vials always ready at hand, there is no hardship in working on this one job continuously during normal working hours, day after day."[70] At St. Olave's, the Industrial Rhythm team promised that the introduction of a "steady rhythmical flow of production" would decrease turnover and boost individual efficiency by 20 to 25 percent, engendering "physical health and contentment" but also an "increased will to work" and an "increased will to co-operate with fellow workers and Management."[71]

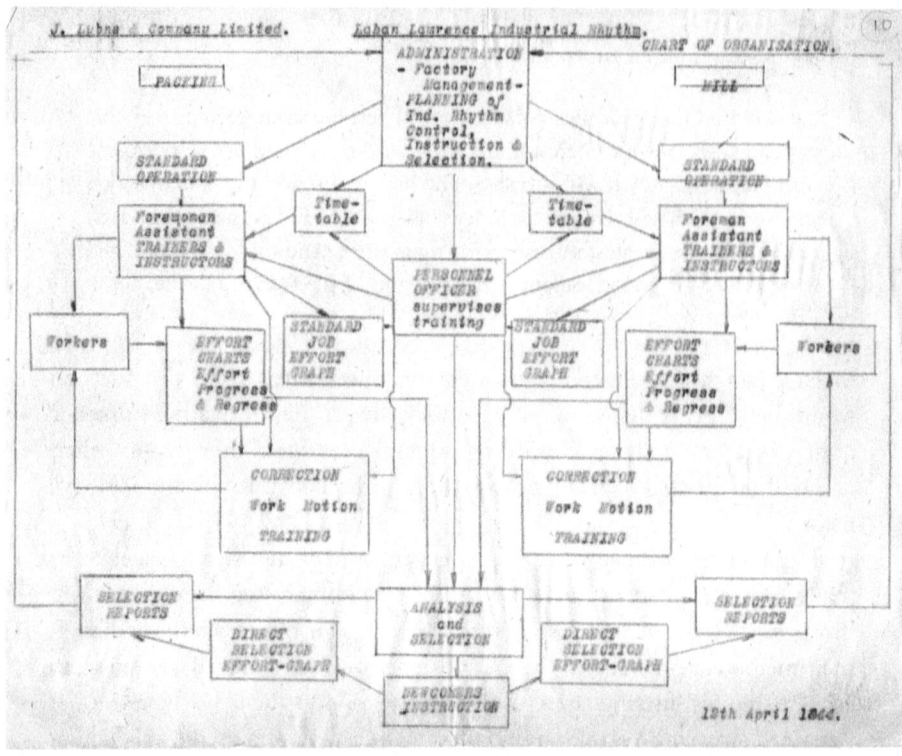

FIGURE 2.4. J. Lyons & Company Limited, Laban-Lawrence Industrial Rhythm, Chart of Organisation, April 18, 1944. Archive Ref No L/E/72/6, Document 10. From the Rudolf Laban Archive, University of Surrey, © University of Surrey. No reuse without permission.

At J. Lyons, Laban and Lawrence again identified the company's central problem as a rhythmical one, this time caused by a mismatch between the factory's humans and its machines. Early in their consultancy, observers from the Industrial Rhythm group noted the frequency with which Lyons's machines were set at speeds too slow or too fast for their operators, resulting in awkward periods of waiting or a frantic scramble. "To be driven or to be compelled to interrupt the flow of personal movement in waiting for the next machine to move is un-nerving," they wrote, "and this is constantly reported throughout the day." The solution they proposed was not an adjustment of the machine's controls, but rather the introduction of working movements that would enable workers to "harmonize their personal rhythm with the machine rhythm."[72] In practice, this might entail the insertion of arm, leg, or hand gestures that did not increase the pace of production but would, in theory, increase worker comfort and pleasure. These kinds of interventions caused the most conflict with Paton Lawrence's existing consultants, who tended to recoil at prescribing anything other than the absolute minimum movement required by a job.[73]

Others were unimpressed with Industrial Rhythm's results. In fact, Laban and Lawrence's very first client, Tyresoles, also proved to be one of their most difficult. A major tire manufacturer and reconditioner whose predominately male staff had been almost wholly replaced by women at the outbreak of war, Tyresoles reached out to Laban and Lawrence in January 1942. At their first meeting Tyresoles's managing director P. G. Hamilton reported that their new female employees succeeded at many crucial aspects of production but continued to struggle with stacking heavy army transport tires, a task he described as "obviously beyond female strength."[74] Laban promised a solution, contending that a carefully conceived training program could compensate for the women's lack of sheer muscular prowess while simultaneously ensuring a steady supply of willing workers.[75]

The initial agreement was as follows: for a total fee of £450, Laban and Lawrence would spend approximately nine weeks at three of Tyresoles's locations.[76] The initial "experimental" week—during which Tyresoles could cancel the remainder of the contract if they were unsatisfied with the results—would take place at the factory at Ormes and would be followed by a month each at Tyresoles's factories at Wembley and "Normand's Garage."[77] As at Mars, Laban and Lawrence evaluated the women's existing movements and prescribed new ones via notation. Workers then participated in movement classes between thirty and forty-five minutes in length, held several times a week. (Laban and Lawrence argued that daily courses would be more effective, but the company balked at the lost working time.) Most classes were attended by between fourteen and eighteen operators and included exercises performed both with and without tires and other tools.[78]

Laban and Lawrence quickly pegged Tyresoles as a success. Output marginally increased, the women seemed to experience less fatigue, and Laban and Lawrence reported that "even the Operators less able to do heavy lifting had been trained, in the end, to handle tyres of 36 X 6 HD—62 lbs, without assistance."[79] There are almost no sources in which workers speak directly about their own experiences of the program, though trainers' log books noted that most of the women seemed happy to attend the classes.[80] One might imagine that—if nothing else—the courses provided a welcome break in a long day of labor. Some women also seem to have reveled in their new physical abilities: an internal Tyresoles newsletter compared the freshly trained workers to the Greek heroine Atalanta and noted that "erring husbands will note with alarm the practiced way in which our 'sylphs' can swing heavy objects."[81] On the other hand, at least one employee seemed skeptical of the program's value, dropping out after one or two class meetings. She later returned—whether of her own accord or at the urging of management is unclear—but apparently remained doubtful, reportedly taking part in the courses in a "manner which is somewhat disturbing the others who take the training very seriously."[82] Logbooks produced at other Industrial Rhythm sites recorded a similar mix of reactions. Some relished the opportunity to move in new ways, others were self-conscious but curious, and a handful were simply uninterested. There were no reports, however, of organized opposition, at least on the part of female war workers; as Laban and Lawrence had predicted, they seemed open to experimentation in ways that their longer-serving male counterparts might not have been.

Tyresoles, however, requested to end the contract earlier than planned, citing the program's negligible impact on the company's bottom line. In the letter informing Laban and Lawrence of his decision, Hamilton first blamed the escalating war in Japan for the "disappointing" turn of events. The remainder of the letter, however, revealed another anxiety. "We do think," he wrote, that "it is very important to show what the actual saving and tangible results are. If the idea of Laban training is to go ahead on a considerable scale, you will have to show results in £ S.D. If it is not going ahead neither you nor we would be interested."[83] For Hamilton, the large-scale remaking of labor relations was not the point; the real proof was in profits and production.

In their response, Laban and Lawrence first addressed Hamilton's concerns head-on, citing favorable production reports.[84] They also, however, refused to waver in their insistence that Hamilton misunderstood Industrial Rhythm's fundamental aim. "It is intended," they patiently explained, "that the application of the Laban Methods shall fit the Women employees for the work they are called upon to do and ensure the optimum production from their efforts, but we wish to submit also that the institution of the Courses will be means whereby the relation between Employer and Employees will be improved and in the present circumstance we feel this is a matter of the greatest importance."[85] The goal was not production alone, but rather a comprehensive remaking of the relationship between workers, their work, and their employers, a process which Laban and Lawrence felt was already well underway. (So convinced were Laban and Lawrence of the efficacy of their methods that they offered limited followup training at no cost even after the contract had expired; privately, Lawrence questioned Tyresoles's motives, suggesting the company might simply be out to get something for nothing.)[86] Tyresoles, however, held firm. Laban and Lawrence remained at the company only through the conclusion of the initial contract.

Sanctifying the Factory

Despite the failure at Tyresoles—and a steady drip of similar complaints from other clients—Laban and Lawrence continued doggedly in their work. On the whole, the companies they worked with were satisfied with what seemed to be improvements in morale and occasional—if small—increases in production. At Mars, for example, the Industrial Rhythm team took credit for a 5 percent increase in output and noted, as well, the new "liveliness" workers brought to their tasks.[87] (In later years, Laban's disciples would forcefully dispute claims that any positive changes were simply a result of the Hawthorne effect.)[88] As the stewards of a growing business, Laban and Lawrence were undoubtedly gratified by this state of affairs. For both men, however, there was always more at stake in Industrial Rhythm than the happiness of any individual worker or the bottom line of any one company, including their own. In the balance instead was the fundamental sustainability of industrial society.

This deeper rationale for Laban and Lawrence's work is suggested by the fragments of a book manuscript they began writing together in the early 1940s. In the text, titled "Man and the Commonwealth," they cataloged the horrors that had plagued the modern West over the past several decades: two world wars, a massive economic depression, ongoing social and political unrest. But rather than blaming these traumas on "the mere will" of certain "demagogic personalities," Laban and Lawrence took aim at underlying economic realities, arguing that the pace of industrialization had resulted in both unprecedented economic inequality and—even more importantly—a mounting and insidious discontent with the experience of work itself. Moreover, while this restlessness was thrown into relief by the peculiarities of wartime production, it was not a momentary anomaly but a symptom of a much deeper ongoing problem.

"People do not grasp how swift, how profound the transition from crafts to industry" was, Laban wrote, but the aftereffects of this transformation could be felt everywhere.[89] Workers who spent day after day on an assembly line making seemingly meaningless, repetitive actions left the factory frustrated, anxious, and spoiling for a fight. The hope that humans would welcome the assistance of the new industrial technologies seemed increasingly naive. As one journalist writing about Laban and Lawrence's work summarized, "the more important and perfect the machine becomes, the more bored and frustrated grows man. This boredom, this sense of being part of a machine and therefore less of a man, is a form of invisible dry rot infecting and invading all levels of the industrial scene."[90]

But while Laban and Lawrence criticized the current state of affairs, they also took a strong stand against any attempt to slow the tide of industrial or technological change. There was no going back, only forward; the notion that the solution to industrial society's woes was a return to "fetishized" forms of medieval craft work was, they held, simply unrealistic. "New forms of labour," they contended, "have come into existence, which must be recognised and accepted by the broad masses, as fundamentals for new joy in the performance of everyone's productive task."[91] Assembly lines and complex machinery were not going away; nor, Laban and Lawrence argued, should they. The underlying problem, they contended, was not industrial development itself but the fact that human beings had not yet devised a way to harmonize their spirits with the rhythms of the machine. Not, at least, until Industrial Rhythm.

Laban and Lawrence's efforts to control movement, therefore, represented an intervention intended to reverberate far beyond the factory itself. In a letter to friends, Laban admitted these "revolutionary tendencies" and highlighted the fact that "movement awareness and training—not in gymnastical form but during work instruction" was designed to aid in the construction of a "new mentality which we feel to be necessary . . . in modern industrial life."[92] Movement, they argued, held the key to the full and final reconciliation of human beings and industrial systems of production. Citing Industrial Rhythm's early successes, they claimed that a comprehensive extension of their program—into factories nationwide, but also into primary schools and the military—would heal the fractures plaguing society and pushing the modern West to the brink of destruction. It would do so, moreover, without any need to fundamentally alter existing technological, economic, or political systems.

In a lecture to the Manchester Association of Engineers, for example, Laban highlighted the case of a pale, sickly girl in the Mars factory, fatigued and depressed, who blossomed under the tutelage of Industrial Rhythm's instructors. She came, in fact, to love her daily labors, entering a pleasant, almost trance-like state as she worked. The story, Laban argued, "should cure all that sentimental nonsense about the dull monotony of a factory worker's life."[93] Elsewhere, he and Lawrence told other tales: of employees finally working together in harmony, swept away by the all-embracing communal rhythms of production; of factories that functioned first like well-integrated "organisms" and then like well-tuned "orchestras."[94]

Laban and Lawrence promised, moreover, that the benefits first realized on the factory floor would eventually spill over, first into workers' personal lives and then into society at large. In a set of notes prepared for a meeting with Tyresoles Laban and Lawrence included a potential script for engaging with workers, claiming that those who embraced the new techniques would "not easily lose your patience, not only at your work, but also in waiting for buses or queuing for food." This practical encouragement, however, was quickly followed by a more opaque exhortation: "And it can be really called a mystery, because it is an almost unexplicable [sic] but definitive positive truth, that you *can* achieve calmness of nerve, endurance, a free view on things and happenings by adequate exercises, which always are best based on the movements you are doing in your work."[95]

In their public lectures and pitches, Laban and Lawrence often presented themselves as quintessential technocrats, armed with a targeted solution to the problems of the contemporary industrial corporation. As the conclusion of their message to Tyresoles's workers suggests, however, Industrial Rhythm remained—as Kinetographie itself had long been—deeply imbued with more mystical elements. For Laban, who never truly abandoned his commitments to Rosicrucianism and Freemasonry, movement and gesture were always conduits to the divine. In the 1926 *Gymnastik und Tanz* (Gymnastics and dance), for example, he outlined how the moving body had been deployed across time immemorial by religious leaders—whether in the "old, erotical prayer exercises" of "primitive" non-European cultures, the "ecstatic temple dances" of the Greeks, or in Catholic ritual built on bowing, kneeling, and gestures of blessing. Such practices, he claimed, were "designed to train the body to obey disciplined movements and thus reunite spirit and body," and movement was thus a "spiritual necessity," uniquely able to uplift the soul, discipline the desires, and connect human beings both to one another and to a larger cosmic order.[96] The truth of these earlier insights, he argued, was often ignored in modern life but needed to be made use of once again.

Laban also believed, however, that in the twentieth century the site for these embodied religious experiences would no longer be the cathedral or the shrine but the industrial workplace. It was the factory where so many spent so much of their days and thus here that movement could be most powerfully deployed. For Laban and Lawrence this was only natural: as they noted in "Man and the Commonwealth," "spiritual life" had "always been borne by the struggle for sustenance of the human race" and would change "with the form of that struggle."[97] In fact they noted that, rightly managed, the "extreme mechanization" characteristic of contemporary work life could ultimately "lead to a deeper understanding of the purely human." Only by accepting this reality, Laban wrote elsewhere, would work "not torture, kill, and exhaust, but invigorate and exalt."[98]

In making this claim, Laban again returned to ideas he had long espoused. In his 1920 book *Die Welt des Tänzers* (The world of the dancer), he first acknowledged that "when the priest advises men to carry God's name in their hearts when doing even the meanest of work, this seems to many to be arrogant. A priest can easily always carry God's name in his heart because his profession constantly reminds him of God, or at least it should. Not every profession reminds of God. Some activity excludes this thought completely during work."[99] The philosophical contemplation of the divine, however, was not what Laban had in mind. He envisioned instead an experiential, embodied form of religious practice modeled on the cultic celebration.[100] Its particular contours were never clearly defined—and perhaps even left strategically vague—but he assured his readers that the right kinds of movements could make the daily toil of any worker as spiritually nourishing as that of "an artist, an actor, a priest, or a scholar." Instead of the Weberian paradigm in which steady toil—and the attendant accumulation of wealth—served as a signifier of divine election, Laban construed labor as a vital tool for experiencing the divine in the here and now.[101] He exhorted his readers: "In you exists joy as a constant inner festival. All you have to do is release it."[102]

But just as a Catholic priest might mediate his congregants' access to God, Laban was convinced that this festive release needed to be carefully managed, ideally from above. The technological mechanism for that control was Industrial Notation. With its neat columns and black-and-white shapes, Industrial Notation could ensure that workers were making the right movements: selecting some, discarding others, monitoring constantly for compliance. The resulting practices would put workers in tune with the fundamental rhythms of the universe but also be regular and easily comprehensible: an ecstatic dance ruled by a bureaucratic technology. As Laban clarified in an early Industrial Rhythm memo, "newcomers must be instructed in order to adapt themselves to the great pattern. Failures and irregularities must be corrected or rejected and—last/not least—the co-ordination can and must be perfected by improving more and more the performance of each single detail."[103] Nothing would be left to chance.

Indeed, Laban and Lawrence explained in a 1943 business prospectus that Industrial Notation's true power derived from its capacity to precisely shape both outer movements and a worker's "inner attitude."[104] Employees need not even be aware of the process for the system to function; as they emphasized elsewhere, "it is not at all necessary or even possible that the workmen become clearly conscious of this effect."[105] These newly managed movements would automatically produce a flood of self-transformations, correct the "mismatch between material progress" and "the masses' mental and spiritual growth," and ensure the steady functioning of modern civilization.[106]

As he established his new identity in England, Laban sometimes publicly downplayed these metaphysical elements, but they never truly disappeared. While Laban and Lawrence spoke about the "joy" or the "happiness" that workers exposed to Industrial Rhythm would experience, they aimed at more than mere temporal satisfaction. A draft of a lecture Laban gave to an engineers' association in 1944 speaks to this fact. The original text reads as follows:

> Human movement in work is first of all witness of the will of man to live. It is a thrust towards happiness, and this will and this thrust have a deep root, which becomes also visible in man's working movements. Movement springs from the desire to fulfill some fundamental rhythm and to be in tune with something, which man conceives of as the infinite or the divine. Man is aware, he knows without proof, that his life is part of a divine conception of eternity, and it is against the background of such harmony, that conscience measures all things.[107]

In a subsequent edit, however, Laban crossed out both references to the spiritual, replacing them with more secular turns of phrase. The "infinite or the divine" to which man hoped to attune himself became simply "his experiences," and the "divine conception of eternity" became the "greater whole." This occasional whitewashing of the stranger, more mystical aspects of Industrial Rhythm was in part a canny marketing strategy, designed to assuage the fears of engineers and executives perhaps already inclined to be skeptical of the dancing consultant who appeared on their doorsteps.

For others, however, Laban and Lawrence's promise of societal redemption through the individual body was the root of its appeal. There was no shortage of anxiety about the individual, cultural, and political consequences of industrial life in midcentury Britain. There was also no shortage of bitter arguments about the appropriate remedies: increasingly powerful trade unions sparred with conservative business interests while the left engaged in rancorous internal battles about the relative merits of incremental reform and revolution.[108] In this climate, Industrial Rhythm filled a particularly useful niche. In its use of the vocabulary of estrangement and alienation, in its criticism of the dehumanizing specialization of industrial production, and in its elevation of labor as central to the human species, Industrial Rhythm drew on Marxist rhetoric. But though Laban and Lawrence did occasionally express support for increasing workers' wages—and even voiced enthusiasm for centralized economic planning—the core of their program located the responsibility for adapting to new conditions of production in the individual's own body. As Laban himself put it, the "great problems" could only be solved by attending to the "the microscopical germ of discontentment" within each individual: "It is in these tiny things where the emotional and intellectual trouble resides."[109] A larger revolution was never on the table.

The most critical responses to Industrial Rhythm painted the program as a tool for making workers content with their own subjugation. In a 1942 piece in the *Derby Daily Telegraph* that spoke to such frustrations, the author addressed the would-be reader directly: "'What,' you will say, 'isn't it bad enough tending a machine all day, or all night? Do they expect us to dance for joy as well?'"[110] And though available sources do not provide a clear window into workers' experience of Industrial Rhythm, one might imagine the article's author groaning at the mantra Laban asked instructors to practice with those undergoing training: "I am entering into a great unit—the greatest and most natural unit of liltfully functioning human beings—I am integrating myself into the creative energy which acts naturally through me."[111]

Laban and Lawrence's disinterest in altering larger structural conditions becomes particularly evident in their attitudes to pay. Though they repeatedly claimed an investment in the "just remuneration of labor," they even more often declared that money—in the end—would come to be beside the point. As Laban wrote in 1944, "There is, I think, no kind of moral exhortation which can replace the gross sportive pleasure in skilled performance. No fines or command or extra remunerations can compete with the enjoyment of an effectively performed movement."[112] This outlook was also communicated to potential clients, as in a proposal that claimed that Industrial Rhythm would "offer balance in the workman's reaction towards incentives," creating a more sensible equilibrium between "pleasure in rhythmical function" and "satisfaction of financial greed or need."[113] Elsewhere Laban admitted that—given the economic realities—the "industrial worker . . . will rarely have the possibility to enjoy the last result of the perfect product," making it all the more necessary that he take pleasure in "the results of his performance of a detail."[114] For Laban and Lawrence, wages, economic restructuring, and top-down political change would not save the twentieth century from its own demons. They proposed instead a kind of religious practice: a renunciation of the satisfactions of the flesh in favor of the deeper comforts of the (embodied) spirit.

In fact, a number of companies may have been enticed by Laban and Lawrence's suggestions that Industrial Rhythm could help keep workers from unionizing. In a confidential report submitted to J. Lyons, for example, the Industrial Rhythm team noted that there was a particularly volatile situation in one of the mills, where—even before Laban and Lawrence arrived—there was "without doubt a kind of agreement between most of the workers in a wish to resist measures and procedures of the administration or management." This dissension, they further suspected, "could probably be traced to common talk, a tendency to form or enter a union, or suchlike organisatory propaganda. This unity or common feeling of the workmen is, in any case, concentrated on negative criticism and not on helpful collaboration."[115] The report, however, assured J. Lyons that a way out was at hand.

First, Laban first dismissed the usual causes of this kind of worker unrest—pay scales, working conditions, fickle foremen. Instead, he reminded company managers of "the well-known fact that the uninterrupted contraction of joint muscles—especially those of the arms and hands—as well as over-violent movements during these contractions, provoke nervous tensions which lead easily to emotional states such as 'anger,' 'impatience,' and even 'fury.'" Because the workers were unaware of these effects, Laban explained, they instead blamed their foul moods "on external circumstances, of which slight deficiencies of administration become naturally the first target."[116] Altering the mill workers' movements, however, could remedy the problem.

Indeed, Laban and Lawrence argued that proof their intervention would work was already available: amid all the grumbling was one man who stood apart, quietly avoiding the simmering rebellion in the rest of the division. Unlike the others, Laban noted, he naturally "performed his work with just that restricted effort which these operations require," and it seemed that his movement qualities had freed him from his colleagues' self-defeating discord—and their concomitant interest in unionization. As the report continued, "it is characteristic that the man working in the right rhythm is the only one who is really aloof from, or immune against, the negative attitude of the team."[117] He would, instead, feel a sense of joyful belonging, knowing that he and his employers were all striving toward the same higher purpose. Laban and Lawrence further suggested that this man be tapped as a trainer for new employees. While the existing workers' bodily habits might be too ingrained for Industrial Rhythm to fully remedy—a situation that was not helped by their indifference to the training procedures—newcomers, "even men or youngsters who are not particularly strong physically," could be properly instructed from the start. In time, they suggested, the situation would fix itself.[118]

This mix of high-flown spiritual rhetoric and hands-on labor control might seem like an unusual combination, but Laban and Lawrence were far from alone in their attempts to build a version of modernity founded on the aestheticized spiritualization of new work processes. Dartington Hall itself—the institution that first brought Laban and Lawrence together—was imagined largely along such lines. As Elmhirst wrote in *Faith and Works at Dartington*, the 1937 pamphlet that became one of the estate's calling cards,

Dartington's synthesis of intellectual, cultural, aesthetic, and economic practices drew on a model previously best exemplified by religious institutions: "Till the Reformation the Church attempted such an aesthetic synthesis by means of the Mass and of its ritual, its mystery play and holy day festivities. It aimed to guide the feelings as well as to steer the reason of its varied flock."[119] In adapting to twentieth-century conditions of production, Elmhirst and Whitney ventured, it was these spiritual techniques that would again become necessary.[120] In the same moment in Hitler's Germany, the Bureau of the Beauty of Labor promoted programs of joyful physical exercise to "reconstitute the soul of the German worker" as part of an attempt to reconcile its vision of traditional, rural arcadia with the demands of an industrial society gearing up for war.[121] In Spain, the Catholic lay organization Opus Dei sought to sanctify technocratic expertise as central to the "work of God."[122]

Laban and Lawrence's work was a kind of crystallization of this larger set of efforts. By paying attention to and elevating bodily work—and making much of its purportedly ancient and mystical connections to the divine—they assured their clients, workers, and Britain at large that industry would not stamp out humankind's spiritual life. At the same time, they promised that the messy, primordial forces the body unleashed would be carefully harnessed for the benefit of capitalist production and social stability. Laban himself sometimes characterized Industrial Rhythm as situated between the twin lodestars of Frederick Winslow Taylor and Isadora Duncan.[123] That one was known for splitting movement into its smallest parts and stripping it of meaning and the other for a form of modern dance that held up movement as the "expression of the life of [the] 'soul'" did not, to him, seem like a paradox.[124]

In its pledge to remove the contradictions between these seeming dualities—reason and spirit, body and soul, past and future—lay Industrial Notation's power. What Laban and Lawrence believed that the traditional Taylorists did not was that financial incentives alone might prove insufficient to stabilize the social order; the creation of a new kind of person would also be crucial to its functioning. Notation, therefore, was not just a tool for solving a geographically or temporally specific problem. It was a technology designed to support the long-term existential maintenance of a world committed to industrial progress.

In fact, at Industrial Rhythm's height a seemingly strange linguistic quirk appeared with unusual frequency in midcentury British news stories about Laban's notation system: the authors referred to the notation not as *script*, but as *scripture*.[125] For at least some of its devotees, it seemed to be just that: a guide to a superlunary realm free of the contradictions between human fulfillment and rampant economic growth, spiritual satiation and production targets. Like all forms of scripture, however, notation could also be used as a cudgel, a technology of indoctrination that elevated industrial production itself into a new kind of deity.

Epilogue: Industrial Futures

Though there was little hard data to support Laban and Lawrence's contention that making work into a dance would transform industrial life, the basic premise seemed too intuitively seductive to reject out of hand. This was particularly true during World War II, when broader anxieties about industry fortuitously combined with a labor market in profound upheaval. In the years following the end of war, however, interest in Industrial Rhythm began to wane, and Laban and Lawrence signed fewer and fewer clients as the 1950s wore on. But even as the actual business of Industrial Rhythm wound down, its techniques and philosophy found a new home in the world of education.

Scattered schools had been making use of Laban's ideas about movement since the 1920s; an article championing Laban's views on educational dance appeared in the *Journal of School Hygiene and Physical Education* in 1924, and thanks to Kurt Jooss Dartington Hall had been devoted to Laban's approach even before his arrival from Germany.[126] Before the 1940s, however, most physical education programs were based either around competitive sports or the highly regimented gymnastics promoted by the nineteenth-century Swedish gymnast and educator Per Henrik Ling.[127] Nevertheless, in 1941 a group of Laban-trained modern dancers organized an official conference under the auspices of the nation's Physical Educational Association to promote what they had begun to call "modern educational dance."

Attended by influential education specialists and members of the Crown's Inspectorate of Schools, the event proved to be a success: Laban's argument that forms of "freer and more expressive" movement should complement or even replace the existing curriculum fell on receptive ears. At its conclusion, the Physical Education Association officially recommended to the Board of Education that modern educational dance be integrated into the national curricula. Interest grew even further in the postwar years as dance became yoked to larger efforts aimed at educational reform and child-centered pedagogy. Eventually, Laban-based programs were introduced into a number of primary schools and training in the system became part of the standard curriculum at Bedford and Chelsea, two major women's physical education colleges. Laban himself, along with Lisa Ullmann, also founded the independent Art of Movement Studio in Manchester 1945, which offered multiyear teacher training courses as well as shorter summer intensives.[128] The courses continued to grow over the succeeding decades, such that by 1967 a report on British primary schooling published by the Department of Education and Science described Laban's theories as among the most significant influences on postwar British physical education.[129] A 1973 *Times* article sang a similar refrain, casually remarking that "after the last war Rudolf Laban changed our whole attitude toward physical education" and likening the extent of his contributions to educational reform to those of Montessori and Piaget.[130]

Though it often went unremarked upon, these new courses were nearly identical to the classes Laban and Lawrence had formerly held for factory workers. While specific instruction in, say, the details of candy-wrapping or tire-hurling was absent, the forms of movement were largely the same. In addition to less structured forms of expression, students were frequently asked to practice working movements, including "sawing, chopping, pulling a rope, hammering, screwing, ironing, scything, digging, or stitching, sewing, and cutting different materials."[131] As in industry, there was a special focus on making students aware of the qualities of their chosen movements as well as the emotional consequences of those choices. The aim was to inculcate the habits students would need later in life. As Laban explained *Modern Educational Dance*, a handbook of practice he first published in 1948, "On the whole, the bodily and mental food for children is fairly well controlled and selected. However, in the important field of the acquisition of [movement] habit by dancing no reliable selection is made. The children are fed either not at all, or with the stale remainders of the spirit of the past, entirely incompatible with the needs and knowledge of our days."[132] The new dance training, in contrast, would "foster the development" of an entirely new sensibility, thus "guaranteeing the appreciation and enjoyment of any, even the simplest action movements," a key skill as students moved on to jobs likely full of endless repetition.[133] Lecturing to the Contemporary Dance Club in London, Laban similarly assured his audience that "educational practice will surely tend to translate and canalize the dark creative forces in man into the blissful domain of human industry" and argued that there was "no better means to bring this modern mood of living near to the growing child then dance-training based on what we call industrial rhythm."[134]

The extent to which Laban's program penetrated the British educational establishment is made clear in a 1974 review of a reissue of Laban and Lawrence's 1947 book on their industrial work, *Effort*. A reviewer in the journal *Ergonomics* largely panned the text, pointing out that its arguments were "not supported by any experimental data" and that the authors were unaware of the problems that accompanied efforts to transfer movements learned in a generalized movement course to a particular work setting. "It is doubtful,"

FIGURE 2.5. Laban/Lawrence Industrial Rhythm exercises demonstrated by first-year students, Manchester, England, 1947. Photo by R. Watkins. Repository GB 1701 Laban Collection, Box LC/A/14/1/8. Trinity Laban Archive Collections, London, UK.

the reviewer wrote, "if this book will be of use to the ergonomist." Still, he grudgingly urged his colleagues to read it nonetheless, as "most of the young persons now entering industry will have experienced 'movement education' based on Laban's ideas and principles."[135] Indeed, *Modern Educational Dance* was reprinted no fewer than six times and served to guide an entire generation of physical education teachers. (This cultural ubiquity would also shape the reception of Laban-derived techniques in the world of British white-collar management, a subject that chapter 4 will explore.)

Thus, even as Industrial Rhythm seemed to disappear, Laban and Lawrence's ideas about movement, selfhood, and the future of industrial production quietly entered the daily lives of a generation of British schoolchildren. Other scholarly accounts have chronicled the full story of these programs, but they have largely ignored the roots of educational dance in Laban and Lawrence's industrial work.[136] As a result, educational dance's public image has been much more closely aligned with the anarchic defiance of Chaplin's Tramp than with the factory against whose regimentation he rebelled.

For Laban and Lawrence, though, the goal was always to bring these two worlds together. They might, in fact, have viewed *Modern Times* as a kind of advertisement for their own projects, insisting that Chaplin's breakdown could have been easily avoided through more thoughtful childhood management of his inborn tendency to caper wildly about. Co-opted in this way, his extensions, leaps, and aesthetic flourishes would become not a repudiation of factory work but the next stage in its evolution. Transformed, the Tramp would be an ideal industrial individual: expressive but controlled and, most importantly, at peace with the machines that increasingly dominated his working life. It was a vision of an improbable future in which there was no conflict between labor and capital, just happy workers dancing beside assembly lines.

3

The Dance Notation Bureau

In June of 1940, John Martin, the dance critic for the *New York Times*, reported in excited tones on the formation of a new local organization. Called the Dance Notation Bureau (DNB), its ambitious moniker reflected its sizable aspirations: to become the "center of information and a kind of national headquarters for all dancers who are interested in the recording of dances according to the principles evolved by Rudolf von Laban."[1] Currently, Martin acknowledged, the bureau was a somewhat minor affair. Housed in the back rooms of a Greenwich Village dance studio, it was run entirely by volunteers and boasted "no fees, no memberships, no formalities," functioning "simply [as] a free association of laboratory workers."[2] He was nonetheless full of praise for its four founders—Ann Hutchinson Guest, Eve Gentry, Janey Price, and Helen Priest Rogers—and certain that their work would soon have a broad impact.[3] "To eavesdrop at a working session of the bureau," he wrote, "is to see hard-headed, serious and alert thinking in operation. These four young women are completely convinced of the value of the work they are undertaking, completely grounded in its fundamentals, and bent and determined to make it practical, accessible and serviceable to as wide a field as possible."[4]

Martin, too, was sure that the bureau's work was urgently necessary. A longtime advocate of notation, Martin was thrilled by the increasing diffusion of Laban's system in the dance world but equally concerned that it had already begun to fragment. "Young as it is," he lamented, "this essentially universal written language of the dance has already begun to develop dialects."[5] One of the bureau's goals was to arrest this process, establishing a standard form to be practiced across the globe, and Martin encouraged "everybody who is desirous to make contacts with fellow scriveners" to "write in to headquarters."[6]

A second goal was broader and a bit less precise: to transform a system that to many still seemed "too concerned with philosophical matters and too little with immediate usability" into a practical tool appealing to "the American field."[7] With the right modifications and presentation, bureau leaders believed, Kinetographie could shed any remaining associations with mystical rituals and German racial theory. Instead, the DNB's efforts would make clear that dance—and movement more generally—was merely information and Kinetographie simply the ideal technology for recording it.

Over the next few decades, the Dance Notation Bureau made significant progress in achieving these aims. Its leaders immediately set to work, for example, on the project of rationalizing and regularizing the system, and in 1959 they participated in the creation of a global standards association, the International Council of Kinetography Laban.[8] They convinced major choreographers—including George Balanchine, Jerome Robbins, Doris Humphrey, Ted Shawn, and Anthony Tudor—to have their work notated and archived, and successfully lobbied the United States Copyright Office to make dance eligible for intellectual property protection. The DNB opened new offices in Union Square and embraced not only fees, memberships, and formalities but also multiyear educational programs, journals, official certificates, and regular meetings and conferences, drawing an average of a hundred students each year. Thanks in large part to the bureau's efforts, notation become part of the curriculum at dozens of educational institutions, including Bennington College, the University of North Carolina, Julliard, Jacob's Pillow, the Boston Conservatory of Music, the 92nd Street Y, and the High School for the Performing Arts; the National Ballet of Canada and the Philadelphia Civic Ballet both maintained resident notators.[9] The bureau even changed the system's name, turning "Kinetographie" into the less-foreign-seeming and easier-to-pronounce "Labanotation."[10]

In some ways the DNB's quest to preserve dance on paper mirrored Laban's early efforts in Germany. Like Laban, the bureau's leaders believed that notation was necessary not just to preserve important works for the future, but to raise dance's overall status as an art form. In other ways, the bureau was very much a product of its own time and place: its members were captivated not by grandiose visions of embodied national community, but by the prospect of noiseless information transfer, international standards, and commercial protection. The DNB's resulting efforts made Labanotation appealing to new constituencies and created a new kind of technical profession—the dance notator—composed almost entirely of women.

This chapter is the story of those efforts and of their sometimes unanticipated results. In particular, it explores how the Dance Notation Bureau's

FIGURE 3.1. Dance Notation Bureau members at the bureau offices at 35 W. 20th Street in New York City, circa 1955. From left to right: Helen Priest Rogers, Judy Bissell, Barbara Hoenig, Irmgard Bartenieff, Maria Nicholson, Lucy Venable, Els Grelinger. Courtesy of the Dance Notation Bureau and the DNB Collection at the Lawrence and Lee Theatre Research Institute at the Ohio State University.

vision of dance as primarily informational raised contentious new questions about the definitions of art and authorship, questions that dancers, choreographers, lawyers, and engineers would all answer differently. It also reveals how these new ideas and practices facilitated movement's entry into the "information age" in a more literal sense, creating lasting ties between Labanotation, engineering, and computing that would prove important to the system's future. As the next pages will explore, however, as art became science and dance became data, the meaning of dancers' thinking, feeling, sweating bodies became increasingly uncertain.

Experimentation and Information

Before establishing the Dance Notation Bureau, its four founders moved in similar circles in the New York City dance world but were not in close contact. All were experts in Kinetographie, though they had trained under different mentors and in different countries: Hutchinson Guest had studied with Laban himself during her days as a student at Dartington Hall, while Priest Rogers worked with Albrecht Knust in Germany after the end of her performing career with the Martha Graham company.[11] Gentry and Price learned Labanotation in New York City as dancers with Hanya Holm's company. (Holm herself had worked with Laban during his time in Switzerland and later joined the company of his protégé, Mary Wigman.)

In the spring of 1940, however, all four found themselves at an informal gathering at Holm's studio. Their discussion turned to Kinetographie and then to the orthographic differences to which their varied national trainings had given rise. In attendance as well was John Martin, the *New York Times* critic, who quickly interceded, suggesting they found an official organization to promote Kinetographie and standardize its form. (As the first dedicated dance critic at any major American newspaper, Martin had a special interest in any tool that promised to give the form its own language.)[12] Hutchinson Guest, Priest Rogers, Gentry, and Price were taken with the idea, and the organization began to come together. Though he neglected to mention his own part in the story, by June of that year Martin was reporting that "nothing that has happened in recent years has about it more potentialities for the development of the art."[13]

In its early years the bureau was indeed a modest operation, relying entirely upon the missionary zeal of its founders, who were often stretched for both time and money. Hutchinson Guest, for example, regularly spent her mornings on DNB business, her afternoons teaching dance classes, and her evenings hoofing it on the Broadway stage in musicals like *One Touch of Venus* (1943) and *Billion Dollar Baby* (1945).[14] Hutchinson Guest's commitment to the organization was nonetheless unshakable, and she quickly emerged as its leader. In time she even moved the DNB's physical headquarters to her own apartment building, where she rented space in an out-of-order freight elevator shaft. "When more people were recruited to the cause," a colleague later recounted, "they expanded to two floors, one above the other in the elevator shaft with a hole in the floor connecting the two floors. A string went through the hole and had a bell attached. Ringing the bell allowed workers to get the attention of people on the other floor."[15]

Perhaps appropriately, it was from inside that piece of mundane building infrastructure that the bureau's members began tinkering with Labanotation's own internal workings. In doing so, they navigated an ambivalent relationship with Laban himself. They exchanged letters regularly, but communications were sometimes friendly, sometimes less-so: as symbols flew back and forth across the Atlantic, Laban and DNB leaders tangled over questions about orthography but also over the ownership, development, and marketing of the system.

Though Hutchinson Guest and the other bureau members were undoubtedly Kinetographie's most avid promoters, they argued that much work was yet to be done in clarifying, augmenting, and standardizing the system for widespread use. By midcentury, Laban had begun to tire of these tasks.[16] Though he took the occasional stubborn stance on matters of money or control—requesting, for example, royalties in exchange for the DNB's use and publication of the system—he increasingly declined to weigh in on the technical disputes that regularly divided his geographically scattered interpreters, most notably Albrecht Knust in Germany and Lisa Ullmann in the UK. As he once wrote to Hutchinson Guest, "I sometimes get somehow contrasting versions which seem to me of equal value, and I think it is not my job to produce final judgments. I feel it is the right way that you fight it out amongst yourselves."[17] When bureau members devised new ways to more precisely notate turns, foot movements, and changes of weight, Laban was similarly disengaged. More interested in the ongoing promotion of Industrial Rhythm,

FIGURE 3.2. Riki Takiff (Kuklick) and Erica Goodman display their notation skirts as Nadia Chilkovsky greets Albrecht Knust at Idlewild Airport. From *Dance Notation Record* 9, nos. 3 & 4 (Fall and Winter 1958), 21. New York Public Library for the Performing Arts, Hanya Holm Papers, (S)*MGZMD 136–410. Courtesy of the Dance Notation Bureau.

the working out of technicalities in the system that had made him famous was no longer high on his priority list.

Thus, even as Kinetographie became Labanotation, the influence of the man himself waned.[18] Bureau members boasted that their devotion to notation was "impersonal," rather than the side effect of a cult of personality, and Hutchinson Guest noted that she was sometimes tempted to simply tell Laban "to go to ___."[19] This is not to say that the bureau disavowed Laban entirely, but rather that his looming shadow slowly receded into the background. As Hutchinson Guest explained in a letter to a friend, Kinetographie was Laban's child, but it was she who had brought it up.[20] Laban's death in England in 1958 made the separation permanent.

By the mid-1950s the bureau was well on its way to establishing an independent identity, a process that a 1955 DNB promotional pamphlet, titled "Dance Notation Conversation," illustrates vividly. Structured as a discussion between a curious potential student and the bureau's leaders, the pamphlet was authored by DNB member Selma Jeanne Cohen and answered such questions as: "First of all, I'd like to know, just what is the value of a dance notation system?" and "Is Labanotation hard to learn?" (The answers: "Providing a permanent record of dance compositions" and "Not at all!").

In response to another mock query about Labanotation's history, Cohen recounted Laban's "first experiments with notation fifty years ago" and boasted that by 1928 "pupils all over the continent were already using the system." She went on, however, to clarify that the version of the system the DNB taught was by no means a carbon copy of the original. Answering the question "So this is the system you are using today?," she explained that, on the contrary, Laban's work was just the beginning: "Over the years, Kinetography Laban (its original full title) has been constantly tested and improved. Today's highly developed version, known as Labanotation, is the result of the Bureau's work in research and experimentation to achieve greater accuracy and broader range. As the system is applied to new movement fields, new needs are discovered and new challenges met."[21]

Cohen then described the DNB as equal parts research center, school, clearinghouse, publisher, library, and recorder. She explained that the bureau offered elementary, intermediate, and advanced courses in notation (as well as specialized courses in score-checking) and that the successful completion of each course entitled the student to a notation certificate. Instruction was provided by "authorized teachers" who had themselves progressed through the bureau's offerings and earned diplomas attesting to their capacity to notate at a professional level. Courses were available either in person in New York City or—perhaps appropriately for an organization that put such stock in writing—by correspondence. Instruction in Labanotation was, of course, ancillary to the key tasks of recording, storing, and disseminating scores: DNB members were hired to notate new choreographic works, the DNB library stored copies of those records, and after obtaining the necessary payments or permissions, DNB's publishing arm released notation exercises, short excerpts, and full scores.

"Dance Notation Conversation" also emphasized the bureau's role as a center of information exchange, noting that its quarterly, the *Dance Notation Record*, was a crucial tool for facilitating communication between notators in the United States, Canada, South America, Europe, and "the Orient," providing a forum to both "discuss problems and to learn of new developments" and encourage the "standardization of procedures."[22] In 1958, the bureau highlighted, for example, Labanotation's use by the dance critic Yamano Hakudai in Chiba-Ken, Japan. Yamano had recently sent a letter to the bureau in which he recounted learning notation entirely from Ann Hutchinson's 1954 textbook *Labanotation* and testified that he had already become skilled enough to teach it to others. For the DNB, this was clear proof of Labanotation's fundamental universality. As the *Dance Notation Record* enthused, "If the acid test for the logic and learning ease of a system of movement notation can be described as the likelihood of its being assimilated entirely from a book by a person whose entire background is utterly different from that of Western culture, in addition to speaking a language which has no similarity whatever to English, then Labanotation has passed with flying colors."[23]

More broadly, the pamphlet reflected the bureau's deep-seated commitment to cultivating a scientific identity, though one somewhat different from Laban's own. While Laban had touted his engagement with turn-of-the-century physiology and physics, DNB members constructed their scientific personas not in relation to any particular theories concerning the body or mind but through their embrace of broader notions of objectivity, replicability, universality, precision, and disinterestedness.[24] In fact, as Hutchinson Guest wrote in a 1952 letter, many in the bureau were eager to fully and publicly "divorce" notation from Laban's underlying theories, and bureau members in the 1940s and '50s almost never mentioned movement's mental or emotional correlates.[25] Instead, publications highlighted Labanotation's utility as an innovative information technology. Though the 1955 brochure mentioned in passing that Labanotation was "based scientifically on the structure of the body," the text was more interested in highlighting the system's "universality, economy, and accuracy" as a method of recording. Any "adequate notation system," it explained, required three traits: it "must be based on elements common to all forms of movement; it must be easy to read and to write"; and "it must be capable of recording even minute actions with great precision."[26]

The DNB also contended that the widespread use of notation would make dance itself more like a science, continually progressing through ongoing research and experimentation. Supporters loudly echoed this claim. In *Dance Observer*, the New York City Ballet's George Balanchine compared Labanotation to Euclidian geometry, while in the *New York Times* John Martin praised the "systematic mentality" of Laban and his followers, their "genius for science and invention, for efficiency and mechanical creativity."[27] Martin and others argued, moreover, that future progress in dance was dependent on the use of such methodologies: "If experiments have been successfully made and the results incorporated into general practice, they must be accurately and scientifically recorded so that every dancer may be able to know what his colleagues are doing, and that he may be spared the wasteful effort of duplicating experiments already made. The dancer, like the mechanic or the physicist, must have a method of preserving in a simple and concise form the fruits of his laboratory work as well as of his creative moods."[28] DNB proponents

FIGURE 3.3. Cartoon from Selma Jeanne Cohen, "Dance Notation Conversation: Facts on the Universal System of Recording Human Movement—Labanotation," 1955, Dance Notation Bureau. Illustrator Doug Anderson. Courtesy of the Dance Notation Bureau.

also claimed that notation instruction would train the mind in more precise observational practices. The "careful observation" notation required would make dancers "more aware, more attentive to detail" while giving choreographers and teachers the "increased perceptiveness" needed to determine "why a dancer fails to achieve a desired effect due to an error in such elements as timing or dynamics."[29]

It was the notators themselves, however, who would be held to the very highest standards of scientific thinking. Hutchinson Guest in particular was constantly reminding DNB members that their efforts should be "mechanical"—two notators observing the same ballet should produce precisely the same set of symbols. And though it was impossible to deny that notation required considerable observational and technical expertise, and that notators might approach the process in different ways, such stylistic differences were never to appear in the notation itself. As Martin put it, notation, if practiced properly, was "as impersonal as music notation and no whit less foolproof."[30] In constructing this new profession, the bureau was relentless in its insistence that notation was more objective than subjective, technical rather than artistic.[31]

The DNB's unwavering commitment to this stance was undoubtedly inflected, at least in part, by the fact that the vast majority of its leaders and members were women. Indeed, women's almost total domination of American dance notation is particularly striking given Laban's own early pronouncements about the field. Though Laban's most famous colleagues, assistants, and students were women—and Kinetographie owed at least one crucial element of its basic structure to Dussia Bereska—Laban had seen his project as deeply gendered, linking written notation with a robust, modernizing masculinity and the "illiterate" oral tradition with femininity.[32]

The women of New York City's Dance Notation Bureau, however, bent notation's purportedly masculine and scientific character to their own ends. In an era in which men dominated high-status public roles as directors or choreographers—the ones in charge of the "girls" on stage—proficiency in notation represented a new way in to previously closed corners of the dance world. Like other female information workers of the Progressive Era and the middle decades of the twentieth century—computers, typists, social workers—the Labanotators strategically carved out a new professional niche by asserting their mastery over the recording and analysis of data.[33] The women of the DNB may not have ever attained the public notoriety of Balanchine or Lincoln Kirstein, but their knowledge gave them a new kind of standing as valued expert consultants. Drawings accompanying the 1955 brochure reflected this aura of ordered technical competence: rapt young women with sensible buns and neat ponytails sit in front of blackboards as slightly older women sketch out, with geometric precision, the features of the notation system. Making dance modern and scientific thus became, in a perhaps unexpected way, "women's work."

Tools of Creativity and Control

Indeed, accomplishing the DNB's lofty goals required a great deal of on-the-ground labor, as the process of notating even a single piece of choreography could take months or, occasionally, even years. In 1965, for example, bureau members Muriel Topaz and Lucy Venable were hired to notate Jerome Robbins's staging of *Les Noces*, a complex twenty-five-minute story ballet set to a Stravinsky score and requiring twenty-six dancers. Combined, the two notators attended roughly seventy-five hours of rehearsals, learning each part along with the dancers and making shorthand notes. Each notator then transcribed her shorthand into careful pencil sketches using the standard Labanotation orthography. Once the rehearsals ended, Topaz spent several months correcting and refining the score, at which point the completed document was sent both to the choreographer and to a "certified checker," charged with spotting inconsistencies, omissions, errors in form, and inefficiencies of expression. After certification the score was returned to the Dance Notation Bureau, where it was hand-inked, a process that took nearly two hundred hours. The final product was a 365-page score resembling, in the words of one newspaper reporter, a "long-lost Aztec codex."[34] In the bureau's final accounting, the full notation process took nearly a thousand hours and cost $4,000 (about $38,000 today), a substantial investment for even the most well-funded choreographer or company.

Until the DNB began its efforts, few publicly complained about existing practices of person-to-person transmission. Dances were stored in the kinesthetic memory of dancers and choreographers, and when works were revived after their initial staging, individuals who had participated in earlier productions were frequently recruited to reconstruct and teach the new cast. The system, of course, was not perfect. If too many people moved, passed away, or simply had fuzzy memories, works could be lost. Nonetheless, oral transmission was functional, drew on long-held traditions, and located epistemic authority in dancers themselves.[35] Why, then, did Labanotation succeed in capturing the attention of major choreographers? Why privilege the esoteric and difficult over the accessible and transparent? What did paper accomplish that memory did not?

Like Laban before them, part of the DNB's commitment to textual representation came from their hope that the existence of a written "language" would counter the charges of stagnancy and empty sensualism that had long bedeviled dance, confining it to the "realm of the minor arts."[36] In publications, meetings, and grant applications, they argued that it was "illiteracy" that kept dance provincial and stymied historical comparison and analysis.

In a letter requesting funding from the Rockefeller Foundation in 1958, DNB board member Muriel Topaz argued that "there is no reason why the experience of great teachers must remain geographically isolated, why there can be no exchange of ideas between these pedagogues, why the dissemination of the actual repertoire of the dance must be carried from one place to another by word of mouth as was prehistoric folklore." She contended that a student who is unable to consult written reference material may have "no idea whether or not he is actually learning ballet or some local dialect. The whole picture is comparable to a world of blind musicians learning by rote and performing from inaccurate memory."[37] A 1950 fundraising letter written by Ann Hutchinson Guest echoed these sentiments, claiming that notation did for dance "what has been a commonplace and accepted service for the other arts—the score for music, the written word for drama, line and color for graphic art, stone and brick for architecture. It records and perpetuates aesthetic bodily movement so that posterity can reproduce, enrich, play variations upon, the artistic creations of the past. It provides for the continuity essential to any art. It builds up a literature in an art that heretofore has been as evanescent as the performers who created it."[38]

For the DNB, dance's illiteracy was archaic, particularly given the state of other twentieth-century information technologies. Echoing the views of contemporaries like Claude Lévi-Strauss, Walter Ong, Marshall McLuhan, and Jack Goody, the DNB saw written recordkeeping as central to the edifice of modernity.[39] Just as, for midcentury anthropologists, writing seemed to have provided the decisive break between "primitive" and "modern" societies, the DNB believed that notation would make dance a truly modern art. In an article in support of the DNB's work, dance critic Bernard Taper reflected this view, noting that "for all its sophistication, ballet really is a prehistoric kind of art. Lacking an accepted written language, it has been able to preserve its masterpieces only by devoted, laborious effort, passing them on from one generation to the next by direct communication, like folk legends. And, like legends, few ballets survive this process unchanged."[40] Choreographer Doris Humphrey agreed. Commenting on Labanotation's success in the mid-1950s, she quipped, "now these dances are no longer legends; they are history."[41] These claims, importantly, were not without political weight. As Susan Manning has discussed, the predominately white world of modern concert dance—where the DNB found particular support among choreographers—was, in the early and middle decades of the twentieth century, in the process of erasing many of its ties to Black dance performance. Embracing notation may have represented another mode of differentiating high-culture modern dance from its "primitive" counterpart.[42] Dance's ephemerality was also often linked to its perceived femininity, another source of the art's relatively low status. As Susan Leigh Foster notes, "Of all the arts, dance, with its concern for bodily display, its evanescent form, and its resistance to the verbal, distinguished itself as overwhelmingly feminine in nature."[43]

The DNB's obsession with written documentation also had more immediate, practical roots. In the years following the atomic bomb, the specter of mushroom clouds hung heavy, and anxieties about cultural preservation—and, in fact, preservation of all kinds—felt particularly vivid. For the first time, Americans were faced with the prospect of national, or even species-wide, annihilation. As news anchor Edward R. Murrow put it at the time: "Seldom, if ever, has a war ended leaving the victors with such a sense of uncertainty and fear, with such a realization that the future is obscure and that survival is not assured."[44] Thus notation, which might have remained a chiefly internal concern about artistic continuity, gained broader cultural traction. Like seed stocks, "native" populations, human genetic material, and antiquities, dance was newly construed as a potentially vulnerable resource in need of protection.[45] "It's a dreadful prospect to consider," one newspaper article noted, "that if all over the world all dancers schooled in classical ballet and all the people associated with them were to be wiped off the surface of the earth, we should have nothing left of classical ballet but the music and a few photographs and stage designs—and, of course, our memories."[46] Tellingly, by 1967 the DNB counted among its members Howard Hamilton of the Atomic Bomb Casualty Commission. Chief of Clinical Laboratories from 1956 until its dissolution in 1975, Hamilton learned the script in an effort to preserve traditional Japanese Noh dances in the face of the deaths of many of the art form's most prominent practitioners in the decades after Hiroshima and Nagasaki.[47]

Still, none of these concerns about preservation *required* Labanotation specifically. Given concerns about the loss of cultural heritage, it is even more surprising that the Dance Notation Bureau made its push for Labanotation just as another reproductive technology—film—was becoming a more accessible alternative. Though by no means straightforward, filming a dance was generally both less taxing and less costly than employing notators, with the added advantage of quickly conveying elements like scenery, costume, and the spark of live stage performance. A filmed dance was also more easily "read" than a notation score, particularly by the many dancers and choreographers who were not interested in devoting the time it would take to learn this complex new "language." Filmed dances—while admittedly presenting their own archival challenges—were also permanent records, and they too could be secreted away in underground bunkers.

Often choreographers couched their objections to film in practical or technical terms. For example, though in theory film captured all the action that occurred on stage, if the camera recorded from only a single angle dancers in the background could be obscured, and elements like spacing might be distorted. Others found the iterative process of watching, rewinding, rewatching, and rewinding again inefficient or exasperating. During World War II, some cited the difficulty of accessing raw film stock as a motivation for employing Labanotation. A fuller analysis, however, reveals that choreographers' preferences only partly stemmed from film's technological limitations. In a 1960 *New York Times* article, John Martin laid out one of choreographers' most important objections to film: "A film records a particular company's performance. It may have been a poor company or an off-night by a good one; or it may be that the choreographer would have preferred other dancers for the various roles but has had to change his ideas to conform to the limitations of the performers at his disposal. A script, on the other hand, records the movements as he has conceived them, without anybody's personal quirks, just as musical notation captures the composer's ideas impersonally and in all their purity."[48] In another publication Martin reiterated this point, emphasizing again that "what the camera records is not a composition but a performance." There is, he said, "no guarantee in any particular film that the dancers in the subordinate roles, let us say, were capable of performing the work as the composer conceived it. There is also a certainty that the personal styles of the performers colored the work itself, adding interpretation to what should ideally be a simple statement."[49] For Martin, the essential problem with film was not what it left out—a gesture here, a subtlety of expression there—but what it left in: dancers. The *real* stuff of dance was the pure choreographic idea, existing only in the mind of its maker.

Hutchinson Guest shared this fundamental assumption. When instructing students, she exhorted them to remember that Labanotation *is* "dance in a written form. When we see a sheet of music we speak of it as 'music,' not as 'music notation.' It is old fashioned to speak of a 'dance notation score'—it is a *dance* score, and Labanotation is *dance* and not a separate, though related subject. . . . [T]he students must experience the movements coming to life at the same time that they learn the written version."[50] Dancers, their particular and unique bodies, their quirks, their interactions with the audience on a particular evening, play little role in this account: dance was, by definition, that which could be captured in writing.

There were, perhaps unsurprisingly, critics of this position. Though George Balanchine was among the first choreographers to notate and copyright his works—and even wrote the forward to the first edition of Ann Hutchinson Guest's Labanotation textbook—his New York City Ballet cofounder Lincoln Kirstein never truly warmed to the system.[51] In 1965, then-DNB president Lucy Venable wrote to Kirstein to ask to speak about why the company's interest in notation appeared to have waned. Kirstein responded with a terse one-line note, stating simply that "I do not feel this system of notation offers much in a practical way over film and/or memory" and ignoring Venable's request for further conversation.[52]

Even skeptics who conceded that Labanotation was indeed an effective means of capturing every last "entrechat or swing" were often troubled by its inability to record individual performances. A 1943 article in *Dance News*, for example, reminded the readers that "even in ballet, the most formal of dance arts, there never have been two dancers who danced the same role alike." Moreover, "in the modern dance, where every composition is a purely

FIGURE 3.4. "Are These Two Pictures the Same to You?" These images accompanied an invitation to a lecture-demonstration of the reconstruction of Antony Tudor's 1971 ballet *Continuo* from the Labanotation score. The second page of the invitation continued: "Let American Ballet Theater and the Dance Notation Bureau show you why they are identical!" Irmgard Bartenieff Papers, Special Collections in Performing Arts, University of Maryland.

personal expression of its creator, notations would produce still less definite results. One can write down Martha Graham's *Frontier*, but what will be left of it when reduced to symbols and subsequently recreated from these symbols? *Frontier* is Martha Graham in a certain artistic mood, not a sequence of steps and poses."[53] These detractors, however, fundamentally misunderstood the DNB's aims. To Hutchinson Guest *Frontier* was, in fact, those steps and poses and *not* a particular Graham performance. And as Labanotation grew in influence, it was redefining what counted as creative product (choreographers' intellectual engagement) and what was mere interference (dancers' corporeal contributions).

Some members of the bureau suggested that Labanotation could remove dancers from the choreographic process entirely. They noted, for example, that "choreographers who are familiar with notation have even found it possible to create by writing out the dance phrases rather than having to try them out each time on the dancers."[54] Such changes, they suggested, might have significant monetary consequences, allowing choreographers to create while minimizing the financial burdens of rehearsal time.

Moreover, once rehearsals began, the DNB contended that Labanotation would facilitate a clear, accurate, and efficient translation of the choreographer's vision. Without any of the muddiness of film, oral communication, or physical demonstration to contend with, the choreographer's ideas could be realized exactly, without fear of revision or distortion. As Muriel Topaz argued in the 1958 Rockefeller grant request, Labanotation was "capable of much greater precision than words and can describe the movement completely. In fact, the consideration of ballet movement in terms of notation forces the teacher into a precise realization of what he wants, and supplies for him and his students an exact and common language."[55] The messiness of the in-person creative interaction—once understood as productive—was here redefined as a problem to be solved.

This was a new mode of thinking about dance-making. Customarily, choreography was understood to be a somewhat collaborative project between choreographer and dancer: the choreographer outlining a schema but expecting that it would be transformed in the process of engaging with the particular bodies and minds who danced it. As George Balanchine initially put it, "I'm not one of those people who can create in the abstract, in some nice quiet room at home. If I didn't have a studio to go to, with dancers waiting for me to give them something to do, I would forget I was a choreographer. I need to have real, living bodies to look at. I see how this one can stretch and that one can jump and another one can turn, and then I begin to get a few ideas."[56] Even for Balanchine—a figure who has often been seen as the paradigmatic solitary male genius—creation came only through the dialogic interaction between choreographer and dancers. In modern dance, interest in improvisational and quasi-improvisational methods was even more explicit.[57] Labanotation, in contrast, posited a different vision of creation, one that drew attention away from dance's rootedness in human bodies.[58]

A 1941 *Time* magazine story on Labanotation captured this framework absurdly well. The article, full of praise for the new system, was accompanied by a photo featuring two dancers leaping sideways through the air, eyes trained fixedly on their notation scripts. The caption read, "Paying no attention to their bodies, Kip Kiernan and Winifred Gregory follow the score of their ballet script. . . . Only experienced dancers can 'read' without watching what they are doing."[59] This was the new ideal—a script that transferred the

FIGURE 3.5. "Paying no attention to their bodies, Kip Kiernan and Winifred Gregory follow the score of their ballet script." From "Toe Writing: Ballet Dancers Learn How to Put Muscles in Black and White," *Friday*, April 11, 1941. Photograph by Hans Reinhart.

choreographer's vision directly to his instrument, a transaction so frictionless that even the dancers lost track of their own forms. Dancers, moreover, might not even notice their demotion. As Martin put it, though notation "has been in the minds of choreographers at least since the time of Feuillet in the Court of Louis XIV," dancers were less invested, "for they are traditionally more interested in increasing their physical prowess than in troubling their pretty heads over hieroglyphics."[60]

Martin's casual misogyny and condescending view of (mostly female) dancers may help explain why their voices were often absent in the public debates about notation. When skepticism did bubble up, however, Martin was incredulous, remarking that "even dancers themselves are frequently on the defeatist side of the argument, setting forth how complicated the human body is in movement, and how laborious it would be to notate each quirk and tremor." He thought little, however, about why this might be the case, dismissing their objections as mere "superstition."[61]

Owning Movement

In spite of these challenges, the DNB achieved a number of landmarks in the decades following its founding. Among the most notable were its interventions in copyright law. In fact, it was in law that the distinctions Labanotation suggested—between head and hand, thinker and doer, dance-maker and dancer, male and female—were reified. Because of its ephemeral nature, dance had long presented a problem for conventional intellectual property regimes. Since 1879, US copyright law has protected particular expressions of an idea, rather than the idea itself.[62] The expression of dance, however, was understood as fleeting and thus not easily dealt with by the Copyright Office's file cabinets and bureaucrats. Until the mid-twentieth century, therefore, dance had never been successfully copyrighted under US law.

The Dance Notation Bureau hoped to change that, arguing that by capturing dance in a written form, Labanotation could open up new modes of legal protection. In early 1948, Hanya Holm was recruited to create the choreography for Cole Porter's new musical *Kiss Me, Kate*. Not only was Holm a fervent advocate of Labanotation, she was also acutely conscious of the financial realities of the arts world; her first company had recently folded due to lack of sufficient funding. Thus, after accepting the job, Holm recruited Ann Hutchinson Guest to notate the musical's complete choreographic score. From the beginning, Holm's motivations were both high-minded and financial. As she recalled in an interview, "I saw the necessity of protection. It was a show which was really running well. It was a show that everybody wanted to do and wanted to repeat."[63] Labanotation served as a defensive maneuver to protect an asset with clear value.

Still, at the time of the musical's premiere it was far from certain that the Labanotation would provide any legal protection whatsoever. Only a decade earlier the Copyright Office had rejected Eugene Loring's Labanotated application for the choreography for *Billy the Kid* on the basis that the Laban system was "not yet recognized as a set method for recording movement."[64] By the early 1950s, however, the DNB's program of advocacy had made inroads: Richard S. MacCartney, Chief of the Reference Division of the Library of Congress, reached out to the DNB and suggested that dance works could now be submitted via Labanotation.[65] In 1952, the *Kiss Me, Kate* score was successfully registered for copyright.[66]

The DNB rejoiced, and intellectual property protection quickly became a key element of its activities. After the completion of a score, two copies of the notation were made, often on microfilm; one was stored in the bureau library and the other immediately sent to the Copyright Office. Not coincidentally, 1952 was also the year that the DNB was officially registered as a nonprofit corporation in the state of New York, likely in anticipation of growing revenue in the coming years. (Agreements with choreographers varied, but individual notators generally received a flat fee for their work, while the DNB itself collected 10 to 20 percent of a choreographer's royalties each time a piece was performed.)

With this change in legal status, the potential benefits to notating a work multiplied. A choreographer with a copyright could, among other things, set compensation rates and determine whether and when a piece would be performed. Holm, for example, immediately refused performance rights to a proposed staging of *Kiss Me, Kate* at Princeton University. The protection, of course, was not absolute, particularly in cases when works were performed overseas, where few enforcement methods existed. Still, Martin noted that the change "gives official recognition to the dance creator as such, which is at least a small step toward the dignity to which he is entitled." He also hoped that it would make younger choreographers and stagers aware of the immorality of uncredited appropriations, thus spurring a "stimulation of fresh creative activity in the summer theatres, as well as a heightened moral code."[67]

Yet Labanotation's role in copyright came with significant ideological baggage. As the bureau's resistance to film demonstrated, the DNB was attracted to Labanotation in part because of how it seemingly captured the unadulterated essence of a choreographic work rather than the expressive details of a particular performance. Copyright legally enshrined this preference and in so doing gave credence to a particularly narrow, intellectual, and disembodied concept of dance authorship. Thus, not only was Labanotation's history now inextricably intertwined with the commoditization of dance, but its widespread use in copyright represented a decision about what kinds of artistic output were worthy of protection—indeed, about the nature and essence of the artwork.

The Holm decision of 1952 expanded the category of acceptable "fixing" methods, but it left other issues unresolved. Since 1909, choreographic works had been included in the standard copyright statute as Class D "dramatic compositions." As such, choreography with significant dramatic content—such as Holm's work for *Kiss Me, Kate*—could, at least in theory, be copyrighted. More "abstract" work, on the other hand—like much midcentury ballet and modern dance—was not eligible for protection.[68]

But in 1955, the US Senate's Subcommittee on Patents, Trademarks, and Copyrights began considering a range of potential revisions to US copyright law in light of the "far-reaching changes . . . in the techniques and methods of reproducing and disseminating the various categories of literary, musical, dramatic, artistic, and other works that are subject to copyright."[69] As part of this process, the committee reopened the question of the protection of nondramatic dance, and in 1959 it produced a formal report on the subject. "Copyright Law Revision, Study No. 28" illustrated the further solidification of a disembodied view of artistic creativity.

For example, although the report acknowledged that film was a legally permissible fixing method, it makes clear that Labanotation was the emerging standard. The document included comments from nine legal and dance authorities, including DNB members Ann Hutchinson Guest, John Martin, Hanya Holm, and Lucile B. Nathanson. Of the remaining experts, only Agnes de Mille lent any support to film, though her advocacy was tempered by concerns about the medium's perishability. Moreover, the report's author, Borge Varmer, explicitly defined the term "choreographic work" as referring "both to the dance itself as the conception of its author to be performed for an audience, and to the graphic representation of the dance in the form of symbols or other writing from which it may be comprehended and performed."[70] Varmer made no reference to either physical or filmed performance, eschewing the traditional legal focus on protecting *expression* in favor of a defense of the pure authorial *conception*.

Information Machines

Even as Labanotators rhapsodized over the perfectly efficient transmission of choreographers' ideas to dancers' bodies, they acknowledged that the production of Labanotation itself was not similarly effortless. In particular, by the 1960s some notators had become increasingly conscious of the time and effort it took to hand-ink pages upon pages of symbols and began wondering if there might be some way to "mechanize" the process.

At least a few bureau members had been interested in using new technologies to speed up the creation of notation for some time. A 1948 DNB meeting, for example, opened with a group discussion of the stenotype machines used at the United Nations for real-time recording of speeches. "Could not such a machine," one attendee asked, "be devised for purposes of writing down movement as it was performed?"[71] Another member quickly threw cold water on the idea, arguing that such a purpose-built machine would likely be prohibitively expensive and would still require a highly trained operator. Instead, the conversation turned to a discussion of mimeograph ("ditto") machines and new developments in xerography, tools that members hoped could eventually make the reproduction of hand-drawn scores easier.

The DNB's efforts to embrace mechanical recording in the years that followed responded to the changing information technology landscape but also reflected growing concerns within the organization that the DNB needed a broader user base. Despite its successes, the bureau's financial footing was never entirely solid; its operating funds came from a mix of licensing income, course fees, membership dues, philanthropic contributions, and foundation grants, but most leaders continued to volunteer their time, and salaries for the few part-time paid staff were low. As a 1965 report noted, "In the 25 years' history of the DNB one financial fact emerges as a certainty. No one identified with the Bureau has ever been adequately rewarded, monetarily, for his work." The fundamental problem, DNB leaders began to suspect, was their "almost exclusiv[e]" focus on professional dance communities, "a segment of the population with limited means."[72] Broadening the board membership, they hoped, would help the DNB connect with larger audiences, and in 1965 they began reaching out to "leaders in related fields of interest," hoping to strengthen their connections with "anthropologists, psychological and medical researchers, physical educators, industrial researchers, etc."[73]

In 1966, these efforts began to pay off, and the DNB acquired a new chair for its board of directors: Earl Ubell, a well-known journalist and the health and science editor for WCBS-TV. Delivering exactly the push some DNB members had hoped for, Ubell quickly began urging the bureau to consider how new technologies might simultaneously ease their workload and make notation more broadly appealing. (In the early 1970s, the sense that it was crucial to "update" Labanotation's production became even more pronounced due to the new availability of consumer videotape recorders, a development which many notators feared would further dilute their market.)[74]

As one of his first acts as Board President, Ubell facilitated a meeting between the DNB and IBM, with the idea that the company's computers might be used to help produce final copies of notation scores, eliminating the laborious work of hand inking.[75] The company's Office of Product Development was intrigued by the idea, though it proved impossible to execute. IBM's room-filling mainframe computers—like the just-released System/360—were designed to store and process numerical data and lacked the memory and display hardware to manipulate graphical information, especially the complex vocabulary of geometric shapes that constituted Labanotation.[76]

As an alternative, the company's engineers suggested developing a product for the company's electromechanical Selectric typewriter. First released in 1961, the Selectric was an immediate hit, famous for its innovative replacement of the traditional basket of individual type bars with a set of interchangeable, golfball-sized steel "typeballs" studded with letters. When a user struck a key, the ball spun and imprinted the page through an inked ribbon. Not only did the new system solve the problem of jammed key bars, but swapping one typeball for another took just seconds and allowed a single typewriter to print in multiple fonts and languages. Promotional films showcased close-up action shots of the typeball as it turned and tilted, luxuriating in the spectacle of the "ingenious printing element that dances across the paper at incredible speeds."[77]

FIGURE 3.6. Photograph from "Lexington Engineers Relate Story Behind Unique Product," *IBM News* 11, no. 1 (January 18, 1974), 4. Reprint courtesy of IBM Corporation © 2024.

IBM set out to develop a typeball for dance itself, and—after several years of back and forth between notators in New York and engineers in Lexington, Kentucky—production began in 1973. As IBM boasted in a press release, the final form of the "L/N Ball" featured eighty-eight symbols that could be arranged to form "a complete vocabulary for recording movement of *any* kind." Striking marketing photographs featured bunned and beribboned ballet dancers leaping and lounging, but the accompanying text made it clear that the element was not intended for dance alone. IBM explained that Labanotation was a "scientific approach to recording movement" and situated the ball as one of a "growing collection of special-purpose typing fonts which IBM has designed for the technical disciplines." In the first year, the company hoped for sales of at least two thousand units at a price of $18 each (about $115 today).[78]

In an article on the collaboration in the *New York Times*, Herbert Kummel, the DNB's executive director, characterized the development of the typing element as a "major breakthrough" for movement notation. Not only, he predicted, would the Selectric facilitate the use of Labanotation at the ninety-plus universities already teaching the method, he believed the L/N Ball's introduction would help the system reach new audiences in "sports, the behavioral sciences, industrial operations and physical therapy." As he explained to the *Times* reporter during a demonstration of the machine, "It might be interesting, for instance, to discover just how Joe Namath (quarterback of the New York Jets) throws those lovely passes. . . . Labanotation affords a depth of perception and accuracy of measurement slow-motion film cannot hope to match."[79] Additional marketing materials illustrated these uses, placing photographs of a football pass, windsurfing, and horse racing alongside the corresponding notation.

Kummel was not the only one to see the Selectric as an inflection point in Labanotation's history; the phrase "Gutenberg leap" was thrown around both in the press and the DNB offices with some frequency. Indeed, many bureau leaders hoped that the symbolic power of the typewriter—long associated with neutrality, universality, and modernity—could help establish the credibility of Labanotation in the wider world.[80] The fact that the script could be produced with the same technology that pumped out press releases, earnings reports, and mathematical symbols would serve as a vivid illustration of an idea that the DNB had long promoted: movement was simply information, and with the right system in place, information that could be objectively recorded. On a more practical level, many notators were excited by the promise that the Selectric would not only significantly speed up the production of scores but also create an even more standardized final product.

FIGURE 3.7A-B. Photographs from an IBM press release announcing the IBM Selectric Labanotation Ball typing element, December 20, 1973. Reprint courtesy of IBM Corporation © 2024.

120 CHAPTER 3

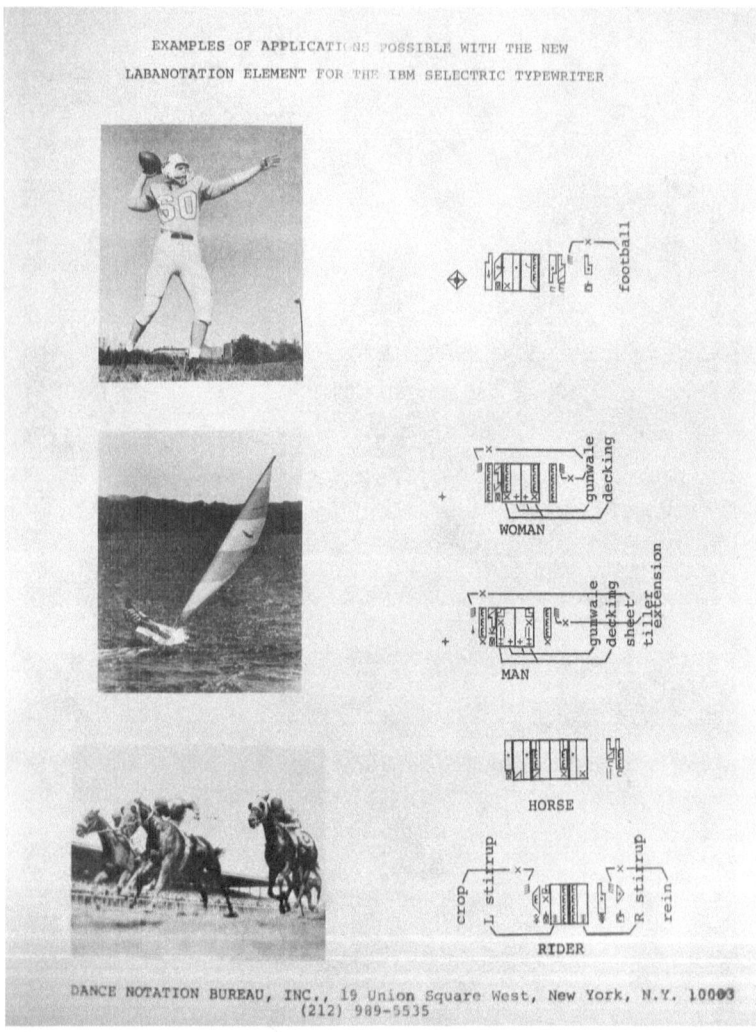

FIGURE 3.8. Possible applications of the Labanotation Element for the IBM Selectric Typewriter. Courtesy of the Dance Notation Bureau.

Just as Labanotation itself had banished the dancer's body from the written record, the Selectric would remove any traces of the notator's particular hand.

In practice, however, attempts to transition from hand-inked to typed scores were rocky. One could not simply unpack the new L/N Ball and go to work, and IBM outsourced training to the DNB, directing purchasers to contact the bureau directly to obtain a thirty-three-page user manual.[81] But even once a potential user had pored over the manual's pages, mastered the basic scheme, and memorized which keys produced which symbols (the manual suggested affixing small paper tabs on top of the standard keyboard as an intermediate step), actually creating notation was a protracted process. Only six of the eleven standard direction symbols, for example, could be produced with a single keystroke; the remaining symbols had to be built piecemeal through elaborate combinations of other keys. Moreover, even those six "easy" symbols could be directly keyed only when they indicated movements of a specific, short duration; more sustained movements required an entirely different procedure.[82]

The process for recording Effort Notation was equally laborious. Recording a single Direct Effort symbol—essentially, a bent line—required completing the following sequence of actions: hitting a key, rolling the platen up one line, hitting another key, backspacing, rolling the platen up one line, hitting a third key, backspacing again, rolling the platen up once more, and hitting a final key.[83] Yet another problem involved the orientation of the page itself, as Labanotation was read from the bottom up, rather than left to right. After some discussion, the DNB decided to address the issue by inserting standard typing paper sideways.[84] In theory, this solved the problem, but it also required that typists compose with the paper at a 90-degree angle to the direction in which the score was read, a process resulting in much awkward

FIGURE 3.9. Handmade key for the Labanotation Element taped to a Selectric typewriter at the Dance Notation Bureau offices. Photo by author.

neck-craning. A considerable number of elements could not be typed at all and had to be hand-inked after the typed score was removed from the machine.

The first page of the manual acknowledged these challenges, explaining to the reader that "the IBM-L/N element is such a new mechanism that you, the purchaser of it, are only slightly behind us in experimentation." It welcomed users to contact the DNB with both problems and creative solutions.[85] Many heeded the invitation, though complaints were more frequent than suggestions. Some were annoyed by how time-consuming the process was or how the typewriter limited nuance, others by the frequency with which the L/N Ball produced not the promised clear symbols but unreadable blobs of dark ink. Still others criticized the aesthetics of the new typed scores and warned that notators should be careful about unthinkingly making changes to suit the typewriter's limitations, as "what we change for the machine will of necessity become a change in the system whether we like it or not."[86] One bureau member argued that—if speed were the truly the issue—it would be better to streamline the hand-drawn scores, rather than "making all kinds of concessions and adjustments for another wretched machine." Evoking the medieval scriptorium, he protested: "We are requiring our 'monks' to continue turning out manuscripts competing with the Book of the Hours. Why can't we make some concessions about hand done scores that would make them less time-consuming, more economical, and scores which are readable and not something that one has to memorize what this dark-shaped blob means."[87]

The fear that the typewriter would alter Labanotation's structure ultimately proved unfounded, though this was in large part a function of the Selectric's limited uptake.[88] Despite some excited coverage in the media, three years after the L/N Ball's release bureau president Ann Hutchinson Guest herself was still complaining that "none of us here is experienced with the typewriter and has never gained any real speed."[89]

By the mid-1970s, however, deploying computers to record, edit, store, display, and print notation seemed increasingly feasible, and the bureau's attention shifted once more. Capitalizing on the excitement about the graphics and memory capabilities of new mainframe and minicomputers, DNB members reached out to a small number of computer researchers about the potential for collaboration.[90] Initially, these efforts were either conceived as explicit alternatives to the Selectric or envisioned using the typewriter in conjunction with the computer as a printing tool. A 1976 Labanotation printing system ("LABA") developed at the University of Pennsylvania's Department of Computer and Information Science, for example, employed a Sperry UNIVAC 70/46 mainframe computer to generate scores on a TEKTRONIX 4010 cathode-ray tube graphical display. Guided by a series of command menus, users employed a device (either a set of positioning wheels or an electronic tablet-and-pen system) to move a cursor across the screen and a keyboard to enter codes representing notation symbols. The computer converted the commands into Labanotation elements and displayed them on the screen. Users could print scores using a photographic hard-copy unit that duplicated the display output one page at a time, but the system's designers hoped eventually to connect the computer directly to a Selectric to produce hard copies more cheaply and easily.[91]

The primary goal of this kind of computerization was the faster production of final scores, allowing notators to make corrections or alterations without the need to redraw whole pages of text. This innovation, the DNB projected, would produce a "a four-fold speed-up in the creation of dance scores as well as an increase in accuracy and a reduction of cost."[92] Yet again, breathless reports predicted that the computer would lead to a "Gutenberg leap" in the world of dance, replicating the promises that had accompanied both the IBM Selectric and handwritten notation itself.[93] Eager boosters envisioned a world with computers in every rehearsal studio and a nationwide "blossoming" of libraries, performing companies, and dance departments all centered around Labanotation. In one memorable formulation, these new users would locate a computerized score, mix it "judiciously with the sweat from dancers true and pure," and produce transcendent art, a process of reconstitution that would supposedly "make the reconstruction of Tang on the moon fade into relative insignificance."[94]

The reality was, once more, more complicated. Though early users of the Penn program marveled at its relative ease of use, the required hardware was still prohibitively expensive and the software difficult to refine or even operate consistently without the assistance of a trained computer professional.[95] (In recognition of this fact, Sperry UNIVAC leased rather than sold the 70/46; the price was $33,000 per month, or $221,000 today.)[96] In 1977, supported by funds from the Mellon and Sloan Foundations, the Dance Notation Bureau was able to purchase a Princeton 801 graphical display terminal and a TEKTRONIX hard-copy unit, but they did not own their own computer, relying instead on the use of a PDP-11 minicomputer at the Columbia University Medical School.[97] As bureau leader Lucy Venable later summarized, "the Bureau periodically received money for a programmer but did not have sufficient access to a computer, or it would be given equipment but did not have funds to hire a programmer for a sufficient length of time."[98]

But as new "microcomputers"—inexpensive devices replicating many of the capacities of mainframes and minicomputers at a much lower price point—began to enter the market in the 1980s, a new crop of Labanotation programs entered development. Unlike earlier computers, these typewriter-sized devices became popular consumer items and spurred increased experimentation in software development. In 1985, for example, researchers at England's Birmingham Polytechnic created KINOTATE for the BBC Micro Model B. Developed as part of the BBC Computer Literacy Project, this affordable personal microcomputer was widely used in British primary schools, and KINOTATE—available on a 5.25-inch floppy disk—was designed as a tool for teaching basic notation to students. The limited memory of the BBC Micro made KINOTATE unsuitable for professional notators, but the program's developers were satisfied that it functioned well enough to be implemented in schools.[99]

Moreover, as computers with increased capabilities became available, software to manipulate graphical images became more robust. Rather than invent a new drawing program to record Labanotation scores, choreographer Andy Adamson at the University of Birmingham created a program named "Calaban" (short for Computer Aided Labanotation), which produced notation using the already popular AutoCAD software used in architecture and engineering.[100] Efforts at the University of California, Los Angeles, produced a similar application, known as LCs LN.[101]

The mid-1980s rise of microcomputer graphical user interfaces (GUIs) also sped the development of notation software. The DNB had established an official "Extension for Education and Research" in the Department of Dance at Ohio State University in 1968; there, with the encouragement of the computer artist and graphics pioneer Charles Csuri, an interdisciplinary group of bureau members, dance faculty, and a succession of computer programmers began a years-long series of experiments that eventually resulted in the 1987 release of a program called LabanWriter. First run on an Apple Computer Macintosh Plus (one of the first microcomputers shipped with a GUI operating system), LabanWriter was intended for users with "minimal computer experience" and allowed operators to use a mouse to select symbols from a menu bar, drag them into place, re-size them as needed, and copy movement segments from across a score.[102] Once complete, the score could either be printed using any compatible printer (including high-resolution laser printers that mimicked hand inking) or stored on a magnetic disk for future output or editing. In 1989, a $40,000 grant from the Mellon Foundation supported further work on the project, and LabanWriter 2.0 (for the Macintosh IIci) was released the following year, provided free of charge to anyone who requested it. By this point Macintosh computers were increasingly commonplace in schools and creative workplaces, and stock units could run notation software without modification. A new IIci could be purchased for just over $6,000 in 1989 (under $14,000 today), or one-fifth the price of a PDP-11 two decades earlier. Most notators considered LabanWriter the first truly practical tool for preparing professional-looking scores.

In an article in the university's alumni newsletter, Lucy Venable—by then director of the Ohio State DNB Extension—celebrated the changes computerization had provided in quasi-Taylorist terms. The process of creating a blank score, she observed, had been sped up by a factor of ten: "To mark off with pencil and ruler, on graph paper, the beats and bar lines for three staffs of four measures, four beats in length, four squares per beat, with three floor plans takes me three minutes and fifty-five seconds. With LabanWriter I can do the same thing in twenty-nine seconds."[103] Other notators were simply happy about their newfound ability to keep their work safe from everyday mishaps. One bureau member, for example, recalled the day when Muriel Topaz, excited about finally completing the score for one of Hanya Holm's Broadway shows, boarded the subway with her cleanest copy and a half-full bottle of gin, intending to celebrate with her fellow notators. "Alas!" her colleague recalled, "in the subway ride, the bottle had seeped open and the hand written pencil copy was soaked! Those were the 'good old days' before LabanWriter and the advances in technology."[104]

Unlike the Selectric or the early Penn system, LabanWriter and its alternatives functioned beyond their symbolic capital. Once the computers became relatively affordable and capable, users could easily store, edit, and print scores and keep them safe from the errant error or spill, whether of ink, coffee, or cocktail. Many notators began to use them. Fundamentally, however, the ability to produce Labanotation via computer did not significantly expand the bureau's reach or increase notation's daily presence in the world of dance. In 1976, Earl Ubell declared that once computers were widely available, any remaining resistance to notation would quickly fade away.[105] Like many such deterministic predictions, however, this one proved overly optimistic: the mere availability of the computer technology did not compensate for Labanotation's lack of ease of use relative, for example, to film or video. While existing notators and choreographers retained their devotion to the system— and Labanotation continued to play an important role in copyright law—the computer itself did little that bureau members had not accomplished decades before.

A 1982 newspaper article about the development of new computer programs for Labanotation reflected these ongoing tensions. The reporter asked a computer engineer whether choreographers and dancers were enthusiastic about the new computer programs he was at work developing. Yes and no, the engineer responded, explaining, "We haven't even gotten to the stage where dance accepts handwriting, much less computers."[106] Again, the issue was not the medium of reproduction but notation itself. Opinions among choreographers were similarly equivocal. While some were intrigued by the possibilities of choreographing via computer, others—including Donald Mahler, the director of the Metropolitan Opera ballet—remained skeptical. Echoing complaints that had been leveled at Labanotation for decades, he noted that any attempt to reproduce movement on a computer screen would be "far removed from the spirit of dance, which is full of qualities other than the steps you are doing."[107] Though at least one newspaper report predicted that "that the computer would finally be the answer to W.B. Yeat's [sic] question of how we separate the dancer from the dance," not everyone was on board with the divorce.[108]

A New Identity

Though the Dance Notation Bureau is still in operation and Labanotation scores continued to be made and used, the organization was never quite able to effect the wholesale revolution it had once envisioned. Its work, however, had other kinds of long-term consequences. Dance scholar Susan Leigh Foster notes that the appellation "choreographer" was rarely used prior to the 1920s, a change she argues was reflective of a new vision of what dance-making meant. In contrast to "making up," "directing," or "staging" a dance, the new title "specified the author of an original work and highlighted the inventive engagement of the artist in crafting movement."[109] Labanotation likely both reflected and accelerated this transformation. As in other areas of midcentury American life where control moved from workers to owners and choreographed order reigned, Labanotation made dance ordered, legible, and commodifiable. Though notation did not ultimately displace other forms of recording and other modes of dance-making, it did enshrine—at least legally—a conception of artistic creativity in which the choreographer appears particularly powerful.

This dismissive attitude toward dancers and dancers' bodies, of course, would not have so easily triumphed in the absence of preexisting—and profoundly gendered and racialized—power relations between dancers and choreographers and widespread beliefs about the relative merits of intellectual and physical labor.[110] In particular, the heavily feminized nature of dance performance may help explain the disparate tracks musical and dance notation followed over the course of the twentieth century. While avant-garde composers like John Cage (a onetime DNB board member) developed new notation systems that foregrounded the indeterminacy of performance and recognized musicians' participation in the act of art-making, dance took the opposite path.[111]

For Labanotators, human movement became an entity that could exist in the absence of human bodies. This transformation permitted the preservation and management of dance works on a scale previously thought impossible, but it simultaneously altered ideas about what dance actually *was*. The act of dance-making was disassociated from the potentially idiosyncratic and unrepeatable somatic contributions of individual dancers. Dancers became mere noise in an otherwise rationalized recording system: a difficulty to be solved, rather than an integral aspect of the piece. They were, to borrow a phrase from information theorist Claude Shannon, simply the "statistical and unpredictable perturbations," distorting the transmission of dance's clear signal.[112]

It would be too simple, however, to read the story of Labanotation at the Dance Notation Bureau as an uncomplicated slide toward control, flattening, and disembodiment. Though the impulse to preserve dance was couched in a narrative of impersonal scientific progress, notators themselves had considerably more complex relationships with their work. For many, the drive to record was an emotionally charged mission to gain respect for an underappreciated, feminized, and all too ephemeral art. Others flocked to the profession as a way of remaining close to dance after their performing careers had ended; Ann Hutchinson Guest herself wrote about how a career in notation helped her deal with the devastating recognition of her limits as a performer.[113] Whatever the final outcome, their professed hope was not to desiccate dance but to care for it.

The process of notation was also undoubtedly an intimate one. Women like Hutchinson Guest often spent hundreds of hours with a single piece of choreography: observing dancers, conferring with choreographers, poring over details, delicately penning symbols. One 1976 guide for checkers reminded them to "remember that certified checking is a devastating experience for every notator," suggesting the profound emotional investments notators made in their craft.[114] Some notators even contended that they deserved a permanent financial stake in the scores they produced, an implicit statement that the act of notation was itself a creative act deserving of remuneration. These claims, however, gained little purchase, undermined in part by the very ideology of mechanical objectivity the bureau itself promoted. It is nonetheless clear that notation was never entirely mechanical or detached, a truth that present-day notators increasingly recognize.[115]

In the public mind, however, the idea of Labanotation as an objective, neutral tool for making sense of movement took hold, and as the bureau increasingly began working directly with scientists and engineers—and attracting attention from medicine and the social sciences—Labanotation's former associations with avant-garde art, esoteric theories, and fascist politics faded further. The transition was so successful that by 1965, bureau president Lucy Venable noted that the organization had begun to reevaluate its mission in light of the fact that there had "almost seem[ed] to be more interest" in nondance uses of notation "than there does in the dance field."[116] Other notators complained that Ubell and Kummel had spent so much energy pushing the new "technology projects" that the bureau had resources for little else. Even Ann Hutchinson Guest, who relocated to London following her 1962 marriage to dance historian Ivor Guest but remained involved with the bureau, wrote a letter of concern to a friend still in New York: "It seems time that DNB meetings were concerned with putting the house in order, not just with computer projects, no matter how fascinating. Can you help? Can you speak up for us absent members?"[117] For Hutchinson Guest, the bureau's new obsession with electronic technology had distracted it from more pressing matters.

Bruno Latour has remarked upon the "extraordinary obsession of scientists with papers, prints, diagrams, archives, abstracts and curves on graph paper. No matter what they talk about, they start talking with some degree of confidence and being believed by colleagues, only once they point at simple geometrized and two-dimensional shapes."[118] Thanks to the bureau's work, in the middle decades of the twentieth century a significant corner of the American dance world began to share this obsession. The motivations of Labanotation's proponents were multiple and complex—a desire to increase dance's prestige by linking it to science, deep-seated concerns about preservation and conservation, hopes of carving out a new professional niche. The outcome, however, was steeped in contradiction: in seeking to preserve a corporeal art, the DNB promoted a new understanding of dance in which the physical body played an attenuated role.

4

Corporate Bodies

When Tom Rath arrived for his interview at New York's United Broadcasting Corporation, he opened the door to find Mr. Gordon Walker, "a fat pale man sitting in a high-backed upholstered chair. . . . He didn't stand up when Tom came in, but he smiled. It was a surprisingly warm, spontaneous smile, as though he had unexpectedly recognized an old friend. 'Thomas Rath?' he said. 'Sit down! Make yourself comfortable! Take off your coat!'" Tom sat, awkwardly draping his coat over his lap, though the room was not particularly warm. Suddenly, Walker "touched a button on the arm of his chair and the back of the chair dropped, allowing him to recline, as though he were in an airplane seat. 'You will excuse me,' Walker said, still smiling. 'The doctor says I must get plenty of rest, and this is the way I do it.'"[1]

Rath, the protagonist of Sloan Wilson's famous 1955 novel *The Man in the Gray Flannel Suit*, had come to United Broadcasting to seek a new, better-paying job, eager to jettison his family's shabby-seeming homestead in suburban Connecticut. But what began as a pragmatic attempt to move up the economic ladder soon became an expedition into the bewildering world of the large corporation, a place governed by unspoken rituals, where smiling seemed to be "a company rule" and personnel men probed deeply into employees' lives. For Rath, the attention was unwelcome: a paratrooper during World War II, he remained haunted by his service and desperate to keep his past in the past.

At its release, Wilson's novel was called "uncannily familiar"[2] and "reportorially exact in its account of the pressures, problems and tribal customs" of the "ambitious commuters who are too young to be either successes or failures but whose time is running out."[3] Within a year, it was adapted into a movie starring Gregory Peck as Rath and Jennifer Jones as his unhappy wife, and "the man in the gray flannel suit" was set on its path to infamy as shorthand for all seemingly dull, conformist strivers.

Wilson, of course, was not the only commentator on the culture of midcentury Anglo-American business. Sociological critiques of the postwar corporate mindset—William H. Whyte's *The Organization Man*, C. Wright Mills's *White Collar*, David Reisman's *The Lonely Crowd*—similarly sketched the emergence of an insidious ideology of loyalty, collectivism, and conformity as well as the development of new technologies intended to enforce this ethic: chief among them personality and aptitude testing. As Whyte reported in *The Organization Man* in 1956, corporate personnel departments increasingly availed themselves of tests designed to evaluate "a man's degree of radicalism versus conservatism, his practical judgment, his social judgment, the amount of perseverance he has, his stability, his contentment index, his hostility to society, his personal sexual behavior," even his sense of humor.[4] (At his interview, Tom Rath is given a typewriter, an hour, and an empty room and asked to convey his life story, another common evaluative technique.) Between 1950 and 1955, the number of blank personality test forms sold had risen 300 percent, and new psychological consulting firms were popping up with shocking rapidity.[5]

As this chapter will demonstrate, however, the evaluation of employees was not confined to written or verbal answers knowingly given to examiners. Many of the men and women who strode the hallways of twentieth-century corporations, commanded underlings from behind their desks, and beseeched superiors for promotions were also evaluated physically. Deconstructed, categorized, and reduced to scribbles on a page, their bodily movements were a means of communication that could be the wellspring of their success or—if used poorly—their professional undoing. As Tom Rath would learn, Gordon Walker's broad gestures, seated posture, and easy recline were not necessarily true signs of friendliness or vulnerability but instead a kind of kinesthetic performance.

In fact, by the time Wilson's novel was published attention to bodily movement in business settings was in full swing, in large part as the result of the efforts of a man named Warren Lamb, a former student of Rudolf Laban. In 1952, Lamb formed Warren Lamb Associates, a consulting group that promised to transform the process of corporate hiring through its signature test, first known as "Aptitude Assessment" and later as "Action Profiling" and "Movement Pattern Analysis."[6] Used for selecting employees ranging from entry-level salespeople to CEOs, the technique claimed to reveal an individual's true character through his unconscious bodily movements. How someone communicated, his decision-making process, whether he would prefer a day of quiet research or a contentious afternoon meeting: all would be revealed with the flick of a finger or a shoulder's shrug. Lamb described it as a revolutionary tool for remaking the future of work, providing badly needed assistance to both employer and employee. Though few CEOs were capable of following the complex logic Lamb employed—or the strange configurations of hash marks that constituted his evidence—they were sufficiently pleased with his results that the company expanded and prospered.

The following pages chronicle Aptitude Assessment's history, exploring how and why this new form of Laban-derived analysis found a congenial home in the midcentury corporation. More broadly, they trace a change in the form and purpose of notation, tracking how a technique formerly used to prescribe movement in specific settings was newly deployed to describe it in everyday life. In the process, individual ways of navigating the physical world—how a person reached for a glass, greeted a friend, brushed their teeth—were reconfigured as sources of invaluable information, resulting in new forms of surveillance and performance that reached far beyond the corporation's walls.

Types and Kinds

Born in 1923 to a middle-class family in Wallasey, near Liverpool, Lamb was set to begin studying for his university entrance exams when Britain officially declared war on Germany in September of 1939. He immediately attempted to enlist but, at sixteen, was too young and decided to find work while he waited. His father, a shipping clerk, suggested that something in banking might make a sensible choice, and Lamb secured an entry-level position in the local branch of the venerable Lloyds Bank. A year later, however, and old enough at last, he joined the Royal Navy, in which he served until the end of the war. Lamb wrote little about his wartime experiences, but he did recount an interest in theater kindled while performing Shakespeare with his fellow sailors on an aircraft carrier's hanger deck in the conflict's waning months.[7]

Upon demobilization, Lamb returned to a job in the Executor and Trustee Department at Lloyds, describing himself as competent but restless. In his free time, he dabbled simultaneously in right-wing politics (he was a member of the Young Conservatives and worked on Ernest Marples's parliamentary campaign) and left-wing drama, taking courses run through Joan Littlewood and Ewan MacColl's progressive Theatre Workshop. In addition to its radical politics, the Theatre Workshop was known for its attention to the role of physical movement in acting, and it was there that Lamb first encountered Laban's theories, where they were taught alongside the work of Stanislavski,

Meyerhold, and Jaques-Dalcroze.[8] Laban, though, loomed particularly large. Not only was he physically present in Manchester, but one of his students and former collaborators, Jean Newlove, had begun teaching weekend movement sessions in 1946 and joined the company as a choreographer, movement instructor, and performer shortly thereafter. Until at least 1960, most days began with an hour-long Laban-based movement course, and students were also sent to study with Laban directly at the Art of Movement Studio.[9] For actors who wished to transform themselves, Littlewood claimed, the ability to analyze and deploy movement was crucial. "With Laban," she would write, "it became possible to divest yourself of your own characteristics, become a new being, live in a different time and place."[10]

Lamb too began transforming himself, though perhaps not in the way anyone anticipated. After developing a fascination with Laban's work—and with eventual personal encouragement from Laban himself—Lamb left Lloyds, using his £62 war gratuity to cover a term and a half of full-time enrollment at the Art of Movement Studio.[11] Within a year he had been recruited to join the Industrial Rhythm program as an apprentice analyst and trainer, a course of events he later speculated was in part due to his intrinsic talent and in part to Laban and Lawrence's desperate need for assistance. Though technically a student, Lamb fast became a key member of the Industrial Rhythm program,

FIGURE 4.1. Lisa Ullmann teaching Meg Tudor Williams, Mary Elding, Valerie Preston, Warren Lamb, and Hettie Loman, Manchester, England, 1947. Photo by R. Watkins. Repository GB 1701 Laban Collection, Box LC/A/14/1/3. Trinity Laban Archive Collections, London, UK.

accompanying Laban on his factory visits, performing analyses, and courting clients. In 1950, he was formally invited to join Lawrence's consulting firm, Paton Lawrence. Once ensconced in the company, however, Lamb's interests began to shift away from shop-floor labor. Taking inherent differences in movement style as a given, Lamb sought to establish that the same assessments of "fit" that Laban and Lawrence had applied to factory workers could be used for managerial roles.

Lamb began developing the technique in his first years at the firm, and Laban and Lawrence were supportive of the effort, seeing it as a natural extension of their existing programs. The two had themselves once considered evaluating managers but ultimately decided it presented too many difficulties because of the "number and kind of effort responses required from a higher executive."[12] They believed, however, that the problem was technical, rather than theoretical, and were happy to give Lamb a chance to solve it. One of the first wide-scale tests of Lamb's methods took place in 1950 at Glaxo Laboratories, where he both organized the packing floor based on Effort capacities and evaluated a number of individuals for supervisory and management roles. To do so, Lamb first created a rough list of the qualities a given managerial position might require, then—drawing on Laban's body of thought—matched those skills with sets of movement characteristics. Finally, he met with each candidate, recorded his or her movements over the course of a thirty- to sixty-minute interview, compared them with the relevant job specifications, and made a recommendation. Though Glaxo ended their contract with Paton Lawrence shortly thereafter—a temperamental Lawrence apparently picked a fight with Glaxo's managing director on an unrelated matter—Lamb's experiment was deemed a success, and managerial assessments soon became part of Paton Lawrence's offerings.

In practice, the process might have unfolded like this. A man, let's say in his early thirties, walks into a corporate office suite in 1955. He is wearing a suit in an appropriately sober shade of gray and a slate-blue tie. He is there for an interview and clutches his briefcase nervously as he mentally reviews his answers to the likely questions. He is ushered into a conference room; the interview commences with the usual pleasantries before diving into a discussion of his background. He and his interviewer are not, however, alone. At the other end of the table sits another figure, who does not speak but continuously scribbles, a steady stream of indecipherable symbols emerging from his pen. When the interview concludes, the figure returns to a small office down the hall and gets to work: making tallies, performing calculations, and eventually producing a set of charts and graphs that claim to provide a perfect assessment of the man's suitability for the position in question, all without considering a single word he has said.

Within a few years, Paton Lawrence had produced a glossy brochure advertising not "Industrial Rhythm" but "The Laban-Lawrence Test for Selection and Placement." Characterizing the test as a "scientific treatment of the 'human factor' in industry," the pamphlet's cover promised that it would quickly and cheaply uncover "a person's aptitudes, how he will get on with his colleagues and his loyalty to his employers," noting that the test had been validated for positions ranging from managing director and junior manager to assistant engineer, salesman, secretary, and designer.[13] The firm also briefly attempted to market the services of the "Laban Lawrence Youth Advice Bureau," a career counseling service for teenagers, which used observations gleaned over the course of a thirty-minute small-group movement session and a short individual interview to suggest suitable future professions.[14]

Paton Lawrence's willingness to back the program—despite the skepticism of some of its staider consultants—was partly the result of Lawrence's and Laban's long-standing relationship. It was also a strategic effort to exploit a new opportunity in a changing business landscape. As Tom Rath had learned, in the years following World War II newly founded personnel departments were challenged with staffing an expanding managerial workforce and so increasingly supplemented existing systems for vetting potential employees with tools like the Meyers-Briggs Type Indicator, the Thematic Apperception Test, and the California Personality Indicator.[15] These well-known tests were joined by an even larger array of techniques either marketed by independent consultants or developed in house, ranging from handwriting analysis to stress interviews to the use of lie detectors.[16] In the immediate postwar years, these practices were far more pervasive in the United States than they were in the United Kingdom. Many in Britain, however, saw their eventual introduction across the Atlantic as a near inevitability, coincident with the widespread idea that the solution to the UK's economic woes was the emulation of American management practices.[17] For Paton Lawrence, the time seemed ripe for a market intervention.

The second factor facilitating movement profiling's emergence was the fact that, by midcentury, the study of nonverbal behavior had become increasingly academically respectable. In the 1940s and '50s, scholars like Ray L. Birdwhistell, Gregory Bateson, Nikolaas Tinbergen, Albert E. Scheflen, George L. Trager, Edward T. Hall, and Adam Kendon had begun studying how human movement was patterned, shaped, and acquired social and cultural meaning. In anthropology, psychology, linguistics, sociology, and ethology—as well as in the relatively new discipline of communications—movement's study had acquired new intellectual cachet.[18] That Paton Lawrence benefited from this moment is surprising, as much of the emerging academic literature took a position in direct opposition to Laban's theories, positing that movement was largely a function of culture rather than a universal, ingrained biological code. Still, the aura of seriousness that newly surrounded nonverbal behavior was enough to bolster claims for the work's validity, at least in the popular media.[19]

In spite of Paton Lawrence's growing reliance on him, Lamb decided to strike out on his own in 1953, driven by annoyance that he had little power to control a technique he played a key role in developing. (Contemporary accounts back up Lamb's sense that he had moved beyond Laban's early work: in 1952, for example, Ann Hutchinson Guest reported that Laban had trouble explaining some of the symbols that appeared on a new Industrial Notation chart, eventually admitting that "the boy who wrote it had invented some things on his own.")[20] The last straw, in Lamb's account, was Laban and Lawrence's decision to rename what had originally been referred to as the "Laban-Lawrence Personal Assessment" the "Laban-Lawrence Test for Selection and Placement" without consulting him. Lamb felt the new branding associated the system too closely with existing personality and intelligence tests and was perhaps also peeved by his own name's omission. Laban and Lamb's relationship never fully recovered from the latter's departure, though the two did stay in contact until Laban's death.[21]

Once fully independent, Lamb rebranded his constellation of assessment tools. Now operating as "Warren Lamb Associates," he promoted his new "Aptitude Assessment" technique, advertising his wares to both potential clients and the general public via speaking engagements, television programs, newspaper and magazine articles, and eventually three books.[22] He also laid out his new and improved analytical categories, the most important of which were "Shape" and "Effort."

Shape, for Lamb, referred to the direction of a person's movements, mapped onto any one of three perpendicular planes in space: a "horizontal" plane, parallel to the ground in all directions and extending outward; a "vertical" plane, centered on the spine and extending laterally through both shoulders; and a "sagittal" plane, also centered on the spine but extending forward and backward. As Lamb explained, someone moving predominately on a horizontal plane "often appears to 'spread himself.' At the front of a queue, for example, he seems to take up a lot of space on each side, making it difficult for those behind to get a clear view. He tends to stick his elbows out on an escalator, poking those who want to get past him. . . . In shaking hands he moves his arm to the side in a circular approach, legs somewhat apart, almost as though pretending to be an aeroplane."[23] In contrast, someone operating on the sagittal plane "almost propels himself through the door and extends his hand straight-forward; he walks with knees and ankles close together and big strides, then gets himself into a position where he can move straight backwards into the chair, or draw the chair forwards towards him. . . . In conversation, he seems frequently to come straight at us, then retire away. He is the sort of person we may find ourselves backing away from."[24] These Shape categories can also be understood as simplified versions of the movement directions recorded in traditional Kinetographie.

Effort, on the other hand, concerned the *way* in which these movements were performed, and Lamb identified four scales on which it could be evaluated. All movements possessed a quality of "Space," which could be either "direct" or" indirect." They also possessed a quality of "Time," which could be "sustained" or "sudden"; "Weight," which could involve "increasing" or "decreasing" pressure; and "Flow," which could be "freeing" or "binding." These too were rough equivalents to the qualities recorded on Industrial Notation's Effort Graphs.

Lamb's major contention was that—though most people exhibited a variety of these characteristics—everyone favored certain Shapes and Efforts over others and these preferences had meaning. He held, for example, that people who moved primarily on the horizontal plane tended to be "investigators," skilled in gathering information but sometimes hesitant in decision-making. "Deciders," on the other hand, favored the vertical plane and could be ideal company leaders.

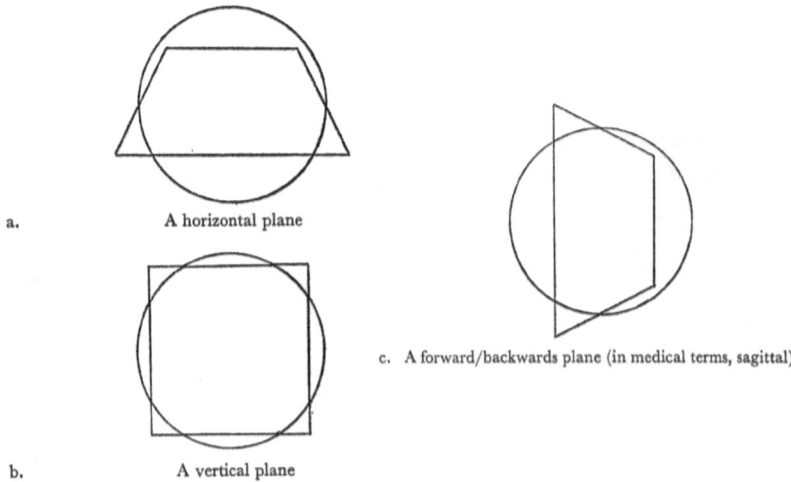

FIGURE 4.2. Shape forms. From Warren Lamb and David Turner, *Management Behaviour* (Gerald Duckworth & Co., 1969), 74.

SUMMARY OF SHAPE AND EFFORT VARIATIONS

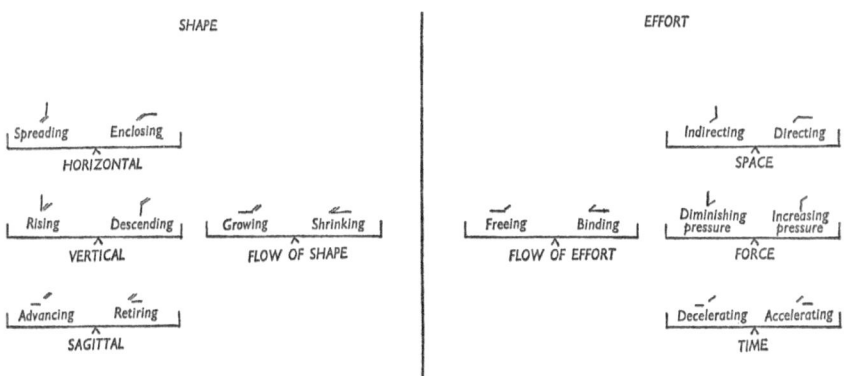

FIGURE 4.3. Notation for Shape and Effort variations. From Warren Lamb, *Posture and Gesture: An Introduction to the Study of Physical Behavior* (Gerald Duckworth & Co., 1965), 62.

Lamb also developed a new set of symbols for quickly recording Shape and Effort (eventually referred to as Effort/Shape notation). Though the revised notation took part of its structure from Industrial Rhythm's Effort Graphs, it was designed not to record *all* qualities of a given movement but rather the single most prominent characteristic an observer noticed. At the conclusion of the evaluation, Lamb would add up the number of symbols in each category and use these totals to determine the subject's movement profile.

In general, Lamb suggested that analysts record at least two hundred movement phrases over the course of an hour-long interview.[25] Not all movements, however, were created equal. According to Lamb, some were "postures," while others were "gestures." As he explained in his 1965 book, a gesture was a movement that was confined to a single part of the body, while a posture involved the body as a whole. Idly scratching one's nose was a gesture; an afternoon yawn that grew into a full-bodied stretch was a posture. The distinction, he admitted, might at first be difficult for the untrained to comprehend: "A bow, for example, might be thought a 'gesture,' yet it involves movement of all body parts, depending on how it is done. A deep curtsey, such as we see performed to the Queen, is a good example; some will do it so that all parts of the body are involved in the movement, i.e., all parts can be observed to do something which contributes to the curtsey." Another group "may succeed in completing a curtsey in which some parts of the body are not involved," by, for example, leaving the head "immoveably fixed to the neck," rather than allowing it to respond naturally to the rest of the body's

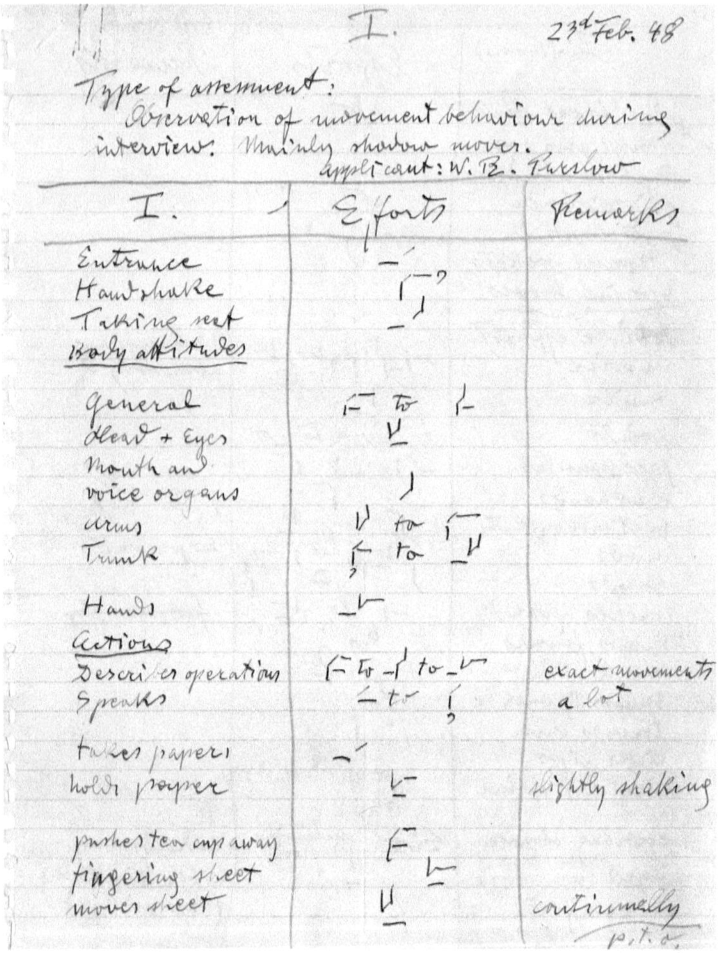

FIGURE 4.4. Personal Effort Assessment. "Observation of movement behaviour during interview," February 23, 1948. Archive Ref No L/E/62/55. From the Rudolf Laban Archive, University of Surrey, © University of Surrey. No reuse without permission.

movements.[26] Lamb further clarified that even if multiple body parts moved simultaneously, they had to do so in coordinated fashion toward a single end for the movement to truly be a posture.

The distinction, Lamb contended, was crucial because the subject's personality was revealed most clearly in the moments in which a gesture was accompanied by a full-bodied postural adjustment, what he called a "Posture-Gesture Merger," or PGM. He urged his analysts, therefore, to focus their movement samples on instances of PGM—ignoring, for example, a moment in which a subject tapped a single finger on the desk but paying close attention

 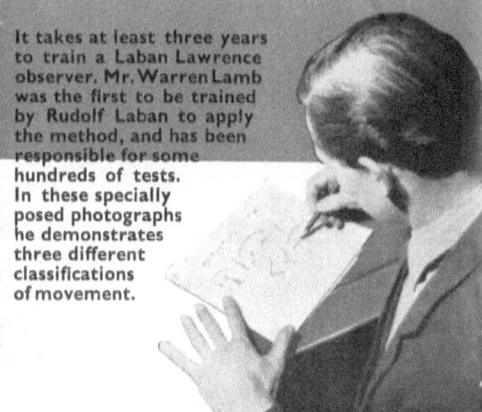

It takes at least three years to train a Laban Lawrence observer. Mr. Warren Lamb was the first to be trained by Rudolf Laban to apply the method, and has been responsible for some hundreds of tests. In these specially posed photographs he demonstrates three different classifications of movement.

ATTITUDINAL MOVEMENT, one which affects every part of the body, such as yawning or stretching.

FUNCTIONAL MOVEMENT, involved in handling an object, as in writing.

SHADOW MOVEMENT the tiny, mostly unconscious movement, such as raising the eyebrows.

Observation of a man's movements in scratching his head are here expressed in notation. The sequence of the symbols is important.

FIGURE 4.5. Warren Lamb demonstrating aspects of movement profiling. From Olive Moore, "Man of the Month: Rudolf Laban," *Scope: Magazine for Industry*, October 1954. Archive Ref No L/E/66/7. From the Rudolf Laban Archive, University of Surrey, © unknown, unable to trace. No reuse without permission.

when he leaned forward to emphasize a point with his entire body. A relative lack of PGMs was its own kind of evidence: because PGMs were the truest expression of the self, their conspicuous absence suggested that the individual under evaluation was less than genuine, or—at the very least—holding back in some significant way.

The Authentic Manager; or, A Plague of Puppets

Given Lamb's incomplete records, it is difficult to determine exactly how many companies employed Aptitude Assessment at a given moment. By his own account, Lamb's first years as an independent consultant were relatively lean; by the early 1960s, though, he seemed to be earning a comfortable living. Between 1960 and 1962 he reported annual income between £3000 and £4300 (amounting to between £73,000 and £97,600 today), derived from a mix of selection consulting, aptitude assessments, career guidance, courses, and lectures.[27] By 1963, he claimed to have performed three thousand assessments of "junior and senior executives in the U.K., the Continent, and the United States," though this total may have included profiles produced during his time at Paton Lawrence.[28] He also received some coverage in the press, including a 1959 appearance on the BBC's *Mainly for Women* television series (the program's other segment was dedicated to instructions for cooking a Hungarian pork filet).[29] Still, the work could be intermittent. As Lamb recalled in an interview years later, he had no active clients in the first three months of 1965, and—after his car "exploded" on a sales trip to the north of England and was left by the side of the road—he returned to London "drearily by train," worried about how he would "pay the rent, or the next term's school fees."[30] Concerned about his business's long-term future, he opened a second company with a marketing consultant named John Reid. Called Executive Search, it focused on headhunting.

Later that year, however, Lamb's fortunes began to improve, largely as the result of the public attention generated by the publication of his first book, *Posture and Gesture*. The book sold eight hundred copies in the first six months, which Lamb happily reported was "by all standards for this type of book ... quite good going."[31] Perhaps more important than the moderate sales figures, however, was the general buzz that surrounded the book, as it garnered appreciative coverage and reviews in, among others, the *Daily Mail*, *Evening Standard*, *Financial Times*, *Business*, and *Sunday Times*.[32] Shortly thereafter, Lamb was featured in segments on *Tomorrow's World* and *Horizon*, two BBC television series covering contemporary developments in science and technology. As the narrator on *Tomorrow's World* described it, Lamb's technique for the "scientific analysis of body movement" was taking the business world by storm. Following in the footsteps of "thousands" of others, he intoned, executives flocked to "a house in Putney to learn more about themselves in a matter of hours than they might learn in months when tested by more conventional methods."[33]

Though *Tomorrow's World* was scientific in its focus—its programming included, for example, segments on heart transplantation, home computer terminals, and the Moog synthesizer—no physiologists or psychologists commented on Aptitude Assessment's validity. The program did, however, reach out to another management consultant, Michael Wood, who was asked to submit to an analysis and provide his perspective on Lamb's techniques. Recalling the experience, Wood noted that the interview had been a comfortable one, and that Lamb's report—ten pages long, provided a week later—was astonishing in its accuracy: "I had known myself for thirty-seven years, my wife had known me for something like eighteen years and there was very little in that report, there was nothing in fact, that we knew about that he did not—and there were one or two things he found out about me that I never knew about." The program's narrator concluded that Lamb's success spoke for itself: "That these hieroglyphics may be a new way to self-knowledge may seem surprising, but many major organisations already consider such assessments as an essential part of personnel selection." "I think," he ended, "I will sit on my hands for a bit!" Wood joined Warren Lamb Associates as an executive search consultant a few years later.[34]

This attitude—that the best proof of Aptitude Assessment's accuracy was its popularity—was not only widely shared, but a key part of Lamb's business strategy. In his books and marketing materials, Lamb invariably emphasized the number of profiles he had performed and the fact that his customers were pleased with the results, using the word "validated" to refer to client satisfaction rather than any independent measure of test accuracy.[35] Lamb himself also freely admitted that was never especially interested in subjecting his techniques to peer-reviewed study, as the audience he hoped to convince was paying clients, not academic researchers.[36] In fact, a number of corporate clients were attracted to Aptitude Assessment for precisely this reason, applauding the fact that Lamb's work had been developed within industry itself, rather than in the rarefied atmosphere of the lab.[37] As a result, there were few sustained efforts to independently check Lamb's system until the 1980s, when the interested parties conducting the research had difficulty substantiating his claims.[38]

Aptitude Assessment thus engaged in a cagey kind of Latourian performance of paperwork—turning movement into data with the help of notation, charts, graphs, and statistics.[39] Those magical "hieroglyphics," mentioned constantly in press coverage, turned a preexisting practice—in this case, the gut-level assessment of body language—into something properly "scientific."[40] Echoing the rhetoric deployed at the Dance Notation Bureau, Lamb described his work as "mathematical" and "dispassionate" and emphasized how his masses of movement "data" could produce results "without resorting to subjective judgments."[41] Lamb even compared Aptitude Assessment favorably to the research on movement emerging from psychology and other social sciences. He described this competing work as well-intentioned but noted that it lacked an objective symbolic vocabulary and was instead forced to rely on "descriptive language which cannot help but be vague and allow a variety of interpretation."[42] (This, incidentally, was a mischaracterization of research on nonverbal behavior. While many researchers did rely on verbal description, others, like Ray Birdwhistell, developed their own symbolic notation systems.)

Lamb's business boomed in the years following the publication of *Posture and Gesture*. By 1967, he was complaining to colleagues that there was simply too much work for him to handle alone; in 1969 he began collaborating with Pamela Ramsden, an Australian student at the Art of Movement Studio with an undergraduate degree in psychology.[43] Together, Lamb and Ramsden began offering multiday training courses for would-be movement profilers across Europe and the United States—including regular offerings at the Dance Notation Bureau—while continuing to take on consulting work. Lamb also released a second book, *Management Behavior*, coauthored with the lawyer-turned-personnel-specialist David Turner. The book's basic content was nearly identical to *Posture and Gesture*, though tailored more explicitly to business audiences. By the end of the decade, fees for a suite of evaluations had risen to as much as £2000, and Lamb's corporate clients included the household goods giant Colgate-Palmolive, textile manufacturer English Calico, confectioner Trebor Sharps, chemical company Laporte Industries, and publisher Sir Isaac Pitman and Sons.[44] He also offered services to individuals who contracted with him directly, and claimed to have produced a total of five thousand profiles. By 1971, the number was seven thousand.[45]

The companies who hired Lamb used Aptitude Assessment in diverse ways. In some cases, profiles were used to choose between top candidates for a job; in others, movement analyses were ordered in response to concerns about particular employees or divisions.[46] In one instance, for example, a profile was drawn up for the thirty-two-year-old son of a corporate group's chairman. The son was managing a subsidiary company but struggled in the

FIGURE 4.6. Warren Lamb studies and charts Michael Slocock. *Sunday Telegraph*, October 20, 1968. Photo: Augustus Rhodes.

role, and Lamb was called in to help him "take a view on his future within the Group" and "advise him on the type of position likely to make the most of his abilities" in advance of the firm's reorganization.[47] Aptitude Assessments were also sometimes deployed as part of a broader assessment of a company by other outside consultants.[48]

Still, given the many other options for assessing workers, why would a firm choose Aptitude Assessment over the MBTI or an annual performance review? For many of Lamb's clients, the most appealing aspect of the system was its claim that its rootedness in the body made it "true" in ways that other evaluations were not. As one reporter, touting Aptitude Assessment's appeal, summarized, "with a little thought and effort man can give (and does give) a completely false picture, both to the world and to himself. He can hide his thoughts behind words or charm; he can smile at his enemy; he can simulate enthusiasm, honesty and loyalty; a spurt of energy when necessary can cover a congenital vacillation; a show of strength when the boss is looking can successfully camouflage the weakling. There is only one thing which no man living can alter or hide—*his basic unconscious movements*."[49] To strategically select the correct answer on a multiple-choice test was one thing, the reporter suggested; to actively conceal one's natural movements for a period of hours was something else entirely.

Lamb contended that even skilled actors were incapable of fully evading the movement analyst's gimlet eye, and that only traumatic injuries or a concerted program of brainwashing could significantly alter a person's movement profile. The most frequently repeated element in newspaper articles about Aptitude Assessment was a story about how Lamb's eye for movement had allowed him to ferret out a "confidence man" who otherwise would have almost certainly been hired for a top corporate job. As one version explained: "Mr. Lamb sat in on a selection board where the unanimous choice was a suave, sophisticated fellow. 'But I didn't like him,' said Mr. Lamb. 'He was just a Gesturer. His body was never involved in his movements.' I made some inquiries—and established he was a confidence trickster."[50] The claim was not only that Aptitude Assessment could pick the best man for the job, but that it could pick out whether a man was truly there for the job at all.

In another case, Lamb was called in to assess a candidate for a general manager position at Air Freight. The subject had already been interviewed by the firm's internal selection committee and was highly rated for competence and experience, but two remaining issues—"the subject's readiness to take a drop in salary and the discovery that he has been looking for a job for at least two and a half years"—were producing ongoing "unease." Upon concluding his analysis, however, Lamb was happy to allay the company's fears. "It is not inconsistent with his aptitudes," his report noted, that the candidate "should have been casually seeking a job" for such an extended period of time. "Nor," he continued, "is a drop in salary likely to disturb a man of his magnanimity, so long as he can keep up his existing standard of living." In fact, Lamb argued that the man's movements indicated that his "application makes sense and his statements during the interview should be accepted. This would include his responses indicating age, health, and general stability."[51] He was hired.

In assessing the Air Freight candidate, Lamb may have introduced himself as a movement analyst, though doing so was by no means a requirement: though most profilees were told what was going on, at least some evaluations were performed without candidates' knowledge. "Men," Lamb remarked, "are now successfully fulfilling top grade positions without realizing they owe their jobs to the fact that their aptitudes were objectively assessed," perhaps simply assuming the quiet figure doodling in the corner was an idiosyncratic division head from another unit.[52] The fact that Aptitude Assessment so closely resembled a "normal" interview and was, as a result, essentially impossible to opt out of was another of its assets. As Lamb boasted ominously, "So long as you move at all—and it is impossible not to—you cannot help revealing yourself." It was a lie detector, but better: all of the insight, none of the wires.[53]

Curiously, Aptitude Assessment was simultaneously promoted as less intrusive than its counterparts. Unlike life histories, personality inventories, or psychological examinations, Aptitude Assessment required no "impertinent and upsetting questioning to bare [an employee's] soul and reveal their anxieties." Instead, the candidates' "true strengths and weaknesses will be evidenced from their movement without the reason for them being discussed."[54] In theory, this process allowed corporations to make decisions based on what one magazine article called "the outward expression of the spirit" while letting job applicants maintain a dignified reserve. Less generously, interviewers using Aptitude Assessment hoped to access applicants' internal mental states without the need to secure their consent, and then use this private information to make hiring decisions.

Aptitude Assessment also served yet another, larger function: it promised to help resolve what many believed to be a growing crisis in British industry. As one scholar put it at the time, the 1960s had seen the production of a "veritable barrage" of reports and research studies calling attention to the country's need to recruit more—and better—managers.[55] The report of the Central Advisory Council for Science and Technology and the Swann Report bemoaned the fact that it was impossible to entice the best graduates into industrial careers, while the Jones Report on British "Brain Drain" highlighted the widespread view that British industry was "bad and rigid" and argued that industry needed to try anew "to make a determined effort to project an attractive image of itself."[56]

An aversion to becoming a dull organization man seemed to be dissuading many of the most promising university graduates from even considering business careers. In 1969, the Cambridge University Management Group conducted a survey of nearly a thousand male Cambridge undergraduates as part of an effort to determine why industry was "not able at present to recruit its full share of available graduate talent." The answer they received was clear. The students feared that managerial careers, while well-paying, would dull their minds, compromise their principles, and stamp out their personalities. These negative views, moreover, were held even by those who were theoretically open to managerial positions. Thirty-five percent of that group, for example, were concerned that a managerial job would lack opportunities to be "creative and original," 58 percent dreaded "involvement in the rat race," and 67 percent were concerned about a "loss of identity." The figures for those who had already decided to pursue another career path were even higher: 62, 76, and 65 percent, respectively.[57] A few years earlier, W. Heffer & Sons had published a compilation of essays written by a different group of Cambridge students that sang the same refrain. One essay described business as "an uninteresting, unchallenging, burdensome mass of menial activity" and despaired of "whether we shall find any glamour in the life or whether it will become dull routine."[58] Another noted that "in industry, brilliance is suspect.... Evidence of leadership, not eccentricity, and commonsense rather than intellect is required. He has to fit in, not antagonize, be quietly competent, persuasive and optimistic."[59] Students made frequent references to texts like Whyte's *Organization Man* and Vance Packard's *The Pyramid Climbers*, touchstones of the working lives they feared awaited them.

The students were concerned, moreover, not just about the present state of British industry, but that the increasing push to adopt American management training methods—personality and aptitude testing key among them—might make the situation even worse. In this specter they saw the withering of the last vestiges of freedom and idiosyncrasy in favor of the dull and compliant running of the "business machine." "American business schools," one author explained, "are gruesome in their efficiency, administering the horrors of group testing, IQ tests and Rorschach Blots with panache. It is only the sentimental English love of the amateur that has spared us so long."[60] Practicing executives similarly worried about the creation of a new expectation that they "surrender . . . their own personalities for the sake of the firm."[61] Company owners, on the other hand, feared that a fully professionalized managerial class might become too independent, operating as a "'third force,' standing between capital and labor, rather than a faithful body of functionaries."[62]

Lamb's Aptitude Assessment undertook to solve all these problems simultaneously. A novel bureaucratic management technology, it promised to "analyse a manager and measure his management skills" with the level of accuracy and precision British industry now demanded. As Lamb put it, because managerial aptitude was increasingly key to any company's long-term success, the manager was "more worthy of the most sophisticated kinds of measurement than any item of plant or equipment or investments."[63] At the same time, Lamb claimed that this process of rating and sorting would produce not alienated automatons but employees who felt engaged, fulfilled, and loyal.

As Lamb explained in *Management Behaviour*, most professional dissatisfaction was not the result of overly intensive management—nor of anything inherently dehumanizing in the pursuit of profit—but of hiring that did not match a person's natural aptitudes and the tasks they were regularly called upon to perform. "There is evidence," he remarked, "to show that men who are, or become, unhappy or discontented in their jobs reach this state simply because they are not having the opportunity to exercise the full range of their abilities."⁶⁴ As he further detailed, "Studies of people at work have shown [that] the more a person is actually using the features which make up his individual pattern in his job the more at ease he appears. If the needs of the job demand movement outside or inconsistent with the individual pattern then he looks 'contrived' in his performance, or stilted, or ill at ease, or forced in his behaviour. He may still be efficient, but there do appear to be serious consequences from behaving out of accord with our physical pattern."⁶⁵ At a lecture given to the London Association of Industrial Medical Officers, Lamb provided a wildly varying list of what these consequences might be, from generalized irritability at one's colleagues to major depression, ulcers, and hypertension.⁶⁶

The alternative, though, was corporate bliss. Using Aptitude Assessment properly, Lamb assured his readers, was part of a larger project of ensuring "not only that the job is performed in a way that will lead to the achievement of goals and successful attainment of objectives, but also that the individual performing the job should himself obtain the highest degree of personal satisfaction—it could almost be called complete fulfillment—from doing his work."⁶⁷ Lamb acknowledged that traditional compensation would continue to matter, as few high-performing executives would take a job without a suitable salary. He hypothesized, however, that money might play less of a role in attracting and retaining employees than the "opportunity for a man constantly to exercise those skills in which he is most successful."⁶⁸

Though Lamb's purported ability to spot a scammer made for a great party trick or newspaper lede, the real appeal of Aptitude Assessment was more mundane. It was not so much designed to spot the professional con artist but rather to identify the men and women who were, in Lamb's eyes, conning themselves, living and working in ways at odds with their essential nature. The finance manager who left the office each day with a sense of bottomless exhaustion, the executive who found himself inexplicably angry: these were the figures that Lamb's system was truly designed to catch—and then to cure. Doing so would, in turn, solve the country's "serious shortage of managerial talent."⁶⁹

But even as Lamb recapitulated Laban and Lawrence's argument that joy in work could be an end in itself, he bent it in a new direction. The world he sought to create was not that of a mass of factory workers coming together in a collective ritualistic symphony but of distinct individuals responding to the dictates of their individual personalities. Lamb took pains to remind his clients that no way of moving or working was inherently "better" or "worse," merely more or less suited for the task at hand. In fact, the only profile he explicitly warned employers against was one in which the subject exhibited no movement preferences, suggesting an individual with no personality at all. The problem was not only that such a person would be a dull and uninspiring leader—the corporate drone that haunted the Cambridge undergraduates—but that he might be inclined to imitate the working styles of others. "Mimicry" of someone else's way of working, Lamb noted, could only "produce a false and unnatural style of management."[70] This formulation, of course, implicitly posited that there *were* modes of management that were both true and natural.

There was almost nothing that Lamb criticized more fervently than what he called "puppet-like behavior," bodily movements that were mainly gestural and emerged from convention rather than serving as a true expression of the inner self. The "cheese-like grin which does not express joyousness," the "routine shaking of hands which does not express welcome"—in Lamb's estimation, the profusion of such actions in the twentieth century threatened to rend the fabric of contemporary society.[71] Like Norbert Elias with a twist, Lamb posited that this change in habitus had likely begun during the Renaissance, but had intensified over the succeeding centuries and become particularly ingrained in the twentieth. "The possibility," he wrote, "that anyone might use parts of his body independently of the whole is a peculiarly modern phenomenon. If we compare barbaric peoples with highly civilized peoples at any point in history up to the modern period it is difficult to find any indication whatever that either moved in a way that segregated Posture from Gesture."[72] "We are now," he proclaimed, living "in a puppet age," characterized by huge numbers of people who moved mechanically, with no coherence, as if an invisible, sadistic puppeteer was haphazardly jerking their strings.[73] The results were not only aesthetically horrifying but psychically damaging as well.

This idiosyncratic perspective sometimes provoked derision. After a lecture at the Association of Industrial Medical Officers, for example, Lamb wrote to a colleague that many of the attendees—used to more staid fare about the importance of good posture—likely wrote him off as "completely balmy."[74] When Lamb assembled a large icosahedron and invited attendees to climb inside during a seminar at the British Institute of Management, the series organizers reportedly threated to cancel his next appearance unless he abandoned the contraption.[75] Still, Lamb was buoyed by the fact that there always at least a few attendees "who come around afterwards and say how impressed they are," confidently noting that "they are themselves usually the most impressive members of the audience."[76] Aptitude Assessment's aura of eccentricity was in fact central to its appeal. Despite the British Institute of

Management's protests, Lamb's seminars there were oversubscribed, filled with managers who were looking for a way to bring life to their daily labors. In comparison to profit-and-loss statements and multiple-choice personality inventories, Lamb's bodily assessments and exercises seemed compellingly strange and ripe with possibility. Those who criticized the system could easily be written off as specimens of the "conventional" thinking of the past, standing in the way of the "exceptional" and "inspired" managers of the future.[77]

On a more practical level, aptitude assessments allowed employers to frame even demotions or firings as a kind of gift, an opportunity for the employee to find the "complete fulfillment" that had, in the opinion of the consultants, thus far eluded him. Lamb explained in *Management Behaviour* that he had identified many cases in which an employee, after performing well in his existing job, was automatically promoted to the next higher status position with little regard for the aptitudes required. Sadly, Lamb remarked, "it is usually the executive himself who takes the consequences of unfortunate moves of this kind." Whether "his life is made so uncomfortable through his own inability to do the job" or he merely felt robbed of the profound satisfaction he previously derived from his work, promotion could be a disaster. Luckily, "these unhappy incidents have usually arisen from an error of judgement by management, and management has in its power the capacity to repair such an error."[78] One of Lamb's eventual partners reported that about 20 percent of her assessments ended in a recommendation that an executive leave the company (or the company remove the executive), but she also framed such decisions as happy opportunities for gaining "self-insight."[79] As one report that suggested denying an internal candidate a promotion summarized, "In his own interests it is doubtful whether he should seek the job."[80] An uncredited cartoon in the Warren Lamb archives indicates that at least some workers were cynical about such claims. In the panel, Satan gives a newcomer to hell his work assignment, remarking, "On Warren Lamb's recommendation, the Sulpher [sic] pit is the best place for you."[81]

Lamb also argued that Aptitude Assessment suggested that short-term contract employment might better serve both employee and company than the traditional model of long-term salaried work. If, for example, an executive's profile indicated that he would be happiest returning an ailing company to health, why not hire him for a limited term, freeing him to find more meaningful work elsewhere once his employer was profitable again?[82] Though Lamb acknowledged a variety of obstacles—ranging from the structure of pension plans to outmoded ideas about the value of long-term corporate tenure—that might make such a change difficult, he was optimistic nonetheless.

"With the increasing sophistication of management techniques and the need for specialists to introduce them into organizations," Lamb wrote, "there are growing possibilities for engaging executive staff on a contract hire basis." He predicted that there might soon be "much greater opportunity to choose between appointing a man to a perpetuating job or to an achievement job."[83] It was a dream of increasingly precarious employment, built on the shaky premise that "flexibility" would only enhance workers' opportunities to find self-expression and fulfillment.[84] Lamb's independent consultancy was itself a kind of model of this new mode of temporary, entrepreneurial, contract-based labor.

One letter in Lamb's archive penned by a former private client attests to the power that this vision seems to have held, at least for some. Jeremy Hemming of Somerset first consulted Lamb in May of 1966 and reached out again in October of the same year. In his letter, Hemming recalled that he had been "very far from satisfied within the tenor of life as a junior-middle management man in a relatively heavy and static industrial concern," and—following Lamb's analysis—decided to make a change, leaving his job to start his own business. Though he was still looking both for a product and for a means of distributing it, Hemming felt confident that his certified aptitudes for "launching and developing a campaign," "awareness of trends and ability to forecast," and "decision-making capacity generally in running the whole of a business" would serve him well.[85]

In his reply, even Lamb seemed a bit worried about Hemming's prospects. While he commended Hemming for his courage and agreed that he "certainly ha[d] the single-mindedness to make a success of it from the aptitude point of view," he also cautioned him about the importance of timing and the risks of starting a business only a few months after the national government had instituted a wage and price freeze.[86] At least briefly, Lamb seemed to acknowledge that aptitude wasn't everything. He ignored entirely, however, another aspect of Hemming's note—Hemming's growing sense that the kind of workplace ennui he experienced might be remarkably common. "This may be obvious enough," Hemming wrote, "but I am struck by the number of well educated and able middle class people who really just don't fit in a large industrial concern."[87] For Lamb to acknowledge this reality, though, might mean that that the problem haunting Britain was not a rash of aptitude mismatches but the more basic conditions of managerial work. Instead, Lamb continued to pitch a world in which the dull-eyed and downtrodden administrator would disappear, replaced by a sea of contented, effective employees with bodies perfectly suited to their positions and minds perfectly devoted to the corporation.

On Observation and Performance

One 1965 article in the *Financial Times* began, "A meeting with Mr. Warren Lamb has left me with a faintly wooden feeling—an effect he has had on quite a number of people." Though the author appreciated Lamb's charm and insight, he felt uneasy in the presence of a man whose life was devoted to close and constant observation. Commenting that he could certainly understand the "suspicion" aimed at Lamb from "people who think he is looking for their weaknesses (he says he is looking for their strengths)," he recalled that "after one anecdote at the luncheon table I kept my hands out of sight for the rest of the meal."[88]

Another piece, published at roughly the same time, echoed these anxieties. "Talking to Warren Lamb," the author noted, "is an unnerving experience. As you tell him what you want to do he is looking at your feet, when you ask him a question he is looking at your hands."[89] If speech had "lost much of its significance," the significance of each movement had increased commensurately, turning formerly insignificant acts like lighting a cigarette or adjusting a lapel into high-stakes performances.[90] As yet another article's headline queried nervously, "Do you give yourself away as you wave goodbye?"[91]

Lamb maintained throughout his career that the act of observing had no effect on the behavior of the people he studied. Movement, at least in his public pronouncements, was simply too ingrained, too fundamental to the self to ever be successfully altered, even if the individual in question actively sought to do so. As Lamb explained to his clients, it was for this reason that Aptitude Assessment outperformed self-reported personality tests and for this reason that it did not matter whether an interviewee knew their movements were being recorded. But accounts like these newspaper articles suggest the situation was far more complicated.

In fact, while little evidence supports Lamb's claim that Aptitude Assessment consistently increased either profit margins or worker satisfaction, there are many indications that movement analysis had other kinds of effects. In particular, Lamb's work helped to generate a new public sense that movement was an important kind of data: that a person's fidgeting held clues to their inner being that could—with the right set of tools—be revealed. Contrary to Lamb's assertions, however, this newfound awareness was less likely to produce people who moved with a new kind of free authenticity than it was to foster the exact forms of consciously self-fashioned, "puppet-like" behavior Lamb claimed to deplore.

The job interview seemed especially ripe for such performances, and a 1965 article in *Ballet Today* envisioned exactly this possibility. In it, the author sketched out a "disturbing" future in which job applicants were obliged to hire professional choreographers to train them before each interview, drilling them in "movements which will suggest to the 'Physical Behaviour Consultant'" that they were "alert, constructive, intuitive and so on." To those who doubted the plausibility of such a scenario, the author reminded them that "something very like it is already happening in America, where applicants for jobs are given the most elaborate psychological tests, and are well advised to produce the right responses."[92] Lamb responded with an annoyed letter to the editor in which he sought to distance Aptitude Assessment from other forms of gameable psychological testing. He cautioned dancers and choreographers not to get "distracted" by outlandish claims about dystopian corporate futures. His point, however, was perhaps undermined by his concluding suggestion that—with just a bit of additional training—dancers might indeed find bright career prospects as movement evaluators.[93]

Indeed, Lamb himself regularly provided detailed instructions for strategically deploying movement to manipulate one's audience. In *Posture and Gesture*, for example, he criticized the conventional advice given to public speakers—to plaster on an "inane grin," to gesture confidently at visual aids, to stand with feet firmly planted. These options created an impression of insincerity likely to do more harm than good. Instead, he suggested that the appearance of authenticity could be cultivated through the conscious deployment of Posture-Gesture Mergers. This technique, he assured, could "be easily learnt," and consisted of: "(1) selecting key points in the speech, deciding what is the appropriate attitude, e.g., appealing, demanding, discussing, eulogising, and noting it down using the present participle of an action verb" and "(2) when these points are reached make sure that you do a Posture adjustment of some sort." These adjustments, whether an "upward stretch," a "forward leaning," or a "throw away jerk," Lamb further clarified, should be performed approximately every two minutes for maximum effect.[94] In *Management Behaviour*, he counseled applying the same techniques in daily business practice, urging managers to "rehearse their physical movement behaviour before a meeting" with the same care with which "they prepare their verbal message."[95] In his profile of the corporate chairman's underperforming son, Lamb suggested that the young man reshape his movements as part of "a deliberate effort to create an 'image' of himself" as resolute and tenacious.[96] It is perhaps telling that Lamb had first encountered Laban movement theory as part of his own acting training.

Lamb also offered hundreds of consulting sessions in which he worked with clients one-on-one. In settings that were more reminiscent of a dance class than a conventional executive coaching appointment, clients practiced

FIGURE 4.7. Warren Lamb demonstrating movement profiles and their associated skillsets and personalities. From Tony Clifton, "Right Arm for the Job," *Sunday Times*, November 16, 1969. The Times / News Licensing. Reproduction courtesy of Rudolf Laban Archive, University of Surrey (Archive Ref No L/E/64/61). No re-se without permission.

tailored sets of movements developed in response to their individual aptitude assessments. The goal of these sessions was twofold. First, clients would learn to bring additional "life," conviction, and clarity into the movements that they naturally preferred and to avoid any inauthentic or assumed "habits and mannerisms."[97] Second, they would work to extend the range of movements with which they were comfortable. If there was a contradiction between these two tasks, Lamb was untroubled by it, explaining that while a person's basic movement profile was unalterable, a moderate expansion of one's range was an achievable desideratum.[98]

Between sessions, clients were given instructions for practicing these same movement qualities throughout their daily lives. The idea was to cultivate a kind of constant, low-level self-surveillance, becoming aware of one's movements whether "dressing in the morning, travelling to work, all the odd occasions of having to wait—as in a waiting room, opening and closing doors, handling a telephone, attending to furniture, operating a car, buying a newspaper."[99] The end result of this new awareness, Lamb promised, would be movement that was both more beautiful and more fully expressive of the self—and presumably also more attractive to the employers for whom Lamb consulted. In a canny business strategy, Lamb sold both the test and the instructions for passing it.

Lamb's movement prescriptions were, moreover, not confined to the workplace and included suggestions for even the most intimate realms of life. *Posture and Gesture*, for example, admonished readers regarding the proper technique for kissing one's paramour: "How much does a Gesture peck of a kiss contribute to interpersonal relations? It will hinder the growth of meaningful interpersonal relations. Better to have fewer kisses, but when we do, to embody them with a consistent and positive physical attitude. It is not suggested that every kiss should be a passionate one. It can be sympathetic, delicate, consoling, teasing, alluring, pacifying, audacious, shy, formal, welcoming—the point is that it should be backed up by a Posture attitude and not confined to a Gesture motion of the lips and head."[100] At times, Lamb argued that *all* movement behavior should be, at least to some extent, consciously engineered. "We go to great lengths to achieve good design in clothes, furniture, homes, cars, bridges, ships, aeroplanes and even machinery," he argued. "Why not aim to design our physical behaviour to be not only functionally efficient, but also as beautiful as we can make it?"[101] Time and again, particular movement profiles aside, Lamb returned to the idea that the modern world required a new form of intense, wholehearted embodiment, free from the inauthentic formalities of the past and "appropriate to the social, political, and educational advances of the present century."[102] The fact that the uninhibited habitus Lamb sought could only be achieved through a constant program of intensive, often corporate-sponsored, self-training—and monitored via a numerical analysis of PGM percentages—seems to have escaped him.

Lamb's efforts to bring movement analysis to a broader public became even more explicit with the 1979 publication of *Body Code: The Meaning in Movement*, coauthored with Elizabeth Watson.[103] Though the book again largely recapitulated the content of Lamb's earlier works, its voice was chattier—Watson was brought on largely to punch up the writing—and the text was complemented with a large number of illustrations by Clare Jarrett.[104] The promotional materials for *Body Code* promised "entertainment" as well as that the book would teach its readers to do a little movement analysis themselves, opening up a "new world where to observe movements can often be more interesting than listening to words."[105] The text bore out this premise, suggesting convenient settings in which to practice movement analysis and a systematic three-step process for "drawing your first action-profile."[106] Both academic and popular reviews were mixed, but the book sold well: Routledge & Kegan Paul issued a paperback edition in 1985, it was picked up and reprinted regularly by the Princeton Book Company in the United States, and in 1981 it was translated into Japanese.

Lamb was quite never satisfied with Routledge's marketing efforts—regularly writing to his editor to complain about the need for more promotional mailers or higher stocks—but an irritated review of the paperback edition in the *Daily Mail* suggests that the book was having at least some of the effects Lamb hoped for. "It started ten years ago," the author complained, "watching the way women cross their arms when squeezing past men in train corridors, the way politicians rubbed their noses when confronted with an awkward question. With almost every other avenue of man's psyche well documented, social anthropologists turned to body watching." While conceding that Lamb did not bear sole responsibility for the "flood" of "semi-intellectual, coffee table manuals" on body language—the popular science writer Desmond Morris alone, for example, had produced three books on nonverbal behavior—the article identified *Body Code* as one of the earliest and most influential guides to the subject.[107]

In the reviewer's estimation, Lamb's work had produced not a society-wide boom in natural gestures but rather a rash of performative physicality, a "splurge of public displays of affection." Princess Diana, the review noted with scorn, "actually kissed the Chairman of British Airways on a visit to Heathrow in April 1984. Maggie Thatcher and Ronald Reagan always begin Transatlantic negotiations with a publicized peck." The less demonstrative were, by contrast, "accused of being psychologically warped if we don't rush up and slobber over a person we've met once before at a dinner party." What's more, the author feared that the public obsession with body language was doing something even worse: "Chipping effectively away at part of our National Heritage: the great British Reserve."[108]

Nonetheless, the reach of Lamb's ideas only grew. In the 1980s and 1990s, a number of students who had completed training courses with Lamb or one of his associates began opening their own consulting practices. In Nottingham, it was the James Thornhill Consultancy; in London, Professional People Development; in New York, Decision Dynamics. Books on the subject also multiplied: Pamela Ramsden's included *Top Team Planning* (1973) and *Action Profiling* (1994), while Carol-Lynne Moore, another one of Lamb's protégés, published *Executives in Action* (1982) and *Movement and Making Decisions* (1994).[109] In 1981 Lamb and Ramsden founded Action Profilers International to serve as an umbrella organization for the growing network of movement profilers, and by 1987 it counted fifty-two practicing members in the United States, Europe, and South Africa.[110] In that same year, promotional materials sent by one Action Profiler listed as former or current AP customers Corning Medical, Kodak, Mars, Hewlett Packard, Sandoz, Colgate-Palmolive, Hoover, Lilly Industries, Bank of America, the Bank of England, Provident Financial Group, London Transport, the National Coal Board, and Saatchi & Saatchi.[111] Lamb's own new clients included a multinational company involved in wind farm construction, a California magnet manufacturer, and SpudULike, a UK chain of fast-food restaurants specializing in baked potatoes.[112] He also continued to promote his work to the broader public, hoping to effect a sea change in how movement was understood and evaluated—and perhaps gain some new clients in the process.

Scripts and Expectations

As the business of Aptitude Assessment (now "Action Profiling") grew in the 1980s, Lamb also altered and expanded the system itself, elevating "Flow" from a subsidiary component to its own category of analysis. Its use provides yet another example of the ways in which such investigations created novel—and in this case, gendered—expectations for movement behavior, as Lamb held that the kind of Flow an individual demonstrated was determined almost entirely by sex. Men, he contended, typically exhibited Binding Flow, while women tended to display Freeing Flow. To illustrate the difference, Lamb provided the following analogy: "Compare also how men and women trapeze artists move differently during their act, including their initial run up to the rope, or the difference in performance between male and female ballet dancers. What is perceived as 'feminine' is the graceful (Freeing) flow of women's movement compared with the more athletic, controlled (Binding) movement of the man."[113]

Ignoring the fact that trapeze artists and ballet dancers are performers in arts with highly formalized gendered conventions, Lamb held that these differences were reflective of universal, binary sex disparities.[114] He furthermore asserted that these movement differences corresponded to differences in professional aptitudes. Men, he argued, were generally more assertive than women because their preference for Binding Flow meant that "a man's investigative skills will not be curbed by sensitivities as those of a woman would tend to be, and a punch will be delivered at full force." On the other hand, women's preference for combining Freeing Flow with Directing Efforts was "useful for teachers or group leaders to use to bring a group of people together.... It is the ideal formula for a heart-to-heart, such as women are good at, completely relaxed but eager for detail!"[115]

There is little need to search for the roots of Lamb's new interest in gender, as he openly situated his fascination with Flow as a response to "a time when there is a lot of talk about equality."[116] As more women entered the labor force, Lamb explained that he sought to counter the "militant feminists" who suggested that women and men were fundamentally alike, asserting instead that "in a number of areas because of their differing movement patterns, men will be more naturally talented than women, and in an equal number of areas women will outshine men."[117] To any regular reader of Lamb, this position would not have been terribly surprising, as his published work had exhibited conservative ideas about gender—and a particular fixation on standards of female bodily comportment—for decades. In *Posture and Gesture*, for example, he complained about the "ugliness" of "some of those products of ladies' colleges who are somewhat loud and aggressive in their use of voice," while praising the graceful habitus of historical figures like Cleopatra and the eighteenth-century English actress and model Lady Hamilton. (Cleopatra in particular, Lamb claimed, was not a conventional beauty but made up for her average looks through her alluring physical behavior.)[118] And while the pen-and-ink illustrations in 1979's *Body Code* include countless men in business attire, women appear primarily as sexual objects, often with nipples distinctly visible through their clothes. A particularly memorable drawing, intended to demonstrate vertically oriented movement, depicts a woman in a tight bustier whose skirt has just fallen off: her arms are anxiously thrown skyward, and her body is entirely naked from the waist down.[119]

This is not to say that Lamb unequivocally objected to women's presence in the business world—he did recommend a number of women for managerial positions. But he had a very particular vision of how they should act once there. He contended, for example, that as a leader, "a woman does better to be inspirational rather than assertive, whereas a man can be assertive about particular goals," and he argued that while "all-women meetings" could be run relatively informally, "when a woman tries to chat to a man in the same way she will be less successful. To try to emulate each other in such behavior is not a solution. It denies their natural talents."[120] In Lamb's view, a woman who was true to herself—who had found her authentic management style—would be graceful and nonconfrontational, both in her movements and her

FIGURE 4.8A–B. Clare Jarrett, drawings from Warren Lamb and Elizabeth Watson, *Body Code: The Meaning in Movement* (Princeton Book Company, 1979), 121, 146.

actions. Any negative reactions to the assertion of female authority, moreover, could be understood as the result of the ineffectiveness of an individual woman's attempt to emulate men rather than a systemic bias against women as a group. In very literal way, Aptitude Assessment provided a new script for the performance of gender.[121]

Lamb had a similar tendency to use movement analysis to impugn liberal political behavior he disliked. In a 1968 essay for *Twentieth Century* magazine, he commented on the recent spate of nonviolent student protests employing a new tactic: the "sit-in." Prefacing his remarks by clarifying that he was not "making any general statement about the value or the ethics of protesting," Lamb noted nonetheless that the protestors' movement behavior often looked "suspiciously insincere." He explained that "if one individual wishes to protest to another he will usually 'stand up to him' both literally and figuratively. To sit down to protest, as is now frequently done, may also achieve some sort of confrontation but also indicates that the protestors are not particularly interested in applying much effort to the situation." "This," he continued, "makes it seem that they are not particularly interested in applying much effort to anything and appears as a contradiction."[122] Here again, "authenticity" in movement was deployed as a weapon in service of particular ideological ends.[123]

There is also evidence that Action Profiling was employed to encourage people of color to move more like their white peers. In a set of remembrances published in 2016, an Action Profiler named Jane Maloney fondly recalled consulting for AT&T in the 1980s.[124] She noted that the company had been making efforts to reduce racism and sexism in its workplace, but that it faced resistance as "many employees felt they were being trained to be politically correct." She then told the story of one employee who approached his manager with concerns that his colleagues "rejected his ideas due to the color of his skin." In response, the manager ordered an Action Profile made, the results of which ostensibly revealed a number of ways in which the employee might modify his behavior to make his coworkers more receptive. As Maloney reported it, the intervention worked, as the employee "shifted his focus from differences in ethnicity to differences in the way people made decisions. In return, his team members noticed the value that he was able to bring to the team by making a few changes to his behavior." Disregarding the fact that the only person asked to change was the already marginalized employee, Maloney described the intervention as "the most powerful diversity training" the company could have offered.[125] Once more, movement analysis produced not a transparent window into the self but a new set of standards to live up to, standards particularly difficult to meet for those outside the dominant movement culture. In this light, one movement profiler's advice to trainees—to "protect oneself from emotional distress coming from the client"—acquires a new kind of meaning.[126]

That profiler, Charlotte Honda, was in fact Action Profilers International's lone Black member, a reality which became uncomfortably evident when the organization proposed to host one of its annual conferences in South Africa in the late 1980s.[127] Concerns were raised by a number of individuals but perhaps most vociferously by Honda, who objected on principle to any travel to the apartheid regime and worried about her own safety were she to attend the meeting. Erik Schmikl, the leader of the South Africa branch, suggested that the organization simply reassure Honda that "she would be treated at most of our major international hotels exactly the same as Whites are treated," arguing that American reporting on apartheid had not provided a sufficiently "balanced perspective" on the country's problems. He also hoped Honda was a "realist" about the need for slow, incremental change on matters of race.[128] Indeed, Schmikl's own movement work reflected this accommodationist stance. Though he professed an interest in diversifying South African business—and created Action Profile–based mentorship programs intended to support non-white South Africans—he did not advocate for radical structural transformation, aiming instead to help Black South Africans better fit into the white business world through a course of "extensive self-development."[129]

Judging by Appearances

From the beginning of his career, Lamb saw the fusion of the individual personality and the corporate mission as an ideal for which to strive. Even in Aptitude Assessment's earliest years, however, others had begun to criticize the troubling implications of such a framework. As C. Wright Mills warned ominously in 1956, "when white-collar people get jobs, they sell not only their time and energy but their personalities as well. They sell by the week or month their smiles and their kindly gestures, and they must practice the prompt repression of resentment and aggression. For these intimate traits are of commercial relevance and required for the more efficient and profitable distribution of goods and services."[130] For Mills, this transformation of deeply personal realms of experience into salable commodities was seriously disquieting.[131]

More recently, scholars like Luc Boltanski and Ève Chiapello have argued that this conflation of the self and the corporate was a key component of the "new spirit" of neoliberal capitalism. While early twentieth-century managers justified their professions in terms of their ostensible social usefulness, by the 1970s companies increasingly promoted managerial work as an intrinsically meaningful site of self-discovery. This shift, Boltanski and Chiapello contend, was a response to radical critiques of capitalism as oppressive and inauthentic, though it ultimately served to legitimate subtler forms of control and exploitation.[132] As corporate leaders portrayed their work as moral and fulfilling and business schools instructed students to be "passionate" and "ecstatic" about their jobs, late twentieth-century business became "an activity transcending the profane task of making money."[133] By claiming that Aptitude Assessment could match an individual with a position that expressed his "inner spirit"—not just his skills and aptitudes but also the basic, natural rhythms of his body—Lamb's work represented a particularly vivid instantiation of this mode of thinking.

Lamb's claims about Aptitude Assessment's ability to reveal the true self behind the public mask eventually moved from the corporate world into law enforcement and national security. In 1974, for example, Lamb and Ramsden organized a one-day seminar in movement analysis for twenty-five members of the Kansas City, Missouri, police department. Describing the course as experimental, Lamb explained that his aim was to find out whether "increased sensitization of patrolmen to non-verbal cues in their own and others' behaviour" would help the officers interact more effectively with the communities in which they worked. By the end of the day, Lamb announced that his hypothesis had been proven correct, as all the officers agreed that understanding movement could help them accurately assess others' attitudes. "Some" were even "prepared to admit that there were occasions on which their non-verbal communication both with members of the community, and with colleagues at peer review panels, was antipathetic to the objectives they wished to obtain."[134]

Lamb then proposed a three-day followup course tailored to the twelve officers responsible for recruiting and training new patrolmen, intended to establish both "a disciplined procedure for training recruits in a knowledge of non-verbal communication . . . whether threatening, supportive, resentful, challenging, etc.," and a "body movement criterion of candidates suitable for Police work."[135] Around the same time, he suggested a similar program to the New York City police and received an enthusiastic response: the relevant officer offered a "whole hearted" endorsement of Lamb's work and a willingness to quickly gather the necessary personnel. Budgetary limitations, however, meant that he was ultimately unable to offer payment beyond a brandy and a "personal tour of this tempestuous town," and the proposal seems to have gone no further.[136]

In more recent years, however, there has been renewed interest in deploying Lamb-derived movement analysis in law enforcement, whether to gauge the truthfulness of people undergoing interrogations or to determine whether an individual's unconscious movements might reveal their intent to commit a crime.[137] The ongoing appeal of such techniques is worrying—not solely because Lamb never produced any robust data indicating the system's validity or reliability but because the assumptions about movement that Aptitude Assessment was grounded on were, as noted, profoundly raced and gendered.[138]

The United States military has also been funding Aptitude Assessment—now known as "Movement Pattern Analysis" (MPA)—research for nearly twenty-five years through a project called "Body Leads." Run out of the US Department of Defense's Office of Net Assessment (an internal Pentagon think tank founded in 1973), Body Leads began in 1996 as a program for assessing foreign leaders through their nonverbal movements. It was led by Brenda Connors, a professor at the Naval War College who identifies herself as a Certified Movement Analyst (CMA), a Certified Advanced Practitioner of MPA, and the recipient of the "Warren Lamb Trust Creativity, Leadership and Innovation Award." In its first twenty years, Body Leads produced studies of dozens of foreign leaders, including Saddam Hussein, Osama bin Laden, and Vladimir Putin, claiming to link their characteristic movement qualities—as displayed in public settings or on video—to personality traits and decision-making styles.[139] The technique, Connors argued, was particularly useful in gaining an understanding of leaders in "centralized and secretive" regimes where conventional methods of information-gathering were limited.[140] She also contended that MPA could help detect deception and verify identity, noting that it had been used in 2003 to distinguish the real Saddam Hussein from a possible body double. "The war on terror," Connors explained, "increasingly demands reliable measures in the area of identity confirmation," making "the remote capacity to identify elusive and lethal figures" a top priority.[141]

While most Body Leads reports are not publicly available and complete financial data is similarly spotty, reporting in 2014 indicated that spending on the program had hovered around $300,000 annually during the five previous years. A significant portion of that amount went to outside consultants who worked alongside Connors, including $230,000 to Richard Rende, then a Brown University psychiatrist and movement pattern analyst, and $113,915 to Timothy Colton, a Harvard University professor of political science. In 2011 Warren Lamb himself—then in his late eighties—received $24,000, most likely in compensation for his participation in a study of Vladimir Putin. Later reporting revealed that Lamb had also directly advised studies in 2004–05 and 2008.[142]

Popular awareness of Body Leads began with a 2014 *USA Today* article and spread to other outlets, putting the program briefly in the public spotlight. Much of the coverage of the project was less than complimentary, as the opening lines of a piece in *Politico* suggest: "Do you like watching Internet videos and then drawing broad, sweeping, pseudoscientific conclusions about the people involved? If so, congratulations, you might be qualified to join the Pentagon's secret team investigating the nonverbal cues of powerful world leaders."[143] Leaders at the Department of Defense worked to distance themselves from the program: a Pentagon spokesman, for example, explained that Defense Secretary Chuck Hagel had neither read nor relied on Connors's reports. He further emphasized that Body Leads was but one of many experimental investigations being undertaken by the Office of Net Assessment.[144]

These criticisms seem to have had little long-term effect. Though it was eventually rebranded under the name ALEADMOVE, Connors's program continued without any substantial alterations. In 2019, for example, ALEADMOVE contracted with the well-known MPA practitioner Carol-Lynne Moore for a number of different services, including new studies of the movement behavior of foreign officials as well as research intended to "operationalize" Lamb's ideas about "Flow" and further investigate its gendered elements.[145] Moore was also charged with providing MPA-based "sensitivity training to faculty and staff associates at the Naval War College," a task with which she was apparently already quite familiar. "In recent consulting assignments," one contracting document explained, "she has profiled leaders in the DOD Office of the Secretary of Defense, CNO Strategic Studies Group and coached them on their findings individually and as teams."[146] In Moore's work, Aptitude Assessment's origins as a tool of corporate management merged seamlessly with its new life as a technology to predict and enact violence.

Throughout its history, Aptitude Assessment appealed because of its claim to expose authentic selves that would otherwise remain hidden. In the office buildings where Lamb spent much of his career, this notion was deployed to make the case that managerial work could provide endless opportunities for the expression of one's unique, individual personality. Decades later, in the halls of the Pentagon, the body similarly became the "ultimate source and container of much strategic information," though here the promise was the conjuring of improbably exact intelligence at a distance.[147] In both these cases, the technique's users clung to the claim that movement was a window into the soul, either forgetting, downplaying, or strategically ignoring the fact that it was always in part a performance, one unquestionably influenced by the act of observation. Nor were they troubled by the lack of rigorous studies underpinning Lamb's theories or predictions. Cloaked in the aesthetics of objective data-collection, Lamb peddled a seductive story, guaranteeing a deep kind of truth and certainty in settings where it was felt to be badly needed.

The effects of movement analysis's presence, however, were far from neutral. Its use was intertwined with coercive ideas about white-collar labor as the path toward self-actualization, and it irreversibly shaped the career paths of untold numbers of workers. For many of those who encountered it, the system also created new forms of physical self-consciousness and new standards of behavior, demands that fell particularly heavily on women and people of color. Aptitude Assessment, moreover, remains with us, quietly influencing military operations and shaping the everyday behavior of white-collar employees, a subject to which the epilogue will return. The next chapter, however, will explore a somewhat different instantiation of this new way of thinking about and recording movement, turning away from the world of business and toward the hospital and therapist's office.

5

Moving On

On a Friday afternoon in October 1952, two women entered the dayroom of the closed ward of a private psychiatric hospital in Glen Oaks, New York. As they pushed the television set and piano against the wall and folded the Ping-Pong table, they observed the "apathetic" mood pervading the space. A few patients watched TV; others huddled in blankets on a couch. Still others stared into space, "apparently doing nothing." There was, one of the women recalled, "no talking among the patients, and little interest was shown by the appearance of the 'dancing ladies,' whose wide-flowing, brightly colored skirts and bare feet may have seemed startlingly frivolous in that drear setting."[1]

Though they received "no sign of greeting or recognition" from those in the room, the new entrants persisted in their plan, filling the room with the sounds of a lively, rhythmic waltz. Initially there was no change, but "as the music continued its insistent lilting beat," they saw that "several patients who had seemed completely detached and uninterested had begun to tap the rhythm gently with their feet and were gradually letting it flow through their bodies. Eyes turned to follow the movement on the floor. The hands of the teachers, stretched out invitingly to be held or to give support, were grasped tightly, and here and there individual patients ventured onto the floor." Patients that had seemed halting and awkward at first slowly gained surety as they were led into rhythmic movements in pairs, then in threes and in fours, until ultimately all joined in a single, unified circle. Within a quarter of an hour, the "whole room was alive with the vigorous swinging of their bodies," and even the patients still skeptically watching from the sidelines "tentatively imitated the movements, stretching out their legs and swaying in their chairs."[2]

The women were dance therapists, and the hospitals they worked in were the mirror images of Warren Lamb's Cold War corporate citadels. For Lamb, midcentury society was populated by men (and some women) who were methodical, vigorous, and full of life. They donned their suits, did their parts, and returned home at the end of each day to serve as pillars of functional communities. They might have struggled to find a career that made them feel fulfilled or felt stymied by an overbearing boss, but their lives were fundamentally contented and predictable. In truth, the world was far more complex, peopled also by haunted war veterans, grieving children, individuals grappling with serious mental illnesses or suffocated by the conventional strictures of home and family life. For every triumphant tale of suburban satisfaction there was a story of anxiety, isolation, and emotional turbulence. For every successful executive there was an ex-serviceman racked by flashbacks and bursts of anger or a Holocaust survivor haunted by grief and trauma.[3]

Indeed, these seemingly disparate identities often coexisted uneasily in the same person. *The Man in the Gray Flannel Suit*'s Tom Rath, for example, was not just a corporate striver but a former Army paratrooper who was tormented by memories—of seventeen felled enemy soldiers, of his role in the accidental death of his best friend, of a child he fathered and abandoned in Italy.[4] His struggles were also neither unique nor confined to the world of fiction. During the 1942–43 Guadalcanal campaign, 40 percent of evacuated nonfatal casualties were psychiatric in nature, and concerns about the long-term mental and emotional consequences of war only intensified in the aftermath of the United States' involvement in Korea and Vietnam.[5] Tellingly, Rath's first task as a public relations specialist is helping the network president compose a speech arguing for the creation of new national mental health services.

Similar historical currents thus shaped the lives of both the company man and the psychiatric patient. Their experiences, however, had more in common than their Cold War milieu. Just as Warren Lamb scrutinized the bodies of middle managers for evidence of their personalities and motivations, a new group of psychological experts saw the moving bodies of people with unsettled minds as a potential source of both diagnosis and cure. Using the same Laban-derived techniques implemented in the corporate world—and some of the same personnel—dance and movement therapists sought to communicate with, treat, and reintegrate the individuals under their care. Though these worlds may have appeared distinct, they were in fact constructed in relation to each other.

Dance therapy as a field did not arise from Laban's work alone. In the years immediately following World War II, psychiatric hospitals increasingly offered their patients new recreational opportunities in sports, crafts, music—and sometimes dance.[6] Initially, exactly what dance therapy was supposed to accomplish was unclear; most of the "therapists" were dance teachers with little or no formal psychological training, and many medical professionals saw dance therapy as not much more than a harmless pastime patients might use to blow off steam.[7] As one contemporary study put it, the practice was justified in a variety of ways: "As a communicative bridge for non-verbal regressed withdrawn patients; as a healthy and pleasurable rehabilitative technique; as a cathartic experience; and as a sublimated, creative expression for primitive feelings."[8] The patients movement therapists treated were similarly diverse—from autistic children to adults with schizophrenia to the intellectually disabled to war veterans in military hospitals. But despite, or perhaps because of, dance therapy's amorphousness, by the mid-1950s more than two dozen hospitals had established dance therapy programs of one kind or another, and the practice had made its way to almost every region of the United States, from Queens, Riverdale, and Manhattan to Tuscaloosa, Topeka, Kalamazoo, and Provo.[9]

Dance and movement therapy attracted even more interest in the 1960s and 1970s as psychiatry shifted its emphasis from providing custodial care for the chronically ill to delivering preventive and rehabilitative services to a more varied patient population.[10] The field acquired a national membership organization, university programs, and new professional registration standards. As it did so, it also increasingly identified itself with both Laban movement theory and the use of Effort/Shape (E/S) notation, a technique derived largely from Lamb's work. This chapter delves into that history, tracing the role of Laban-based practices in making movement therapy make sense in the America of the 1950s, '60s, and '70s. It focuses in particular on the work of two especially influential therapists, Irmgard Bartenieff and Judith Kestenberg, and begins by exploring Effort/Shape's use as a tool of professionalization. This effort was particularly important in a new field, led almost entirely by women, which often struggled to position itself in the shadowy terrain between art and science. Symptomatic of this tension, many of the practitioners whose careers this history covers referred to "dance therapy" and "movement therapy" interchangeably, and I will follow this practice throughout the chapter. Those who did prefer "movement therapy" tended to argue that it suggested a more neutral, medical, or scientific body of knowledge; others defended "dance therapy" precisely because of its artistic connotations.[11]

Laban-based movement therapy's appeal, however, also stemmed from its promise to assuage broader anxieties about the human body's capacity to silently incubate trauma, fear, and aggression. E/S users positioned the notation as a revolutionary tool for accessing the inner lives of patients: even when they could not or would not utter a word, specific gestural patterns might signal schizophrenia, depression, early childhood trauma, or repressed anger, bringing to the surface that which could not be spoken. Therapists argued, moreover, that this endeavor had not just individual stakes, but collective ones, claiming that movement therapy could play a role perpetuating stability in a world threatened by political and social breakdown.[12] In particular, they saw Effort/Shape as a tool for managing one key tension at the heart of movement therapy: a desire to allow patients a forum for free-form psychological release, but to ensure at the same time that the resulting catharsis would be appropriately managed and measured. Much as it had in Germany decades before, though with drastically different political aims, notated dance was enlisted to mediate the relationship between the self and the social whole. Indeed, in a deep irony, many of movement therapy's leaders were themselves survivors of the Holocaust's terror. They had lost work, fled their places of origin, and mourned family and friends murdered in the death camps. They turned, nonetheless, to the work of a former Nazi collaborator, hoping that by remaking the body they might help cure the traumas, both obvious and hidden, that threatened life in the second half of the twentieth century.

Creating a Profession

Though she would eventually be known for her work in hospitals, Irmgard Bartenieff's path to movement therapy began on the stage. Born Irmgard Dombois in Berlin in 1900, she spent her childhood in the city, attended a few semesters of university in Freiburg as a premedical student, and in 1922 moved to Munich with her first husband, a historian. It was there that Bartenieff developed a serious interest in dance and first encountered Laban's work. In 1925, she sought him out in person, traveling to his Wurtzburg school before following him back to his Choreographic Institute in Berlin, where in 1926 she received her Laban Diploma in Dance and Notation. By then divorced, she remarried Michal Bartenieff, a fellow dancer and a Jewish refugee whose family had settled in Germany after fleeing Russian pogroms in 1906.[13] The two opened a school and company together in 1932, but 1933 Nazi racial laws prevented them from teaching or performing publicly. Though it was clear that the political situation was worsening, the Bartenieffs tried to continue their lives in Berlin, choreographing alone in their empty studio, finding in dance, as Irmgard put it in a letter, a realm in which "we could really forget the 'sickness of the world.'"[14]

Sadly, the respite dance provided was only fleeting. In June 1936 Irmgard and Michal traveled to the United States on visitors' visas in hopes of establishing connections that would enable them to make a permanent home across the Atlantic. Drawing on her existing contacts in the arts world, Irmgard gave lectures on Labanotation at Bennington College, the New School, and Columbia University's Teachers College. The following year, aided by letters of support from university officials and members of the local dance community—including the *New York Times*'s John Martin—Irmgard and Michal received visas allowing them to settle permanently in the United States.[15]

They had, however, left two children behind. Igor, born in 1929, and George, born in 1933, stayed with Irmgard's sister as the couple attempted to establish themselves in their new country. Irmgard's first attempt to retrieve the children was a failure. Arriving in Germany on November 10, 1938, she was met with the news of the previous day's Kristallnacht and found that her family had not prepared the children's immigration paperwork as they had promised. Her mother wanted the children to remain in Germany and urged Irmgard to bring them to Berlin's Kaiser Wilhelm Institute of Anthropology, Heredity, and Eugenics to have them anthropometrically reclassified as Aryan.[16] Unable to remain in Germany herself and worried that her family's denial would put the children at risk, Bartenieff moved George and Igor to an anthroposophic residential school in the Bavarian Alps, where she hoped they would be better protected while she made other plans. The children ultimately arrived in the United States in August of 1939, leaving Germany on the last passenger ship to depart before the war began.[17]

While they waited for George and Igor's arrival, the Bartenieffs took courses in Swedish massage in New York City and eventually—with the help of some local doctors—took over a small physical therapy practice in Pittsfield, Massachusetts. Irmgard also gave children's dance lessons, hosted playgroups on weekends, and taught dance in local summer camps. In 1943, she moved back to New York to enroll in the inaugural program in Physical Therapy and Physical Rehabilitation at New York University, bringing the children but leaving Michal behind. (The two lived separately for the bulk of the remainder of their marriage and divorced in the early 1960s.)

Bartenieff's clinical training was completed at Bellevue Hospital with Dr. George Deaver, one of the founders of rehabilitation medicine, and after graduation she found employment at New York City's Willard Parker Hospital. There she spent seven years working with polio patients, combining standard physical therapy practices with insights derived from Laban's movement theories, taking advantage of a therapeutic climate newly open to experimentation.[18] As she recounted in a 1947 letter, "Working under a very dynamic progressive American doctor, I was allowed to use any approach to movement as long as I would make them walk or making [sic] them use their hands and arms for daily needs and professional work."[19] She further credited her Laban background—along with her new medical training—with honing her ability to "discern by eye and touch very fine deviations" from normal physical functioning and creatively synthesize appropriate remedies.[20]

This kind of therapeutic work seemed like the vocation Bartenieff had long sought. Writing to her former teacher Laban, she explained that "As you probably remember, this 'insulated' business of what we used to call '*Kunsttanz*' [art dance] has never fully given satisfaction to me—I am much rather an artisan with good tools and alert senses to perfect and understand movement in its many manifestations and work with many different people. And for that here in America is ample opportunity."[21]

Bartenieff was indeed quite successful as a physical therapist. After her time at Willard Parker she obtained a position as the therapeutic director of Blythedale Children's Hospital, an orthopedic facility in Valhalla, New York. She also took on a number of private patients, receiving frequent referrals from the city's doctors, including David Gurewitsch, Eleanor Roosevelt's close friend and personal physician.[22]

But even as Bartenieff manipulated limbs and massaged tender muscles, she never lost her conviction that human movement was more than mere mechanism. As she wrote to Laban, her work in physical therapy had only made her "crave" dance movement in "a deeper sense." "Once one has experienced with a deepening awe what it does to a human being when the language of the limbs has become blurred or distorted," she explained, it was impossible not to believe that movement was intimately connected with psychological life.[23] She observed, for example, that the children under her care frequently experienced emotional damage from the experience of hospitalization itself. In addition to suffering from a wide array of physical maladies, the children were "listless" and "regressed," "frustrated" by their removal from the cadences of everyday life. In coordination with an art therapist, a psychiatrist, and a number of social workers, Bartenieff sought to disrupt these negative patterns, arguing that true rehabilitation was "not always just a matter of teaching them body mechanics. There is a definite need for rhythmic and expressive movements."[24] Her task, she wrote, was to "find ways of keeping alive the movement impulse—the root of all development of a thinking, feeling, acting human being" in a hospital climate characterized by "stasis and regression."[25] In its most basic form, she argued, it was not so different from Laban's earliest ambitions to use dance to awaken the slumbering soul. Writing to Laban in England, Bartenieff asked, "Can I call myself still your pupil? It means a lot to me."[26] If she felt any anger or anguish at Laban's former role in the Nazi state apparatus, she did not express it.

Bartenieff's more formal investigations into dance therapy began in the early 1950s when she became Laban's student again more literally, traveling to Germany for the first postwar Dancers' Congress and then to England for a summer course at Dartington Hall. With 250 other students she participated in a variety of courses led by teachers including Laban and Lisa Ullmann, all focused around the new "effort approach" then being deployed in British industry and education. Many of her fellow participants were dance and physical education teachers, including two government physical education inspectors. Other students included lawyers, musicians, physical therapists, psychiatrists, and an occupational therapist "working in one of the big mental institutions near London."[27] In a letter to the Dance Notation Bureau, where she had held leadership roles since the early 1940s, Bartenieff remarked on the way this new material "had the effect that one discovered a great variety of shades in one's own movement." She mentioned too that she had become newly conscious of the ties linking everyday movement and dance.[28]

At Dartington Bartenieff also met Warren Lamb for the first time. While Lamb's main focus was the corporate world, he had also begun to dabble in therapy.[29] As he told a group of London psychiatrists in 1952, the transition was relatively seamless: he simply used the same movement assessments and training he had previously applied to "quite normal people" with patients in private psychiatric hospitals, justifying his work with the growing psychological consensus that "mental balance" was "all a matter of degree."[30] The results, he claimed, were quite remarkable, leading patients toward greater happiness, harmony with others, and "mastery" of themselves. Asserting that most psychiatrists already accepted the basic premise that bodily movements could influence the mind, he presented his intervention as primarily technical, rooted in his expertise with advanced methods of observation and recording. "What we do," he claimed, "was to discover the precise sort of exercise needed for a particular individual, and teach it in such a way that the sense of awareness of the movement is developed."[31] While this therapeutic work never received the majority of Lamb's attention, he was passionate about its utility and keen to pass on his insights to others.

Bartenieff was similarly eager to learn, and the two struck up an ongoing correspondence. As her interest in movement as a diagnostic and therapeutic tool grew throughout the 1950s, Bartenieff would reach out to Lamb when she had questions about how to refine her powers of observation or what a particular constellation of gestures might signify. In the summer of 1958 she traveled to England to study with him in person. For a fee of £75, they arranged to meet two or three days a week for a three-month stretch, Lamb promising that by the fall Bartenieff would be equipped "with sufficient facility in the observation and analysis of movement to apply it in your professional work and gain results."[32] By and large, his prediction came to pass, though Bartenieff acknowledged as late as 1959 that she still occasionally struggled to discern certain movement qualities, catching only the occasional glimpse of their presence, "like a little sun glitter on a stream of water."[33]

Once back in the United States Bartenieff began deploying her new knowledge in a position as the staff dance therapist at the Jacobi Day Hospital, then affiliated with Yeshiva University and the Albert Einstein College of Medicine. Her early efforts were experimental, focused around getting the hospital's outpatients—diverse in both diagnoses and backgrounds—to interact with one another and their bodies in new ways. In reports to administrators, Bartenieff described using dance to enliven patients' days, to increase their sense of connection through collective movement, and as a prompt for expressing a range of emotions from joy to sorrow.[34]

Initially Bartenieff recorded her notes on her dance therapy sessions in conventional language for the convenience of the supervising physicians. She believed, however, that her new ability to "think" in Effort terms was at the root of her observational prowess. Soon, "just for practice," she began to notate doctors' movements during staff meetings, interested in what her records might reveal both about their individual temperaments and their intra-group relations. Noticing her scribbling, the physicians were intrigued and asked Bartenieff to try notating meetings between patients, physicians, and their families, watching through a one-way observation screen. At a moment when family and systems therapy was increasingly ascendant, the doctors were particularly interested in what movement might reveal about unspoken group dynamics.[35]

The hospital's staff was also curious about whether patients' movements reinforced or contradicted what they stated verbally—as Bartenieff put it, whether movement was "covering up true feeling or expressing it."[36] In one case, to test what movement alone might reveal Bartenieff and several physicians observed a newly admitted patient as she joined a group dance therapy session. From behind a screen with the sound turned off, they watched the woman's progress, noting that she initially appeared "more lively and dynamic than most newcomers to a group." Relatively quickly, however, Bartenieff began to observe "peculiar" and "repetitive" patterns in the phrasing of the patient's movement, causing her to suspect the patient was suicidal. When the sound system was turned back on, Bartenieff's insight was confirmed: the patient was in the midst of confessing that she wanted to poison herself.[37]

When the physicians were pleased with the results she produced, Bartenieff began observing talk therapy sessions on a regular basis, eventually obtaining a half-time position primarily devoted to producing movement records.[38] Among her tasks was assessing whether patients' movement patterns changed over the course of their usually two-month treatment—an indication of whether their mental state had in fact improved at the time of their discharge. Her early results were mixed. While she noted that most patients leaving the program exhibited an increase in the frequency of postural changes—suggesting a greater and more authentic engagement with their environments—she also observed that they frequently displayed an increase in "exaggerated bound flow." This led Bartenieff to conclude that "the apparent improvement is not so great." In such cases, she explained, "the patient still struggles with his conflicts, he is not flowing in his discharge of expression or functional activity—he is holding back."[39] In conjunction with Martha Davis, a NYU psychology student who trained under Bartenieff and eventually became an important therapist and nonverbal behavior researcher in her own right, Bartenieff also gradually developed movement profiles for various psychiatric states, including depression, schizophrenia, and anorexia.[40]

Bartenieff's work at Jacobi was supported by Dr. Israel Zwerling, the hospital's director of psychiatry and the chair of the newly created division of social and community psychiatry at Yeshiva University. A reformer by nature, like many in the 1960s Zwerling was an advocate of community-based alternatives to hospitalization, arguing that day programs, clinics, and family therapy served patients more effectively and more humanely than long-term inpatient treatment.[41] He was also an advocate for creative arts therapies, arguing that music, drawing, and dance provoked responses "precisely at the level at which psychotherapists seek to engage the patients, more directly and more immediately than do any of the more traditional verbal therapies." The feelings that were then "aroused and expressed" became "available to the therapist to identify, to develop, and to change."[42] Zwerling acknowledged that the precise mechanism by which such therapies worked remained unclear but argued that he had seen too many examples of their efficacy to doubt their power. He felt too that arts therapists were more likely to approach the patient as a full person rather than as a disease to be cured, an orientation he found lacking in psychiatry as a whole.

In 1966 Zwerling was appointed director of the Bronx State Hospital, a new thousand-bed psychiatric institution opened in response to the recommendations of the President's 1963 Joint Commission Report on Mental Illness. Bartenieff—and movement therapy—came with him. Not only did Bartenieff and a growing group of new trainees continue to work with both long-term residents and day patients, but they established footholds in novel areas of practice. In 1970, for example, movement therapy became a key component of the treatment offered by the hospital's Cognitive Development Service, a new department designed for developmentally disabled adults with psychiatric illnesses. Previously these individuals, many of whom were nonverbal, had been left to languish on the regular wards, often for months. Existing verbally oriented therapeutic programs were unsuited for their needs, and staff frequently resorted to physical restraint as a means of control.[43] The new unit, led by Dr. Yasuhiko Taketomo, was formed to provide more humane and effective alternative treatments, as well as to test out therapies that might eventually be employed in the broader community. Upon intake patients were evaluated by a neuropsychologist, a clinical psychologist, and a linguist, as well as by a "senior kinesthetist," who was tasked with analyzing their "nonverbal communications, body movements, and expressions within a cultural context." During hospitalization their strengths and weaknesses would be further assessed via direct observation as well as through "repeated filmings of interactions among patients and with staff."[44]

Patients were also offered dance therapy, both individually and in larger groups. For some, the process was focused on "breaking into the isolation that has engulfed patients for many years," using group rhythmic movement to spur them to collective participation in "an intellectually, physically and socially stimulating activity."[45] For others, movement therapy focused on remaking their relationships to their own bodies and minds. "Josephine," for example, was a forty-three-year-old woman who had been recently transferred from another ward where she was regularly "tied to a wheelchair because of her history of seizures, occasional hyperactivity and the fear that she might fall on her face."[46] When she arrived at the Cognitive Development Service unit, Josephine was completely nonreactive and needed to be assisted with even the smallest basic movements. After months of individualized dance therapy, however, Taketomo reported that Josephine had been transformed: she engaged in voluntary ambulation, she could throw a ball, and perhaps most importantly, she had begun to interact with other patients and staff. As Taketomo explained at a 1971 meeting of the American Academy of Psychoanalysis, the combination of Josephine's organic disorder and her previous movement restriction had produced a "disturbance in the experience of motor identity and the awareness of the totality of self" that movement therapy seemed to remedy. He acknowledged that—given the other interventions Josephine also received—he could not attribute her development to dance therapy alone, but he argued that it was undeniably significant that "these non-verbal approaches" appeared to provide greater access to the "primordial levels of the 'self system.'"[47] Taketomo hoped that additional research on the subject would further clarify the mechanism of action, providing insights that would be valuable across patient populations.[48]

Throughout this period Bartenieff was a stalwart campaigner for dance therapy's professionalization, a goal she saw as dependent on the widespread adoption of Effort/Shape notation. Although by the late 1960s dance therapy was gaining increasing acceptance in medical circles, Bartenieff feared that the field as a whole continued to suffer in comparison to more established practices with standardized procedures, vocabularies, and entrance requirements. In contrast, she complained, "the models of training" in dance therapy "seem either be centered around a particular gifted, creative individual who combines creative expression with some psychological terminology or psychology methodology, or a very generalized training in the accepted elements of dance education which were developed in American modern dance."[49] Arguing that a profession built around the "personal intuitions, hunches and ESP" of a small number of individuals was fundamentally unsustainable, Bartenieff believed that Effort/Shape could bring necessary order and coherence to what was still a messy mélange of theories and techniques.[50] As she explained in a proposal for a movement therapy training program in the early 1970s, the use of "a system of graphic notation . . . will standardize language, observation, and recording techniques" and "will result in a body of knowledge required to produce a scientific base to the development and use of Dance Therapy. Without such development in notation, validation of observation, diagnosis and treatment remain primitive and unreliable."[51] At the same time, notation would continue to permit some degree of therapeutic pluralism. Bartenieff emphasized that she did not consider Effort/Shape a particular method of therapy, but rather "a tool to come to grips with differentiating shades and textures of movement in its visible and experiential aspects—in the sense that any craftsman in the arts and crafts traditions develops specific terms to deal with sensory-feeling aspects of his craft."[52] It could also be used by every kind of practitioner—not just in the hospital, but also in the growing sector of private dance therapy.

Moreover, in a field that was deeply self-conscious about its status—one practitioner wrote about her constant fight against the perception that dance therapists "are women and dumb"—Effort/Shape seemed positioned to help its users communicate the seriousness of their endeavor to the doctors and psychologists who might consider employing their services.[53] That had certainly been Bartenieff's own experience: it was her mastery of Effort/Shape that had first attracted the attention of Zwerling and the other physicians at the Jacobi Day Hospital.[54] Others similarly saw the potential of talking about movement in a way seemed consonant with the existing medical patois. As one therapist explained, the use of Effort/Shape automatically "gives some credence to the therapist's interpretations of patient behavior," helping "communicate not only with colleagues but with other persons professionally involved with patients."[55] As another put it, "LMA legitimates my experience."[56] In fact, as Effort/Shape's public profile grew, it found users beyond dance therapy proper, including among those medical professionals who had always felt movement was somehow significant but lacked accepted techniques for formally incorporating its analysis into their practices. "Effort/Shape terminology makes my intuitive hunches 'legal,'" as one psychologist explained to Bartenieff.[57] Therapists further hoped that the existence of a clear, common vocabulary could help differentiate them from the growing number of "alternative" providers of therapeutic services, from Christian Scientists to adherents of the Human Potential Movement.[58]

Though personally self-effacing, Bartenieff was an unstoppable force in American dance therapy, and as the field began to institutionalize in the mid-1960s both she and Effort/Shape notation played prominent roles. Bartenieff helped found the American Dance Therapy Association (ADTA), which held its first conference in 1966; that same year she spearheaded the creation of a certificate program in Effort/Shape analysis at the Dance Notation Bureau.[59] By the mid-1970s it was graduating nearly fifty students a year; it would eventually evolve into a separate bureau division and then in 1978 into its own organization, the Laban Institute of Movement Studies (LIMS).[60] Bartenieff further raised dance therapy's profile by making connections with prominent academics in the social sciences interested in movement notation, most notably Margaret Mead, who became one of the first members of the LIMS advisory board. Other particularly well-known advisory board members would include the writer Joseph Campbell and the psychiatrists Albert Scheflen and Ian Alger. In 1980, Bartenieff published *Body Movement: Coping with the Environment*, an overview of her contributions that became a touchstone in the field.[61]

Bartenieff's work at Bronx State Hospital also resulted in the creation of a long-running collaborative movement therapy training program there. By 1979, nearly ninety movement therapists had spent some part of their training at Bronx State, a number which was then between 20 and 25 percent of the total population of movement therapists in the United States.[62] Many of Bartenieff's students also went on to teach themselves, including three who were responsible for the creation of the first US master's degree program in dance therapy, founded in 1971 at New York City's Hunter College and funded by a grant from the National Institute of Mental Health. In 1975 the ADTA

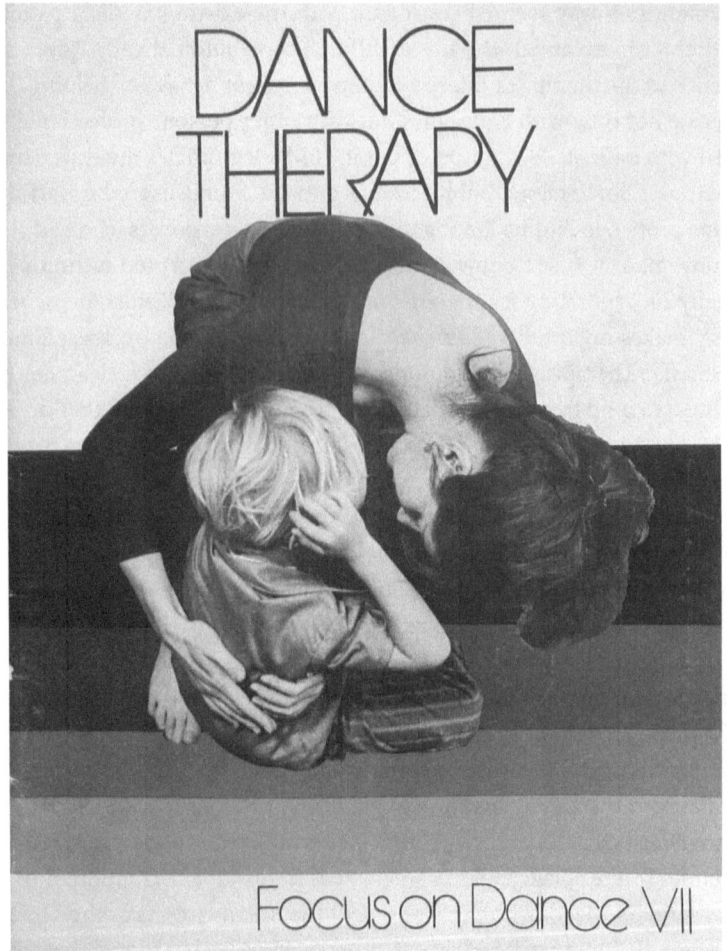

FIGURE 5.1. Cover of *Dance Therapy: Focus on Dance VII* (American Association for Health, Physical Education, and Recreation, 1974). Permission granted by SHAPE America. © 1974, American Association for Health, Physical Education, Recreation, and Dance (AAHPERD), https://www.shapeamerica.org/. All rights reserved.

dedicated its annual conference to Bartenieff; in 1980, the Laban Institute of Movement Studies was renamed the Laban/Bartenieff Institute of Movement Studies in honor of her work.

This is not to say that Bartenieff's leadership crowded out all other approaches to the field—dance and movement therapy remained heterogeneous in a number of ways—but rather that she exerted a profound influence on the shape of its development. Most importantly, Bartenieff provided the vocabulary that would help bind its practitioners together. As she would explain in 1980, "Labanalysis provides a means of perceiving and a vocabulary for describing movement—quantitatively and qualitatively—that is applicable to any body movement research even when there may be differences in interpretation of function and communication."[63] Or, as another therapist put it several years earlier, because "effort-shape deals with the process of movement and employs consistent and scientific principles," it could be used to develop an independent theory of dance therapy, "much as mathematics has been used to develop theories about physical reality."[64] Notation helped make dance therapy into a discipline.

A Science of the Unspeakable

One of the figures who benefited most from the ground Bartenieff laid was Judith Kestenberg (née Silberpfennig). A psychoanalyst specializing in child development, Kestenberg was born in Tarnów, Poland, in 1910 and trained in medicine, neurology and psychiatry at the University of Vienna. After completing her doctorate in 1934 she started additional training at the Vienna Psychoanalytic Society, though her Jewish identity and socialist politics soon made Austria a hostile home. Kestenberg emigrated to the United States in 1937, where she finished her education at the New York Psychoanalytic Institute with Hermann Nunberg and then began work alongside the neurologist and psychoanalyst Paul Schilder at Bellevue Hospital in New York City.

Schilder had also trained in Vienna, and he was known for his attempts to integrate psychiatry and psychoanalytic theory as well as for his focus on the physical body. In his 1935 book *The Image and Appearance of the Human Body*, Schilder developed the concept of the "body image," an entity born out of the interactions between the physiological state of the individual, his or her social experience of the body, and the felt reality of daily life. The body image was not a steady state but a dynamic one, full of fluctuation and change.[65] Schilder had also studied disorders of the central nervous system and believed they were rooted in interactions between the physiological and psychological.[66]

This effort to combine analyses of the mind and the body attracted Kestenberg. Though she operated primarily from a psychoanalytic perspective, her training in neurology had left her convinced that the operations of the psyche were intertwined with other kinds of organic experiences, including bodily movement. When Kestenberg encountered Bartenieff in New York, therefore, she was fascinated and asked Bartenieff to teach her the basics of Labanotation. The two women quickly began collaborating, and

Kestenberg too sought out additional training in Effort/Shape from Warren Lamb in the 1950s.[67]

Kestenberg thus absorbed the fundamental tenets of Labanalysis and always retained strong ties to the Laban community, teaching classes at the Dance Notation Bureau and at the Laban Institute of Movement Studies, where she served as a member of the board of trustees. She also developed her own form of movement notation derived from Laban's work. Called the Kestenberg Movement Profile (KMP), it required that the therapist be trained in Laban-based observation but asked that the observer concentrate primarily on the "rise and fall of flow." As she put it, this form of notation "consists of the recorder's freehand drawing of the increase and decrease of muscle tension during movement" and was based primarily "on the observer's kinesthetic mirroring identification with the observed subject."[68] The resulting diagrams resembled waves of varying amplitude and frequency.

Using the KMP, Kestenberg began recording and analyzing movement data in her role as a faculty member at New York University's medical school and the Long Island Jewish Hospital, integrating Laban-based methods with her own psychoanalytic training and focusing primarily on infants and young children. She posited, for example, that Freudian developmental stages were characterized by specific kinds of movements, distinguishing the "anal-erotic" phase from the "anal-sadistic" in terms of the kinds of bodily rhythms the child most often utilized. Both stages, she held, were characterized by intervals of holding and releasing, but one was significantly slower than the other.[69] At the same time, Kestenberg echoed Lamb's contention that certain movement traits—and thus personality characteristics—were essentially inborn and difficult to alter. Writing of one toddler she studied, Kestenberg noted that "Charlie" seemed "to prefer to have objects presented in space so that he could reach for them by moving forward or laterally. . . . This does not mean that Charlie could not move in all directions. He tended to choose his preferred directions over others when he also had to do new things." Indeed, "despite all the efforts of those about him to bend Charlie's rhythm to fit theirs, he seemed to remain basically unchanged in temperament." "Even in the nursery," she wrote, "Charlie gave the impression of being an important citizen. He looked and felt like a heavy viscous mass."[70] For Kestenberg, these observations had diagnostic value, as they allowed her to detect significant personality traits, emotional disturbances, or developmental delays even among very young children who were unable to communicate verbally.

Kestenberg also argued that the KMP could be a therapeutic tool, contending that the roots of any number of psychological disorders lie in a mismatch between parents' movement preferences and those of their children. She explained, for instance, that infants predisposed to anal-sadistic rhythms "become easily frustrated when they are not allowed to do things their own way, in their own rhythm, and in the posture of their choosing. If mothers interfere too much and too early, babies, who have less pronounced preferences for anal rhythms, will also react with prematurely increased aggression, which may lead to obsessive development."[71] Similar problems could crop up at any stage of development or simply as the result of a personality mismatch between parent and child. While these conflicts might at first seem minor, Kestenberg held that ignoring them was leading to what she saw as a profusion of otherwise preventable psychiatric illnesses. This work also neatly dovetailed with a broader postwar focus on the psychological consequences of disruptions in the parent-child relationship. In Britain, much of this research grew out of concerns about wartime family separations and resulted in a renewed stress on the undisturbed mother-child bond. In the United States, new kinds of residential institutions sprang up to treat "emotionally disturbed" children, many of which operated from the premise that dysfunctional family systems were key contributors to children's maladjustment.[72]

In 1972 Kestenberg founded her own independent institute devoted to child development in Sands Point, Long Island, called the Center for Parents and Children. There she worked with groups of "typical" (exhibiting no obvious pathology) middle-class families, who came regularly to the center for sessions of play, dance, art, music, reading, and meals. As parents interacted with their young children, Kestenberg and her colleagues observed their movements, noting each family member's tendencies and how they did nor did not harmonize. The center then focused on retraining parents—mostly, though not exclusively, mothers—to move in ways that were more responsive to their children's natural patterns, whether playing, bathing, or at mealtimes. For Kestenberg this reeducation was a crucial form of primary prevention, necessary for ensuring that "personality disorders can be prevented and psychological problems corrected before they have become established."[73] Being ever-vigilant about their own unconscious movements demanded a great deal from parents, but to Kestenberg this was a small price to pay.[74] The alternative was permitting problems to fester unseen, planting seeds that might erupt years later with potentially devastating consequences. And though Kestenberg's test population was well-off suburbanites, she hoped that the models she developed could eventually be deployed across the country, including in "underprivileged" populations.[75]

Kestenberg's work in Sands Point was largely about the "normal" suffering of infancy and childhood, but she also turned her attention to the long-term effects of larger collective traumas. As with Bartenieff, the Holocaust hung heavy over Kestenberg's life. She had fled Austria just before the Anschluss, and her father was killed at Auschwitz. Her husband, the lawyer Milton Kestenberg, whom she married in 1942, had similarly left Poland in 1939; his mother and brother were murdered in the Treblinka extermination camps. After the end of the war, Milton devoted much of his time to representing Holocaust survivors seeking reparations in the German courts.[76]

Judith Kestenberg frequently encountered child survivors of the Holocaust as part of her professional practice and over time began to notice commonalities in their physical presentation. Focusing particularly on children who had been hidden or otherwise confined, she set out to study how the physical experience of restriction might lead to permanent psychological changes, asking:

> When the adaptive patterns Laban referred to as "efforts" cannot be used because of confinement in space or/and the need to avoid noise, in order not to be discovered, do these patterns freeze to be applied again when the restrictions are removed? At the same time, movement patterns which serve relationships to people, which we refer to as "shaping planes," must freeze as well. Is it incumbent upon the endangered individual then to regress and rely wholly on tension changes which allow us to express our anxiety and also to serve the rhythmic discharge of drive impulses? If so, will this individual, when not restricted any more, tend to become anxious and less relaxed whenever there is a need to restrict motion?[77]

Eventually, Kestenberg developed a more wide-ranging fascination with psychic pain's somatic consequences, the ways in which traumatic experiences left invisible scars upon the body. On Manhattan's Upper West Side, where she lived for many years, she developed an unusual tendency to suddenly fixate on a seemingly random stranger passing by. Intently training her gaze on her quarry, she would follow the person of interest for a few blocks before approaching and asking to speak. "Tell me," she would say, "about your experience in the Holocaust." The intervention was undoubtedly jarring for most of those Kestenberg waylaid, especially as they had no way of knowing how this woman had picked them out of a crowd and why she was so certain of the contours of their private histories. In most cases, they had not been overheard speaking, nor did they have visible tattoos or other marks that would have tipped her off. For Kestenberg, though, the movements of victims of Nazi terror were impossible to conceal or fake, a language survivors spoke even when silent. She prowled the avenues looking for trauma's telltale grammar.

As one of Judith's colleagues later recounted, "With her very direct, straightforward manner, she managed to obtain these interviews. Although sometimes this appeared intrusive to me, after the interviews the interviewees were clearly grateful."[78] Whether or not that assessment was correct, Kestenberg's doggedness reflected her own growing sense that keeping past traumas buried wreaked its own kind of damage. Indeed, beginning in the late 1960s Kestenberg started exploring the ways in which Holocaust survivors' suffering was transmitted to the bodies of their children, even when the children had not been directly exposed to Nazi persecution.[79] "There is something here that is almost universal," she observed. When survivors' children "reach the age when they themselves had been persecuted, they begin to act out, and the children too begin to act out the parent's Holocaust experience," losing weight, suffering from insomnia, or developing otherwise unexplainable physical symptoms.[80]

By the 1960s and 1970s, the American psychological establishment increasingly shared Kestenberg's interest in trauma's persistent effects. As psychiatrists grappled with the pain of returning Vietnam veterans, feminists explored the lasting psychological wounds of rape and other forms of sexual violence, and psychoanalysts sought to amend Freud's theories to account for the new kinds of symptoms they were encountering in their patients.[81] As Allan Young has demonstrated, the moment was characterized by a new kind of widespread agreement that "intensely frightening or disturbing experiences could produce memories that are concealed in automatic behaviors, repetitive acts over which the affected person exercised no conscious control." "Without the intervention of an expert," Young explains, "the owner of a 'parasitic' memory remained unaware of its content and ignorant that it influenced aspects of his life."[82] In treating such patients, therefore, the psychological professional's aim was to excavate these memories, in the hopes that doing so would eventually allow patients to move through and past them.

Kestenberg reflected this approach in her work with individual patients, using movement analysis as a tool for surfacing memories they could not yet express verbally. She also, however, sought to use her movement expertise as an alternative method for revealing the reality of the collective trauma induced by the Holocaust to the wider world. Such an effort seemed particularly urgent in the 1970s as public memory of the Nazi regime had begun to fade. Even psychoanalysts, Kestenberg noticed, often failed to connect their patients' present-day suffering to their wartime trauma, and the first widespread stirrings of Holocaust denialism—always present, but previously somewhat contained—had started to emerge in the United States and Europe.[83] Horrified by these new attempts to deny or minimize the suffering of survivors, Kestenberg increasingly conceptualized the KMP as a tool for connecting dispersed individuals joined by a common history. By allowing therapists to identify Holocaust survivors and register physical evidence of their trauma, it would provide definitive evidence of the truth of their anguish.

This kind of evidence had a clear symbolic value, but as Milton Kestenberg's work in the German courts eventually made clear, it could also serve immediate, practical ends. In the early 1950s, the new West German government set up a fund for providing individual Holocaust survivors with monetary reparations. At first, the only people eligible for payments were those who could establish that they had experienced a disabling physical injury as the result of Nazi persecution; in 1965 the system opened to claims from individuals who had suffered psychological damage. Establishing the existence of a qualifying emotional or mental injury, however, proved to be more difficult than it might have initially appeared.

Immediately, survivors confronted a plethora of obstacles. They were required to produce precise timelines of their movements from camp to camp and to locate witnesses who could attest to their whereabouts and experiences. Disability claims required an in-depth—and often distressing—examination by a physician, and all medical documentation had to be submitted in German, a language which many of the claimants did not speak. Any inaccuracy, however minor or irrelevant to the claim's substance, was immediate grounds for a suit's disqualification. Initial denials could be appealed but necessitated that the victim obtain a lawyer and take the case to court, where they would be subject to examination by a phalanx of general physicians, psychiatrists, bureaucrats, and court judges. Claims of psychiatric injury proved particularly difficult to establish, as adjudicators frequently argued that there was insufficient proof of a direct link between the claimant's current mental state and persecution decades past.[84] Some survivors were told that had they been psychologically healthy to begin with even horrifying events would not have produced long-term effects. Their pain, claimed some psychoanalysts, must have instead sprung from other early childhood experiences: "Maybe their parents' marriage had not been happy; maybe they were oversensitive."[85] As the analyst Kurt Eissler put it in the title of a contemptuous essay in 1963, "The murder of how many of one's children must one be able to survive asymptomatically in order to be deemed to have a normal constitution?"[86]

These discussions occurred, moreover, in a nation in which many of the medical and legal professionals responsible for adjudicating claims had themselves participated in the Nazi project, and political and public hostility toward the very idea of reparations ran rampant. Skeptical physicians accused sympathetic doctors of bias and subjectivity, arguing that they were swayed by emotion and had no definitive evidentiary basis for supporting survivors' claims.[87]

In representing clients, Milton Kestenberg encountered these impediments again and again. He told, for example, the wrenching story of one man who had made a small error in recounting his wartime history. Told by his caseworker that his claim had been denied because he had "'cheated' the authorities by making an inaccurate statement, he tore his shirt and screamed, 'Kill me, I cannot face my family,'" afraid that no one, including his children, would now believe the horrors he had experienced.[88] Others were informed that spending their infancy or toddlerhood in concentration camps or in hiding—or even having a family member killed in front of them—was not grounds for compensation, as very young children would "not remember the details of that suffering and cannot be permanently damaged."[89]

The problem of proof, as Judith later explained, was an intractable one, in part because a lack of memory was often a symptom of the very trauma survivors were trying to establish. As she noted in 1994, "it became increasingly apparent that the survivors, not only the child survivors, had memory problems and could not always recall the dates of the persecution and their whereabouts at certain times." Children in particular "suffered from a double jeopardy: their memory was adversely affected by persecution, and many were too young to remember when, where, and what happened to them."[90] As Milton engaged with his clients, therefore, he began to deploy methods inspired by Judith's work, searching for bodily cues that might gain him admittance into new realms of memory. Milton's attention to the somatic also accorded well with a strain of long-standing German medical doctrine that held that persistent—as opposed to temporary—post-traumatic psychological damage would only occur in cases where the patient also suffered physical harm.[91]

A woman he called Janine, for example, experienced knee pain that impaired her ability to work, but the original examining physician contended that her suffering had no organic cause and was unrelated to her childhood experiences. When Milton interviewed her, however, he prodded Janine to talk about the moment in which her father was taken away by the Gestapo. In tears, she recalled throwing herself on her father's body. At first, Judith recounted, "the attention was focused on the father, not on the child herself. Only when Milton asked her what happened next did she recall that a Gestapo officer kicked her in the leg with his heavy boot. Milton then asked Janine where she felt the pain of the kick before she could identify the spot on her knee that hurt her now." Focusing on visceral childhood experiences, Judith explained, "allowed Janine to describe her own actions and feelings, which enabled this young survivor to establish the connection between the Nazi-inflicted trauma and the present-day affliction."

Milton also used movement to help survivors identify perpetrators, asking the men in his lineups not just to stand, but to speak and gesture. He argued that—whatever other efforts defendants might make to disguise themselves—their basic movement patterns would give them away. A tendency to "use strength easily," for example, would be difficult to mask, as "the subject cannot be in full control of his postural movement, even if he would want to."[92] In other words, just as survivors' movements bore the scars of their past trauma, perpetrators' movements carried indelible markers of their own.

As German court records are sealed, it is difficult to determine whether Milton Kestenberg was able to directly utilize evidence drawn from movement analyses in his legal complaints. What is clear, however, is that Judith and Milton deeply influenced one another's thinking about the body, trauma, and the sources of healing. They shared subjects, methodologies, and eventually a long-running research project: the International Study of Organized Persecution of Children, founded in 1981 and devoted to—among other causes—the investigation of the "nonverbal expressions of normal and disrupted child development."[93] They also arranged and archived interviews with more than 1500 child survivors of the Holocaust as well as a smaller number of the children of Nazi perpetrators; these interviews are now held by the Oral History Division of the Avraham Harman Institute of Contemporary Jewry at the Hebrew University of Jerusalem.

For Kestenberg, therefore, movement analysis ultimately became not just a technology of scientific inquiry but a tool for voicing truths otherwise unspeakable. In a coauthored 1996 book, *The Last Witness: The Child Survivor of the Holocaust*, Kestenberg wrote that the reader "will note" that "we offer no photographs, tables, graphs, maps, blueprints of the gas chambers, copies of incriminating Nazi documents, or extensive statistics."[94] This choice, she clarified, was not because these forms of evidence did not exist. It was instead because they were only a pale suggestion of the true consequences of the genocide, forms of pain "so profound that present-day language does not express them clearly." Only movement allowed a glimpse behind the verbal facade of postwar normalcy, producing a new language of evidence and—in the process—a new vision of history. It was, as one of Kestenberg's interviewees later put it, a "bridge of living tissues" linking the past to the present.[95]

Taming Aggression, Constructing Community

But while Kestenberg's work with Holocaust survivors used movement as a window into the past, another group of dance and movement therapists saw their primary goal as shaping the future. Most urgently, they became interested in using movement to model novel types of selves and communities, creating in microcosm the kind of democratic societies that would ensure the horrors of the past decades would not be repeated. Such a goal would continue to involve patients hospitalized with persistent mental illnesses, but it also implied that movement therapy might be of use to an even wider spectrum of people. As Bartenieff wrote, "the problems of healing and restoring mental, emotional health are not confined to pathological extremes. There are differences in degree of mental health, rather than kind. The current explorations in healing oriented toward 'restorations' of wholeness—some ancient and some apparently new—are reflections of the wide-spread concern with problems of personal and community wholeness and survival."[96] Though the immediate threat posed by Nazi Germany had ended, dance therapists still lived in a world that felt precarious, and they hoped that their expertise might aid in setting it on a better path.

Movement therapists were not, of course, the only ones haunted by the fear that the current peace was only temporary. As the Cold War ramped up, the twin specters of totalitarianism and communism hovered ominously over the American landscape, leaving many in the country casting anxious glances over their shoulders. In 1950, moreover, the publication of *The Authoritarian Personality*, coauthored by Theodor Adorno, Else Frenkel-Brunswik, Daniel Levinson, and Nevitt Sanford, brought to public attention the idea that the rise of National Socialism might have been at least in part attributable to the proliferation of fascistic personalities in Germany. Such individuals, the authors argued, were characterized by reverence for authority, rigid adherence to conventional moral standards, a distaste for originality and intellectualism, and pronounced aggression toward those perceived as outsiders. Significantly, these traits were not inborn but rather shaped by early childhood experiences, in particular by parents who were excessively domineering or punitive.[97] For many American social and behavioral scientists, therefore, the challenge became figuring out how to cultivate the opposite kinds of citizens: "democratic" personalities who were creative, tolerant, flexible, and autonomous, simultaneously capable of individual thought and eager to exist in harmonious society with others.[98]

Though not always in explicit conversation with these scholars, movement therapists shared their goals. Kestenberg saw her contribution to the cause as rooted in prevention—keeping the most dangerous kinds of personalities from ever emerging. Many of the movement interventions she taught at the Center for Parents and Children focused on identifying and countering early "aggressive patterns," ensuring that they did not metastasize perniciously.[99] Directly echoing the Freudian arguments advanced in *The Authoritarian Personality*, she maintained that "what the one-time persecutor and the one-time persecuted carry with them are traces of their previous roles for which they were trained in childhood," and she sought to proactively adjust this instruction.[100] For other movement therapists operating outside the psychoanalytic tradition the sources of aggression may have been less definite, but the problem was no less pressing. They similarly worried about hidden fonts of anger and rage and agreed with Kestenberg that art could be used to "provide channels for impulses and wishes which have to be tamed to rear civilized human beings."[101]

Even Warren Lamb expressed concern about concealed aggression, particularly in returning veterans. After attending a 1944 meeting of the Research Board for the Correlation of Medical Science and Physical Education, for example, he criticized the short medical inspections that determined discharged soldiers' emotional fitness, arguing that those who had been certified "well" were often anything but. As he put it, "I have seen shadow-moves which contradict the final findings" of doctors, a state of affairs which will "surely lead to disaster." In his view, the most significant problem was that existing rehabilitative measures were oriented only toward returning military personnel to a state of aggressive combat readiness, an approach that might impede their reintegration into regular civilian life.[102]

In Bartenieff's view, dance was *the* ideal medium for negotiating these tricky waters. As she articulated in a 1972 open letter to the American Dance Therapy Association, dance had in fact always served a social purpose: "Its oldest role has always been to order, to renew, to re-affirm, to delineate degree of closeness and distance in human relationships."[103] Though the public often perceived dance as a medium for completely free-flowing, individualistic expression—the epitome of "art for art's sake"—the truth was much more complicated. Movement, Bartenieff explained, often did prompt participants to reach "reach heights of excitement, in fact complete oblivion, and abandon, as well as utmost serenity and calm." It was also, however, a form often practiced in community, one that required that the explosive emotions it provoked be channeled in service of a larger design. As she put it quite bluntly in a report on her work at the Jacobi Day Hospital in the early 1960s, "Dance is the art of regulating behavior in a very literal sense."[104]

Movement therapists' commitment to the dual goals of personal expression and social engineering are clearly evident in accounts of their practice. On the one hand, the vast majority of dance therapy sessions were characterized by a great deal of openness and flexibility. Whether working in psychiatric hospitals—or, as they increasingly did by the late 1960s and '70s, in special education facilities, correctional institutions, halfway houses, and private studios—movement therapists sought to encourage play, creativity, and individuality.[105] At least in theory, patients were never forced to attend dance therapy sessions, and meetings were designed to respond to the specific needs and desires of the individuals participating on a particular day. Group leaders suggested movement phrases and structures, but "repetitive, mechanical rote-learning" was discouraged, and patients were prompted to engage dynamically with their environments.[106] Franziska Boas, for example, practiced dance therapy with New York City children and encouraged her charges to follow their movement instincts, whether pretending to be animals, climbing on one another like jungle gyms, or literally bouncing off the walls. It allowed them, she argued, to "stimulate and find expression for primitive and deeply buried fantasies" and express their inner selves in a too-often restrictive society.[107] Prospective therapists were themselves evaluated not just for their technical skill but for their degree of "personal spontaneity."[108]

Alongside this encouragement of individual expressiveness, however, ran a strong conviction that previously suppressed feelings and desires had to be carefully managed once surfaced. Movement, therapists believed, could help patients recognize the presence of their aggressive or destructive impulses, but it could also be utilized to keep them "within reasonable bounds." As one therapist explained, it was only when such emotions were "repressed, denied, ignored or feared, or conversely indulged in to excess that they form a volcanic element which is liable to become very destructive."[109]

There was, therefore, considerable discussion among dance therapists regarding the precise mixture of free expression and structured movement that served patients—and society—best. As Bartenieff summarized, the questions up for debate included whether structure "inhibited or released freedom" as well as "even larger questions about the nature of freedom, the relationship of permissiveness to freedom, the differences between freedom and license, the degree, if any, of manipulation that is necessary or acceptable in educational methods, parental controls or lack of them, the therapist's responsibilities, and others."[110] For her part, Bartenieff saw at least some degree of control as profoundly important. "Too often," she wrote in a retrospective account of her career, "there is a misconception that freedom should be identified with any outburst of feeling at a given moment." She emphasized that dance

FIGURE 5.2. Photograph for the National Institute of Mental Health. In *Writings on Body Movement and Communication*, American Dance Therapy Association, Monograph no. 3 (1973–74): 22.

therapy should not be overly "structured by mechanical measurements" but also maintained that it could not be allowed to "deteriorate into amorphously indulgent self-expressiveness." Instead, "the delicate balance between structure and permissiveness must be gently maintained because either extreme becomes destructive."[111] Though it might not be perceptible to a casual observer, the skilled movement therapist was perpetually scrutinizing her patients, soundlessly shaping their choices.

One example of what this careful negotiation might look like—as well as a case that highlights the crucial role Laban-based movement analysis played in the process—can be found in a story Bartenieff told regarding a recently discharged hospital patient who had returned for an individual outpatient session. As she recalled:

> On this particular day, she seemed unresponsive and resistive to movement. All of a sudden—and it may not have been all of a sudden, perhaps I just failed to notice the signs—she seized a billiard cue from a table next to her and tried to attack me. I managed to seize the stick with both hands; we became locked into each other. I began to respond to her pushing, pressing quality by a Sustained (Time Effort), Strong (Weight Effort), Indirect (Space Effort) rhythm, with slight fluctuations in the components. Several minutes passed in this locked, strong encounter. I had no intention of yanking the stick out of her hands. In a dim way, I held on to the intent of dancing with her. She finally dropped the stick very suddenly on the floor, exclaiming with a broad smile, "That was some workout. I am tired. I feel fine." We parted on friendly terms.[112]

In the aftermath, Bartenieff noted how the interaction had taught her "something crucial" about "transforming an aggressive tension by gradually dissolving its Effort elements from fighting to indulging qualities," explaining that "this kind of transformation is at the heart of dance."[113] Viewed at the right angle, the ultimate end of free expression might be a kind of control.

Other therapists wrote similarly about the violent impulses that arose in their sessions, highlighting the level of attention and expertise required to redirect them properly. Audrey Wethered, a Laban-trained therapist who practiced in England, noted in 1968 that she had once given patients cushions to help them "deal with aggression" until "on two occasions it led to people using them in such a violent way, seriously bodily harm could have been inflicted if it had been directed at a person." She also related a story about a former patient who returned years later and thanked Wethered for helping her first confront and then control her anger. As the patient explained, "I was always afraid to let go of my aggression, I thought if I did, I would have to kill." Only through movement therapy, she claimed, did she "realize that I had a choice, I didn't have to kill."[114] Using all the tools at her disposal, it was the therapist's job to ensure that this transmutation occurred in a carefully choreographed way. "Very dangerous situations can arise," Wethered cautioned, "unless there is de-climax at the crucial moment."[115]

Concern about the possible consequences of uninhibited emotional release even accompanied practitioners to their own conferences. The 1972 ADTA annual meeting, for example, featured a number of workshops in which, as a pedagogical technique, experienced therapists guided their fellow attendees through mock sessions. In the aftermath of the conference some worried that the workshops—many of which attracted dozens of participants—had been irresponsibly large. One leader noted that even in settings like the conference, in which participants were presumably both relatively healthy and well-prepared for what they might experience, "the power of the experience of movement" had to be respected. When "the darkness which hides repressed feelings, the camouflage of confusion which often surrounds conflicts" were "removed for the first time," a "plethora of emotions and reactions" could burst forth. Packing a ballroom to capacity, she argued, made it impossible for the therapist to conscientiously attend to these individual needs, and she advocated limiting future pedagogical sessions to no more than fifteen participants.[116]

Despite the dangers of working with very large groups, communal movement was crucial to the ideology of movement therapy. In fact, most movement therapy was practiced in group settings: while therapists sometimes worked one-on-one with patients, collective sessions were the practice's emblematic form. The reasons for this preference for group activity were in part financial, as after the explosion of the patient population in the years following World War II, group therapies provided an economical means of providing treatment. For many dance therapists, however, there was also a more philosophical justification: collective movement, in their view, functioned as the perfect model of a democratic society, a setting in which each individual was encouraged to express his or her particular self but also required to adjust in lively negotiation with others. As one therapist explained, dance provides a "group situation" in which "the individual has to find his proper place within a unit of which he is only a part. Through the need for group awareness and sensitivity, for group action and formation, the individual personality is pulled out of the boundaries of his independent self to play a role in a happening which follows its own laws."[117] By enacting these relationships through movement, participants would literally rehearse the social and political roles therapists hoped they would embody in the future.

The idea that collective movement practice could produce immediate political dividends was also reflected in the increasing range of settings in which dance therapy was practiced. In the early 1970s, for example, movement therapist Joyhope Taylor contracted with two public schools in the Los Angeles area to provide dance therapy to students in kindergarten through second grade. Noting that learning to manage aggression and cooperate with others was a key task of early childhood socialization, she criticized the "lack of district-endorsed methods to release hostility" available prior to her arrival.

Without such an outlet, she explained, the school day was characterized by "continual 'crises'" produced by children who hit others, were unable to wait their turns in line or to speak, competed constantly, and had difficulty authentically sharing their feelings. Pitching movement therapy as a preventive mental health intervention, Taylor explained that she taught "dance therapy structures which allow the child to find himself, express himself, relate to others, and discover his wholeness: I'ness, You'ness, We'ness, Us'ness, All'ness."[118] Notably, the two schools in which Taylor worked were quite different from one another: one was in tony, predominately white Beverly Hills, while the other was in the economically depressed and largely Black neighborhood of Watts, still scarred from the Watts Rebellion just a handful of years before. For Taylor, though, the need for intervention was the same. (It is, however, perhaps telling that only the Watts project received external funding, receiving support from the Division of Preventive Mental Health at the Charles R. Drew Postgraduate Medical School of Martin Luther King Jr. Hospital, itself established in the immediate aftermath of the 1965 riots.)

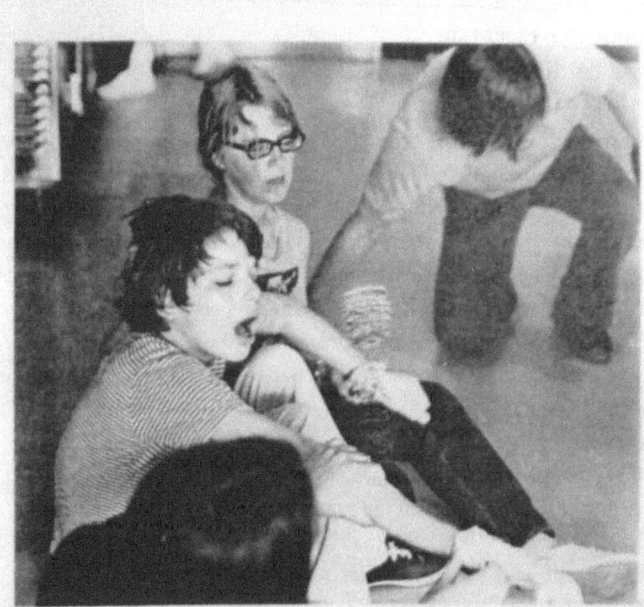

Sometimes it is hard to be part of the group.

FIGURE 5.3. Photograph for the National Institute of Mental Health. In *Writings on Body Movement and Communication*, American Dance Therapy Association, Monograph no. 3 (1973–74): 63.

In practice, therapists used a variety of techniques create a sense of group cohesion. Some urged participants to come early and prepare together, and most prohibited nonparticipants from observing, fearing it would lead to self-consciousness. Other tactics were employed during the sessions themselves. Bartenieff, for example, almost always began her group sessions in a circle, waiting until a "ground rhythm" organically emerged from one or two individuals. As others joined in, the rhythm would often be changed or transformed, but the circle formation generally persisted: it was, Bartenieff noted, "an almost organic expression of non-aggressive relationship. Everyone sees everyone else and the joining of hands relates people and aids the experience of streams of movement energy spreading from one to the other."[119] As a result, the experience of dance was kept from being entirely self-centered. "Since the primary function of the therapist is," Bartenieff wrote, "to help a patient find an acceptable identity and satisfying mode of behavior for himself and in his society, it is important to maintain a context of flow between the internal and external as active as possible." Individual expressiveness was important but never at the expense of the group.

This sometimes fraught relationship between expression and self-effacement, creativity and control, is vividly illustrated by an early account of a problematic movement therapy patient, "E.M.," treated at Hillside Hospital in the 1950s. As the therapist recounted:

> Her improvisations were an expression of a creative urge, but unfortunately one which was too undisciplined and too uncontrolled to be fully utilized. . . . Because the sessions were primarily directed toward providing a group experience for as many patients as possible, this patient's highly individualized needs were felt to have been only partially met. Because of her tendency to get overstimulated by the material, unlimited permissiveness of her self-expression could not be encouraged. Furthermore, within the group situation, the patient's overactive response inhibited other patients who were in need of greater support and more direction. As a result, the patient's full creative potentiality in this medium could not be fully explored. The limitations on complete freedom of expression were felt to be necessary, not only to preserve the integrity of the group, but also to control the regressive behavior of the patient.[120]

E.M. was thus allowed to express her impulses, "but always within the controlling limits of the reality situation." Moreover, "in spite of the constraint of these imposed restrictions," the therapist noted that she was still "able to experience exhilaration and joy, not only in the sense of physical release which the dance afforded her, but in the opportunity it provided for creative expression."[121] Patients in dance therapy classes were told to explore space, to create, and to fully experience their emotions, but they were also cautioned about creativity's excesses. Therapists sought to stimulate patients' originality—and through it their selfhood—but these efforts were nearly always tempered by an attention to the individual's place *within* the larger group. As Bartenieff constantly reminded her students, "The 'holy madness' of the individual artist and the emphasis on the spectacular aspects of theatrical display constitute merely one side of the dance. The other side is composed of the regulatory, interactional forces." These forces, she held, had historically played a primary role in "forming and regulating group and individual relationships," though an appropriate reverence for their power had been lost in modern, technologically oriented "Western" societies.[122]

This use of movement to create new forms of community echoed Laban's work in Germany in ways both unsurprising and profoundly ironic. Even as Bartenieff and Kestenberg grappled—both personally and professionally—with the lingering horrors of the Holocaust, they embraced tools and ideas whose history was intimately entwined with the Nazi state. Bartenieff freely referred to Laban as a predecessor, and in the 1970s she published a series of translations of Laban's work in the *Effort/Shaper* newsletter, highlighting how Laban had understood the "process of preparing and presenting choric movement works" as part of a process intended to "awaken . . . new values."[123] The *Bewegungschor* remained, in a very real way, a key model for movement

FIGURE 5.4. Movement therapy workshop. Irmgard Bartenieff Papers, Special Collections in Performing Arts, University of Maryland.

therapy. In private, however, at least some movement therapists had reservations. In discussions surrounding the naming of the organization that eventually became LIMS, for example, a number of board members brought up the "bad press" that might come with highlighting their affiliation with Laban's thinking. To outsiders, he might seem too "mystical, too complex, fascist, communist," and the therapists feared being cast as "an evangelical cult of women devoted to a male figure." Still, their concerns were largely about perception, and the name was ultimately retained.[124]

There were also, however, significant differences between the kind of collective movement experience Laban envisioned and those engineered by movement therapists in the 1950s, '60s, and '70s. These disparities are highlighted by the juxtaposition of Laban's preferred notation for group movement—Kinetographie—with therapists' utilization of Effort/Shape. While both technologies were undeniably prescriptive, Kinetographie was intended to dictate each and every movement a participant made, while Effort/Shape provided somewhat more flexible structure, suggesting a series of movement qualities and relations rather than imposing them exactly.

As Bartenieff explained the contrast with Labanotation in 1962, "EFFORT Notation deals with relationships and proportions rather than with absolutes and can thus better deal with the ultimate subtleties of human behavior."[125] Movement therapists also prided themselves on incorporating individual patients' preferences into the collective movements they prescribed on any given day, explaining that "notating the dynamics of and sampling some improvs of the group members" helped therapists "arrive at the final structural framework, at the same time leaving freedom for the participants to individually fit themselves into the group structure."[126] Each session, therefore, represented an embodied microcosm of the idealized social world: one in which citizens were autonomous, but regulated; unique individuals, but fully devoted to the social whole. This was a view of mental health highly contingent on a particular set of Cold War American ideals.

Therapeutic Life

It is, unfortunately, difficult to access direct accounts of the experiences of movement therapy patients. When therapy was conducted within the hospital, documentation was filtered through the physician or therapist's perspective and considered part of patients' confidential medical files; independent therapists were not required to retain records at all. There are also almost no publicly available descriptions produced by participants themselves, whether because of an understandable desire for privacy or because so much of the practice was premised on the inadequacy of words.[127] Nonverbal patients were even less likely to produce lasting documentation.

In therapists' notes and recollections, patients' reactions to the interventions were almost uniformly positive. They included records of patients who expressed how much "the experience in the group had helped them" whether in terms of their capacity for self-expression or in their ability to relate to others. One participant was reported to have said that the process had "put life back into my body," while a therapist recalled that "there was no 'them' and 'us'; just a group of people creating movement and life together.'"[128] When patients did struggle, any challenges or unpleasant emotions were usually depicted as necessary steps on the road to growth. A therapist working in California, for instance, wrote to Bartenieff about a young female patient who "allowed herself to experience strong aggressive movement for the first time" and in doing so "found a new voice inside her." In subsequent sessions, she looked and sounded like "a different person," a transformation that left her both "ecstatic" and "a little frightened." "We still have much work," the therapist noted, "but I believe she will be able to claim herself and I'm delighted."[129] Accounts of patients' initial skepticism or reluctance were also relatively common, but nearly all concluded with the person in question being gradually won over. Presumably, however, at least some participants disliked the experience enough that they either stopped attending sessions or declined to take part in the first place.

We also know little about how the experience of movement therapy varied across lines of race, class, gender, sexuality, or ability. While many in the Laban-based movement therapy community saw themselves as politically progressive and sensitive to the ways in which movement—at least in some cases—could be culturally shaped, they had also been trained within a tradition in which movement was understood to be governed by immutable, universal laws. Perhaps as a result, movement therapists only sporadically recorded the racial or ethnic identities of their patients, although—particularly in New York City's public hospitals—they encountered enormously diverse populations. There is evidence, however, that therapists at least sometimes saw their patients' movements through a racialized lens.

In treating a four-year-old Black child at Bellevue Hospital in the early 1960s, Bartenieff, for example, identified the boy's "overindulgence in free FLOW" as a key factor in his physical and emotional problems. This tendency, Bartenieff argued, could be attributed in part to his specific medical challenges (likely cerebral palsy), in part to his "deprived environment," but also in part to the fact that "Negro movement" was generally more "abundant in FLOW," a characteristic that could be seen in African dance as well as "the working movements of the manual laborer in this civilization."[130] To be clear, Bartenieff did not necessarily pathologize supposedly "Negro" movement in and of itself but rather argued that like all movement styles, its excesses needed to be curbed. In other cases, the conflicts were more explicit. One observer recalled a "painful" dance therapy session in which a "goodhearted" but uninformed Czech therapist tried to impose her European movement style on a racially mixed group of patients in the American South. She labeled a "lean Baltimore dude" antisocial for declining to partake in the prescribed circle dance and instead amusing himself with a "snakehips" dance.[131]

It is thus not difficult to imagine other tensions between the predominantly white, female, and middle- to upper-class therapists and many of the populations they encountered. At the 1972 ADTA conference Martha Davis forthrightly acknowledged that "there is embodied within the Laban system some notions about what is healthy and what isn't. It's true. What is good and what is bad. . . . [T]here are certain assumptions, certain value judgments."[132] Movement therapists working in the psychoanalytic tradition were perhaps particularly prone to these kinds of judgments, attempting, for example, to correct boys whose movements they assessed as excessively "effeminate."[133] Early in her career, Kestenberg herself sought Lamb's movement expertise before hiring a male nanny whom she suspected might be homosexual.[134] But even those therapists who saw Effort/Shape primarily as a recording tool rather than an all-encompassing theoretical framework sometimes found it hard to get away from its—and their—underlying assumptions.

Unlike Laban or Lamb, however, many movement therapists did attempt, albeit imperfectly, to grapple with this bias. In the late 1960s and '70s, a number of therapists expressed discomfort with the homogeneity of their profession and actively sought to recruit more people of color into its ranks.[135] Their efforts, though, seem to have borne little fruit; even decades later, movement therapy remained largely—though not exclusively—white and female.[136] There are undoubtedly multiple reasons for this inertia, ranging from outright racism to more subtle forms of dissuasion. One practitioner suggests that the field's embrace of a systems perspective in the 1960s and '70s made earlier therapists more attentive to their roles as political actors, an awareness that seems to have diminished over time.[137] Movement therapy's static complexion might also be explained, at least in part, by broader transformations in the medical landscape of the 1980s that stymied the field's overall growth.

As it turned out, the 1970s represented movement therapy's apogee as a profession. Following the founding of the ADTA in 1966 and Hunter College's dance therapy master's program in 1971, the field was newly institutionalized and seemed ripe for further expansion, and a significant number of master's and other training programs were established over the following decade. Bartenieff was inundated with letters requesting information about the field and ADTA membership grew steadily. Its practitioners continued to disagree about many things, but—in part with the help of Effort/Shape notation—they had indeed formed an active, identifiable community.

Dance therapy came into its own, however, "just as the ideological and material support" for creative arts therapies started to disappear.[138] The new funds appropriated for community mental health care in the 1960s had begun to run out by the late 1970s, and the increasing closure of inpatient treatment facilities eliminated many of the sites in which movement therapists had practiced.[139] Psychiatry, moreover, was in the midst of an attempt to redefine itself as a medical specialty, one more concerned with the internal operations of the brain than with the psychiatrist's role in a larger social world. As such, while psychiatrists like Zwerling and Schilder had once eagerly sought out the expertise of movement therapists, a new generation of practitioners confronted a field that had less and less interest in their work. As the New York community psychiatrist C. C. Beels summarized in 1989, small-scale, community-focused therapies like dance had little chance of attracting widespread attention given the money and prestige that research in genetics or psychopharmacology offered.[140] Even more importantly, despite their concerted efforts movement therapists were largely unable to get either private medical insurance companies or Medicare and Medicaid to cover their services, a fact which drastically limited their client pool.[141]

None of which is to say that dance and movement therapy went away—the ADTA lists over a thousand active therapists on its website, as well as seven approved master's degree programs—but rather that the field never fully established itself as an intrinsic part of mainstream medical or psychological practice.[142] For some therapists this was disappointing. Others, though, were content to exist in a unique liminal space, relishing the freedom that separation from the institutional strictures of science or medicine provided. For them, movement therapy was somewhere between an art and a science and denying the importance of either would have irreversibly impoverished the form.[143]

In what some of this latter group saw as a cautionary tale, in the mid-1970s researchers at both the Massachusetts Institute of Technology and Harvard University's Aiken Computation Laboratory made contact with members of the DNB's Effort/Shape division. Both groups seemed interested in somehow computerizing movement analysis, using the therapists' insights and Effort/Shape notation to produce programs that would allow hospitals to automatically screen their patients for a variety of conditions, effectively replacing the notator with a machine.[144] While DNB President Earl Ubell was thrilled by the possibility, Bartenieff and others excoriated the project, criticizing its "mechanistic view of people and simplistic understanding of movement." They insisted that movement therapy was fundamentally rooted in human relationships and that even diagnosis required a human observer to "feel" the "moments of change."[145]

On the other hand, movement therapists continued to make clear that their work was not the entirely unstructured, free-form practice that the public so often assumed it to be. To be practiced responsibly, they argued, dance therapy required finely honed observational skills as well as the capacity to document one's observations in standardized form and communicate them to adjacent professions. In taking on this task therapists not only endeavored to unearth truths that others were literally unable to see but made a broader case for the importance of movement in negotiating modern social and political life. The human body, they contended, was itself a kind of recording device, its flesh impressed with all the conflicts and traumas of the twentieth century.

It was for this reason that movement therapy required both care and careful planning. Emotions were uncovered but also managed; patients' self-expression was encouraged but only to a point. The goal was a new kind of synthesis between creativity and control, individuality and belonging. As one movement therapist elegantly summarized, "to understand and master movement is a very human way of achieving a balance between the claims of society and individuality. It requires the adoption of a new outlook on life, a new way of interacting with other people."[146] It was political philosophy in embodied form, a vision of society one could feel in one's bones. Indeed, for Bartenieff, Kestenberg, and their colleagues, the study of movement was rooted not just in a simple curiosity about what was unspoken, but in a fear of what ignorance of such truths had produced in the past—and might produce again in the future.

6

From *Volk* to Folk

It was 1965, televisions were blaring, and Alan Lomax was worried. As *Gilligan's Island*, *The Andy Griffith Show*, *Green Acres*, and *Bewitched* capered into tangerine family rooms, the famous American folklorist expressed his horror. TV had once, he proclaimed, "promised to be a marvelous telescope that could bring the whole world into our rooms—a periscope through which we could peer, unseen and unabashed, into other lives." Instead, "it has erected an electronic curtain, composed of our own prejudices and preconceptions, through which the outside world can only be dimly perceived."[1]

Lomax, perhaps the most prominent folklorist in United States history, is best known for his efforts to uncover and protect the country's musical heritage. Beginning in the 1930s, he spent decades traveling the back roads of the US South, seeking out the songs of prisoners and sharecroppers and recording thousands of them for the archives of the Library of Congress. His "discovery" of figures like Lead Belly, Jelly Roll Morton, Muddy Waters, and Woody Guthrie helped make their careers, and his fanatical advocacy for previously ignored or maligned genres and performers left an indelible imprint on the American musical imagination.[2] For Lomax, this work was both artistic and political, of a single piece with his activism for racial and cultural equity and progressive economic change.

Unlike the Soviet "iron curtain," Lomax's "electronic curtain" was not a physical border across which people and information were not free to cross. It was instead a perceptual barrier, born of the growing power of the American news and entertainment industry to drown out local media outlets both within United States and across the globe. There were concrete consequences to this transformation—national news coverage, for example, that neglected the concerns of the distant and the poor—but something more ineffable was also shifting. Lomax had begun to notice eerie changes in the way people across the planet moved their bodies, observing in particular the seeming omnipresence of the "the head-back, chest-out, erect posture of the North European elite." Noting that scholars in the new field of kinesics had found that human beings "respond below the level of awareness to the movement patterns they encounter," he warned that the global media was in the process of invading humanity's very bones and sinews. Soon, he feared, even the denizens of the world's most remote forests would be striding like London bankers.

In the histories of folklore, music, and sound, Lomax is a figure both towering and controversial. He is as often lauded for his extraordinary efforts to preserve the music of marginalized populations as he is criticized for the ways in which his collecting depended on the exploitation of his subjects or collaborators—sometimes in a single breath.[3] Existing accounts of Lomax's career, however, make only passing reference to the fact that music was not his sole concern. In the late 1960s he turned his attention toward dance, arguing that it too was an art form with profound political implications—as well as an untapped resource for understanding humanity's history.

The result was a project called "Choreometrics," an attempt to collect filmed samples of dance from each and every one of the world's cultural groups and then use them for a massive comparative study of movement patterns. To complete this mammoth project, Lomax needed a technique for making data out of movement's cacophony and ultimately turned to Laban-based forms of analysis. With the help of Irmgard Bartenieff and Forrestine Paulay (who began the work while one of Bartenieff's students), Lomax gathered filmed examples of dance from nearly two thousand communities, carefully classifying and evaluating each one. He used this information to theorize about historical migration patterns, structures of production, and the mechanism of behavioral evolution. By the project's conclusion Lomax had produced several films for public broadcast, an unpublished book manuscript of more than a thousand pages, an unprecedented collection of dance on film, and boxes upon boxes of coding sheets.

This chapter is the story of their production. It is not, however, a simple account of endless acquisition. For Lomax Choreometrics was always more than either a conventional project of salvage anthropology or a novel scholarly method.[4] At its heart, the undertaking was rooted in a quintessentially modernist faith in the power of the right kind of databank to remake the world. The ultimate telos of these films, coding sheets, and publications, Lomax hoped, was a wholesale "recalibration" of the human perceptual and kinesthetic apparatus, remaking the American mind and body to support a new kind of multicultural polity. In studying Choreometrics, therefore, this book comes full circle. Though the vision of the community Lomax hoped to create stands in stark contrast to the one Laban himself envisioned, their beliefs about the moving body's role in achieving those goals were remarkably similar.

The account that follows reconstructs Choreometrics's history, beginning with an examination of how Lomax came to understand bodily movement as a site of important anthropological data. It then continues with a description of Choreometrics on the ground: the contested process of gathering filmed dance, categorizing it, performing statistical analysis, and presenting the findings to the public. Finally, the chapter concludes with an analysis of the radical motivations for Lomax's work and considers how his revolutionary dreams remained elusive—and what those limitations had to do with the structure of movement notation itself.

Preludes

An FBI file opened on Lomax in 1941 and periodically updated for the next four decades described him as possessed of an "erratic, artistic temperament," a "bohemian attitude," and a tenuous grasp of the typical standards of personal grooming. Some informants attributed these traits to his single-minded devotion to folk music, others to his close associations with the "hillbillies" he studied.[5] Either way, his training started early. Lomax's father John was himself a folklorist—as well as a sometime academic administrator and bond salesman—with a special interest in the cowboy ballads of the American South and West. Alan, born in 1915 in Austin, Texas, thus came into a household already closely attuned to the sounds of the rural countryside. He also experienced them firsthand when, starting at the age of ten, he was sent to spend the summers working on a ranch in Comanche, Texas. (His father hoped the labor would toughen his sensitive and not particularly athletic son.) Though often ill in his youth, Lomax excelled academically, gaining acceptance first to the Choate School in Connecticut in 1929 and then to Harvard in 1931. He never felt quite at home in Cambridge, however, and a combination of illness, distraction, and the family's unstable finances eventually took him back to Texas, where he concluded his studies at the University of Texas at Austin.[6]

Even while still an undergraduate, Lomax began assisting his father with his collecting, lecturing, and writing, to which John had turned full-time after the Great Depression cut off his other sources of income. With the aid of some minimal resources from the Library of Congress, the two traveled to prisons, farms, porches, churches, and work camps across Louisiana,

Alabama, Tennessee, Mississippi, and Georgia and coauthored two books: *American Ballads and Folk Songs* in 1934 and *Negro Folk Songs as Sung by Lead Belly* in 1936.[7] Alan also began venturing out independently, including a trip across Florida with the writer and anthropologist Zora Neale Hurston in the summer of 1935. Eventually, supported by a small congressional appropriation in 1937, Alan was appointed to a position as the "assistant-in-charge" of the new Archive of American Folk Song of the Library of Congress, a post he held until 1942.[8]

During World War II Lomax worked first in the Office of War Information, later with CBS and the BBC, and finally as a draftee in the Army Signal Corps and the Armed Forces Radio Service. By the time the war was over, his job at the Library of Congress had evaporated, and he began piecing together a new kind of career. Over the following decade he would achieve recognition as a folk DJ; publish a number of songbooks and records including the critically lauded *Mister Jelly Roll* in 1950; give lectures; conduct research with the support of major foundations; organize concert series; consult with movie studios; and work as an editor for Decca Records.[9] In 1948, as a member of People's Songs, Inc., he composed music for the 1948 Progressive Party presidential campaign of Henry Wallace. In 1950 he was contracted to create a series of radio "ballad dramas" to persuade ethnic minority groups to seek blood tests for syphilis and gonorrhea in response to the new availability of penicillin drugs.[10]

Lomax's politics eventually attracted the attention of the FBI, whose attention he first caught after an anonymous source reported hearing John Lomax make an offhand remark about his son's Communist sympathies. That complaint was relatively quickly dismissed, but in 1950 Lomax again came to the bureau's attention after cosponsoring a Civil Rights Congress dinner honoring a group of lawyers who had defended alleged Communists. Six months later, the privately published pamphlet *Red Channels: The Report of Communist Influence in Radio and Television* listed Lomax and 150 other actors, writers, musicians, and performers as suspected Communist sympathizers, and Lomax decamped to Europe. He would remain overseas until 1958, continuing his collecting, traveling, writing, and producing, though now with a larger geographic purview.

By the time he returned to the United States Lomax had developed a new project focused on the comparative analysis of folk culture. His travels had led him to suspect that there were direct connections between social organization and elements of folk song style, and he began to try to map their relations. Supported by Margaret Mead, he presented an early version of this research at the 1959 annual meeting of the American Anthropological Association, suggesting, for example, that high-pitched singing was characteristic of societies in which women were repressed.[11] He argued too that scholars needed to develop new tools for evaluating and categorizing music, as Western musical notation had proved incapable of accurately capturing the global panoply of sound.[12] Though many at the meeting were skeptical of Lomax's specific conclusions, they were sympathetic to his ambitions, particularly as a growing number of anthropologists shared his conviction that the analysis of visual, oral, and corporeal culture needed to become more central to the field.[13]

In 1960 the American Council of Learned Societies awarded Lomax a nine-month, $6,000 grant to continue this research, which he had begun to call "Cantometrics." Though he had no formal academic affiliation and no doctoral degree, Mead helped arrange for the funds to be administered through the Department of Anthropology at Columbia University, where Lomax would maintain an unpaid affiliation for another quarter century. With the help of collaborators including musicologist Victor Grauer, anthropologist Conrad Arensberg, and statistician Norman Berkowitz, the project consumed much of Lomax's attention for years to come.

Urged by ACLS staff to seek graduate training, Lomax enrolled in a special seminar led by the anthropologist Ray Birdwhistell at Temple University; it was at this point that movement joined music as a key subject of Lomax's concern. Widely acknowledged as the founder of the field of "kinesics," Birdwhistell famously contended that only 30 to 35 percent of the social meaning of a conversation was carried by its words and that nonverbal cues were crucial for establishing both individual and cultural identity.[14] Lomax had hoped that Birdwhistell's approach would give him insight into song as a form of communication, but the senior scholar suggested that dance might represent an even better quarry. Birdwhistell explained that dance was more "primal and preverbal than song, and . . . more directly connected to everyday work and the social movements of the body."[15] He pointed out, moreover, that dance's formalized and repetitive qualities would make it an easier object of analysis than everyday movement.

Choreometrics: Collecting and Coding

Lomax did not abandon his project on song style, but he soon conceived a new study, Choreometrics, focused entirely on dance.[16] Dance analysis, however, required another novel set of tools. Though Birdwhistell used a personal shorthand notation for his own movement research, he was enthusiastic about Laban-derived techniques and directed Lomax to Bartenieff and her student Forrestine Paulay.

Bartenieff and Paulay were enthusiastic about participating—perhaps intrigued by the commonalities between their therapeutic and conservation work and Lomax's own political aims—though the project's limited resources meant that they could only be paid part-time.[17] (Lomax received a number of prestigious grants for Cantometrics and Choreometrics in the years that followed, including awards from the Rockefeller and Wenner-Gren Foundations and a six-year grant from the National Institute of Mental Health. Still, as he had no permanent academic appointment, securing continuous funding was always a concern.) Nevertheless, Bartenieff and Paulay's contributions were significant: not only did they provide technical expertise on dance, but they also collaborated on nearly all other aspects of project development.

Data gathering was the first step. Late in 1965, therefore, Lomax, Bartenieff, and Paulay set out on a seemingly impossible mission: to obtain at least one sample of dance from every world culture. Using George Murdock's Human Relations Area Files at Yale University as a guide, Lomax's initial list included approximately 1900 distinct cultural and ethnic groups.[18] In earlier historical periods, the sheer size of this group would have represented an insurmountable obstacle. Lomax may have criticized previous dance scholarship for being "virtually empty of hypothesis," but he also readily admitted that "our predecessors in this field did not have the privilege of observing the whole range of choreography."[19] Even "the great Curt Sachs," he recalled, "based his seminal [1937] book on world dance largely on written accounts, sculptures, paintings and photographs. He had seen little dancing with his own eyes outside of Europe and North America."[20] Lomax and his compatriots, on the other hand, had the advantage of an increasingly large, albeit decentralized, store of global film.

Lomax reached out to anyone and everyone who might have filmed dance, from scholars at Paris's Musée de l'Homme to the physician and medical researcher Carleton Gadjusek, adventurers' clubs, a retired vascular surgeon, and the US military.[21] The process was long and sometimes contentious: letters negotiating fees, rental arrangements, and mailings fill nearly a dozen boxes of the Lomax archive. While some institutions—such as the Institute für den Wissenschaftlichen Film in Göttingen—were eager to participate, others—including the Russian cultural ministry—required delicate diplomatic negotiations. Wealthy adventurers were often the most difficult to work with, as they frequently used their collections as centerpieces of lecture circuits, "thus realizing the 'cash' value of their traveling and filming." As one collaborator cautioned in a letter to Lomax, "they must be handled with kid gloves."[22] The team even unsuccessfully attempted to locate lost footage made by Orson Welles.

Lomax recognized that many of the films were not intended for scholarly analysis, but he remained confident about their value. While an inexperienced or biased cameraman might, he acknowledged, shape the data in minor ways, Lomax believed that movement style was so deeply rooted in the body that its basic elements could not help but emerge.[23] He argued, for instance, that it was entirely possible "for a couple of tourists who casually point their cameras as they walk past some Samoans to bring us some usable material on the Samoans."[24] Despite Lomax's vociferous criticisms of Disney, US military filmmakers, and the adventurers who disturbed the peace of remote communities, Choreometrics relied heavily on the artifacts they produced. As Jonathan Sterne has pointed out in the case of nineteenth-century recordings of Native American music, the "work of anthropological cultural stewardship coincided with the decimation that necessitated the stewardship in the first place."[25]

Indeed, Lomax saw the world's "vast, endlessly provocative, prejudice-laden, existing sea of documentary footage" as perhaps the "the richest and most unequivocal storehouse of information about humanity."[26] He spoke frequently of his vision for a new kind of Library of Alexandria, where "all important cinematic documents would be stored, catalogued, and analyzed," preserving, for the first time, a "living, moving record of all human behavior." Such a "temple of knowledge," he claimed, would "cost no more than an atomic submarine," but its influence would quickly outstrip "all the libraries that ever existed."[27] Choreometrics was only an initial step toward this comprehensive collection but, Lomax hoped, a conceptually important one.[28]

Ultimately, Choreometrics drew upon approximately 250,000 feet of raw footage, selected from the far larger body of film that had been collected. This collecting enterprise, of course, was only the beginning. Once all the films had arrived at Lomax's offices at Columbia, Lomax, Bartenieff, and Paulay began to watch the films and, more importantly, to code them, looking for the "principles that unify and differentiate the movement styles of the species."[29]

Lomax, Bartenieff, and Paulay first considered using traditional Labanotation to make and record these observations. Soon, however, they realized the technique was not necessarily well-suited to their needs. Labanotation was enormously detail-oriented, designed to record choreographic works in their entirety. Each and every movement a dancer made was painstakingly recorded, and the process was repeated for each and every dancer.

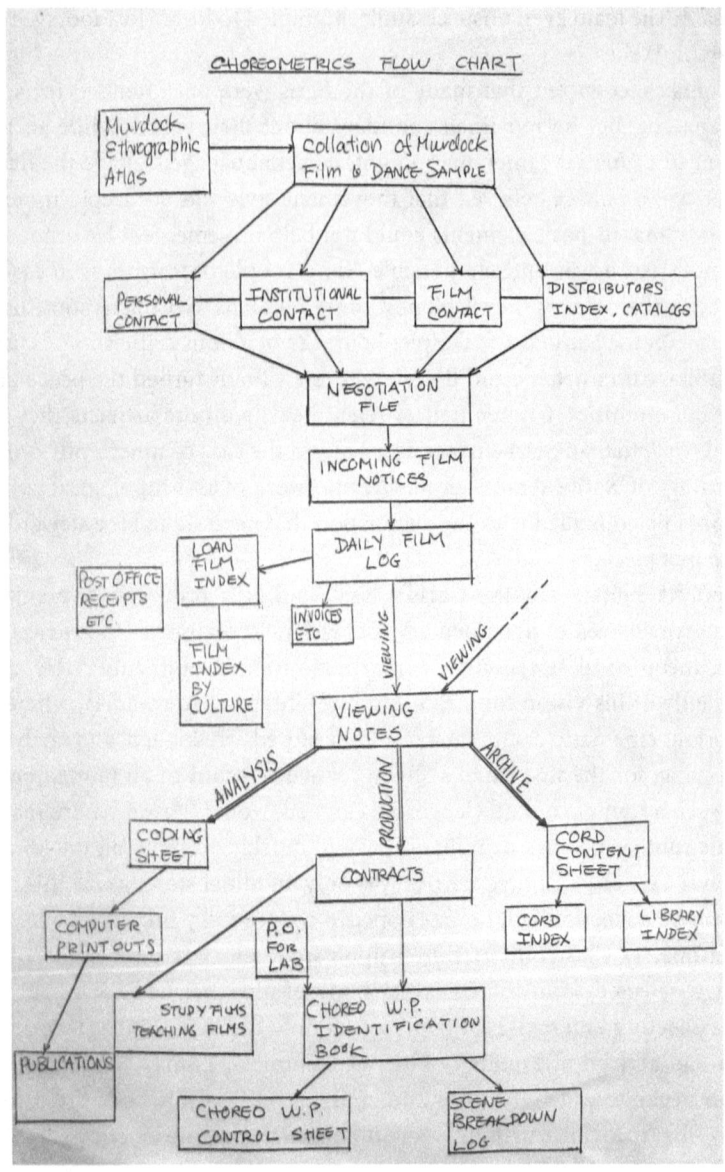

FIGURE 6.1. Choreometrics Flow Chart, 1968. Alan Lomax Collection (AFC 2004/004), American Folklife Center, Library of Congress.

The Choreometrics team, in contrast, was more interested in the overall characteristics of movement than in specific choreography. As such, Bartenieff and Paulay decided to draw on the analytic skills they had acquired as notators but not to attempt to use Labanotation to capture complete dances, turning instead toward Effort/Shape analysis. As Lomax put it, the methods Choreometrics ultimately deployed combined "linguistic and para-linguistic analysis, the Kinesic study of verbal communication by Birdwhistell, the Laban effort-shape analysis, Murdock's ethnographic survey, and the structural approach to culture as formulated by Arensberg, combined with extensive use of the computer to handle large data banks and multi-scale descriptive systems."[30] But even with all these tools at their disposal, evaluating dance in a way that facilitated meaningful analysis seemed overwhelming, and the team decided that their first task would be to limit the range of movement elements under consideration. Thus, as they made a first pass at the films, they resolved to write down traits that jumped out especially forcefully, or that seemed particularly characteristic of one area. In an ideal situation, the Choreometrics team would then locate a trait's "opposite"—smooth movement, for example, as opposed to jerky—in another region. While movement qualities that seemed globally ubiquitous were excluded from future coding, traits that "proved to be useful in differentiating large world regions from each other" were incorporated into the system.[31]

Lomax, Bartenieff, and Paulay worked in this way for about two years before arriving at a set of approximately eighty-five key movement traits. After

FIGURE 6.2. Irmgard Bartenieff, Forrestine Paulay, and Alan Lomax at work on the Choreometrics project, c. 1983. Alan Lomax Collection (AFC 2004/004), American Folklife Center, Library of Congress. Courtesy of the Association for Cultural Equity.

a "three-way collaboration, in which each word and phrase was discussed and voted on," these characteristics were set down and defined in the official Choreometric Coding Book.³² The Coding Book was not intended solely for the project's leaders but also for the new, untrained raters—graduate students from a variety of disciplines—enlisted to analyze the remaining dance samples. The coding process could be enormously complex. Take, for example, a set of proposed instructions for coding just one variable, "body vertical": "Here the coder records the kind of change in and out of the vertical, if any change occurs. At the center of this line, he notes that the dancer maintains a normal upright position. At the far right, he records a violent plunging and rolling about on the earth and, on the far left, narrow and highly controlled shifts of body attitude take place in the upright position scale."³³ Nine other potential ratings for "body vertical" included:

- Continuous sectional change. In this case, controlled action takes place at one level for some time; then, there is a change and there is a long engagement in activity at that level; then another and another, etc.
- Smooth change. Action in which the level of verticality changes gradually and smoothly under great control: dipping, alternating change of verticality, side to side, or sagittally. The shift is marked, the vertical distance traveled being at least 6 inches.
- Neutral. The body held at one level as the dancer remains in place. If there is forward movement and vibration is not coded, the vertical distance traveled should be less than 2 inches. Here, a normal, smoothly controlled walk would be coded, for instance.
- Sporadic leap/discontinuous. Here some other kind of activity is punctuated by leaps or a by a series of leaps.
- Leaps/total trust. Here the mover crouches low to the ground and then, straightening his knees, leaps to his full height. This is the maximal extension possible.³⁴

Predictably, even with these instructions, Lomax, Bartenieff, and Paulay worried about consistency, and they eventually made the decision to drop twenty additional characteristics about which raters had difficulty agreeing. "We must," they noted, "leave out these variables in the protocol which do not yield order, comparisons, and classification."³⁵ The remaining variables, Lomax boasted, generated an average inter-rater agreement rate of 80 percent. Once a dance was fully coded, technicians transferred the data from coding sheets to punch cards and then fed them into Columbia University's IBM 7094 computer. There, with the help of Columbia statisticians, the Choreometrics team began to look for meaningful patterns.

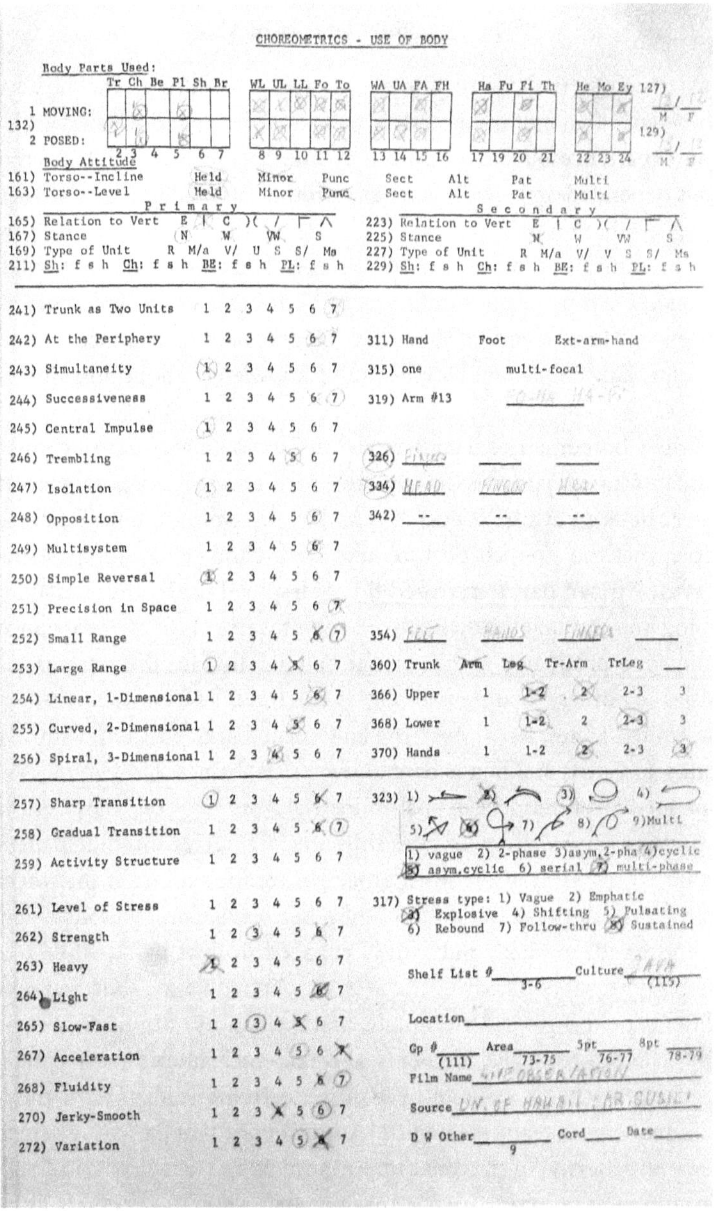

CHOREOMETRICS - USE OF BODY

Body Parts Used:

	Tr	Ch	Be	Pl	Sh	Br	WL	UL	LL	Fo	To	WA	UA	FA	FH	Ha	Fu	Fi	Th	He	Mo	Ey	127)
1 MOVING: 132)																							—/— M/F
2 POSED:																							129) —/— M/F
	2 3 4	5	6	7	8	9	10	11	12	13	14	15	16	17	19	20	21	22	23	24		

Body Attitude
161) Torso--Incline   Held   Minor   Punc   Sect   Alt   Pat   Multi
163) Torso--Level   Held   Minor   Punc   Sect   Alt   Pat   Multi
        P r i m a r y   S e c o n d a r y
165) Relation to Vert E | C)(/ ⌐ ∧   223) Relation to Vert E | C)(/ ⌐ ∧
167) Stance   (N   W   VW   S   225) Stance   N   W   VW   S
169) Type of Unit   R M/a V/ U S S/ Ms 227) Type of Unit   R M/a V/ U S S/ Ms
211) Sh: f s h   Ch: f s h   BE: f s h PL: f s h   229) Sh: f s h   Ch: f s h   BE: f s h   PL: f s h

241) Trunk as Two Units   1 2 3 4 5 6 7
242) At the Periphery   1 2 3 4 5 6 7   311) Hand   Foot   Ext-arm-hand
243) Simultaneity   (1) 2 3 4 5 6 7   315) one   multi-focal
244) Successiveness   1 2 3 4 5 6 (7)   319) Arm #13
245) Central Impulse   (1) 2 3 4 5 6 7
246) Trembling   1 2 3 4 (5) 6 7   326) _____
247) Isolation   (1) 2 3 4 5 6 7   334) HEAD
248) Opposition   1 2 3 4 5 (6) 7   342) --   --   --
249) Multisystem   1 2 3 4 5 (6) 7
250) Simple Reversal   (1) 2 3 4 5 6 7
251) Precision in Space   1 2 3 4 5 6 (7)
252) Small Range   1 2 3 4 5 (6)(7)   354) FEET   HANDS   FINGERS   ____
253) Large Range   (1) 2 3 4 (5) 6 7   360) Trunk Arm   Leg   Tr-Arm   TrLeg   ___
254) Linear, 1-Dimensional   1 2 3 4 5 (6) 7   366) Upper   1   1-2   (2)   2-3   3
255) Curved, 2-Dimensional   1 2 3 4 (5) 6 7   368) Lower   1   1-2   2   2-3   3
256) Spiral, 3-Dimensional   1 2 3 (4) 5 6 7   370) Hands   1   1-2   (2)   2-3   (3)

257) Sharp Transition   (1) 2 3 4 5 (6) 7   323) 1) ⊢── 2) ── 3) ── 4) ──
258) Gradual Transition   1 2 3 4 5 (6)(7)   5) ── 6) ── 7) ── 8) ── 9) Multi
259) Activity Structure   1 2 3 4 5 6 7   1) vague 2) 2-phase 3)asym,2-pha 4)cyclic
                                                       5) asym,cyclic 6) serial (7) multi-phase

261) Level of Stress   1 2 3 4 5 6 7   317) Stress type: 1) Vague   2) Emphatic
262) Strength   1 2 (3) 4 5 6 7   3) Explosive 4) Shifting 5) Pulsating
263) Heavy   (1) 2 3 4 5 6 7   6) Rebound 7) Follow-thru (8) Sustained
264) Light   1 2 3 4 5 (6) 7   Shelf List 0_____ Culture _APA_
265) Slow-Fast   1 (2) 3 4 (5) 6 7   3-6   (115)
267) Acceleration   1 2 3 4 (5) 6 (7)   Location _____
268) Fluidity   1 2 3 4 5 (6)(7)   Gp 0____ Area ____ 5pt ____ 8pt ____
270) Jerky-Smooth   1 2 (3) 4 5 (6) 7   (111) 73-75   76-77   78-79
272) Variation   1 2 3 4 (5)(6) 7   Film Name _OBSERVATION_
                                                       Source _UN. OF HAWAII - AR SUBJ_
                                                       D W Other _____ Cord ____ Date ____
                                                                  9

FIGURE 6.3. Draft Choreometric coding sheet. Alan Lomax Collection (AFC 2004/004), American Folklife Center, Library of Congress.

"Dance as the Measure of Man"

Lomax, of course, had entered the project with some idea of what he hoped to find: evidence that cultural practices were not mere window dressing on the human experience but rather crucial to human survival. While nonhuman animals depended on genetic change to produce new adaptive behaviors, Lomax argued that humans passed on knowledge about how to thrive through symbolic cultural codes.[36] In this view, music, dance, art, and literature all had direct relationships to the fulfillment of both basic needs and species-level persistence. More specifically, Lomax hypothesized that dance functioned as an especially important storehouse of knowledge about adaptive bodily practices.

As the Choreometrics data dribbled in Lomax felt validated, contending that the movement patterns that characterized a society's dances seemed also to appear in its repertoire of everyday and working movements. He noted, for example, that the stooped posture and "deep shoulder rotation" associated with West African dance mirrored the widespread use of the short-handled grubbing hoe and that the distinctive hand movements of carnival dancers in a champagne-producing French village had parallels in "the vintners, pacing the aisles of bottles, rotating two bottles at a time."[37] Of contemporary dance in the United States, he wrote: "The half-cocked arm position, with vaguely pumping forearms and hands bent at the wrists, brings to my mind's eye the occupations most frequent in post-industrial America—the sales clerk wrapping packages and writing up sales slips, the bureaucrat shuffling papers, the secretary typing away at the word processor, and the driver at the wheel of a delivery van—all these and a thousand other service occupations where the arms and hands are laxly but swiftly engaged in light work, while the rest of the body remains inactive."[38] Lomax also linked dance style and climatic conditions, pointing out, for example, that the dances of Inuit peoples prominently featured explosive elements and that such movements represented "one effective way to generate heat in the extreme cold."[39] With this data, the Choreometrics team seemed to have evidence that dance enshrined the movement patterns "of maximal importance to the actual physical survival of the culture," preserving ways of moving that were "as necessary as breath and food."[40]

These colorful examples also served Lomax's more systematic aims, which were shaped by the emergence of neo-evolutionist anthropology. In the late nineteenth and early twentieth centuries, the study of the world's cultures was interwoven with an evolutionary model of development. Founded on the premise that all societies progressed through the same series of cultural stages, from the primitive to the increasingly complex, this perspective was linked to systematic racism toward "non-Western" peoples. With the ascent of the Boasian school in the United States, however, midcentury anthropologists increasingly sought to understand individual cultures in their particularity, moving away from questions of development or hierarchy.[41] Nevertheless, by the 1960s, some believed that the flight from the scholarly study of "temporal change in social structure" had gone too far. Neo-evolutionists like Leslie White and Julian Steward believed they could reconcile Boasian cultural relativism with a renewed attention to the structural, rather than biological, determinants of cultural change. Their aim was to "put historical 'progress' back on the agenda of social thought" without falling prey to racial or cultural hierarchies, social Darwinism, or modernization theory.[42]

White, for example, contended that human culture "could indeed be examined as a single whole, having a dynamic of change and growth quite apart from human consciousness of it, a process moving *directionally* toward more large-scale, organizationally encompassing forms of association and state power, paced by growing quantities per capita of physical energy in social use."[43] Steward was less convinced by unidirectional accounts of cultural evolution but nevertheless believed it possible to locate common patterns of social development via comparative ethnographic evidence. Both maintained that such studies did not have to rely on discarded racial tropes and could incorporate new theories of cultural exchange, interaction, and fluidity.

Lomax's thinking clearly drew on such approaches. In the introduction to an unpublished manuscript about Choreometrics, Lomax acknowledged that there had been a "very considerable argument, pro and con, as to whether human culture has evolved or, whether, once invented, it has simply changed its shape according to circumstance, expanding or collapsing, winning and losing capabilities in the struggle to satisfy human needs in earth's varied environment."[44] Lomax was certain of the correct perspective. "On one point," he notes, "there is little doubt. The productivity of human subsistence systems, for which archeology and the study of living cultures provide the record, has progressively increased from simple extractive systems to systems with an ever-increasing ability to transform large portions of the environment into food and energy."[45]

Lomax hoped that Choreometrics would provide additional support for this thesis. Though evolutionary theorists had generally ignored dance—seeing it as decorative, rather than functional—Lomax believed that certain habits of bodily comportment were prerequisites for the development of increasingly productive cultures. Dance as a field had long suffered from the curse of invisibility, overlooked in part because it was relatively difficult to record and second because in the modern "West" it was so often associated with women.[46] Lomax hoped that Choreometrics would alter this state of affairs—not only transforming the picture of *how* culture worked but of *whose* labor propelled it.[47]

This instinct seemed to be borne out in the coded data. In his early analyses, Lomax noticed a correlation between certain clusters of movement traits and particular subsistence systems: "Some movement features appear in cultures that represent an early stage of socio-economic development, others in the middle range, yet others in the more complex productive systems." He argued that high levels of bodily "articulation," in particular, appeared characteristic of the most "advanced" systems.[48]

These findings only increased Lomax's confidence that dance was nothing less than "the behavioral representation of man's increasing control of his environment through rational manipulation." Though he admitted that his conclusions were only preliminary, it seemed more and more likely to Lomax that movement style played a role in a population's ability to "carry out the ever more differentiated tasks of production and social relations that support an ever larger human population on the planet."[49] Like culture itself, dance evolved into increasingly complex forms, forms that Lomax now believed he could accurately track.

Still, like the neo-evolutionists, Lomax worked to reconcile this sense of teleological progress with a belief in the essential equality of all cultures. He frequently issued reminders that—although some were more complex than others—all movement styles were equally adaptive. He was eager to point out, for instance, that "a lot of people using the shuffling step that blacks use when they dance thought this was all about laziness. But it's not, it's connected with the fact that blacks have always worked as hoe agriculturists, and they have a broad stance, and they moved like this when they were working. . . . We see that all throughout the tropics." For Lomax, there was almost no better counter to accusations of indolence than the fact that "these are the people who pioneered the whole tropical part of the planet during the last two or three thousand years. A part of the world where we couldn't survive, we Europeans."[50] Lomax even came to think that movement style might represent a less problematic way of assessing human difference than race, contending that it was both "far more reliable" and "surely more interesting since it deals with behavior and communication rather than physical traits over which we have no control."[51] In this his work reflected a broader postwar scientific skepticism about race as a biologically meaningful category, as geneticists, for example, focused their attention on statistical variations across population groups, rather than within socially recognized racial categories.[52]

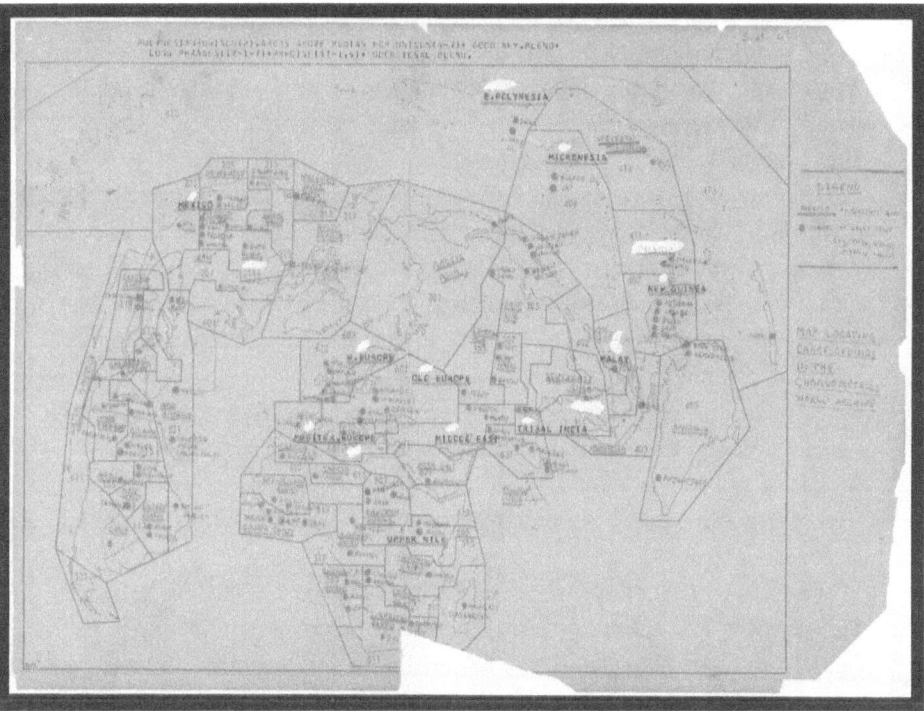

FIGURE 6.4. Dymaxion map depicting dance codings in the Choreometrics World Archive, September 1969. Buckminster Fuller developed the Dymaxion map in 1943 to correct for the size distortion of land masses in the Mercator projection as well as to emphasize the connectedness of the earth's inhabitants by suggesting past migration routes. The map was created by projecting the globe onto the surface of an icosahedron which could then be unfolded into a variety of configurations. Alan Lomax Collection (AFC 2004/004), American Folklife Center, Library of Congress.

Recalibrations

The instructions for raters at the beginning of the Choreometric Coding Book included the following caution: "The rater is advised not to attempt to count the frequency of a feature by breaking down the action or scene into similar parts or units and then summing up his impressions in numerical terms. If a quality is not strongly and emphatically present throughout the whole scene, or else if it is not markedly emphasized in some way in the scene, it should be given a low score."[53] This flight from numbers seems, at first, at odds with a project that generally emphasized its quantitative credentials.[54]

A core assumption of Choreometrics's methodology, however, was Lomax's contention that the system itself would render numerical comparisons unnecessary. As a rater viewed more and more films, coding sheets in hand, he or she would gradually absorb the schema until it became second nature.

"Training in Choreometrics," the team contended, "consists, fundamentally, of the recalibration of the observer's standards of tempo, etc., to the full human range."[55] Raters thus participated in the project in two ways: by gathering data and by offering themselves up as test subjects for the system's power to alter perception.[56]

The trial, it seemed, was successful. As the project went on, Lomax, Bartenieff, and Paulay placed greater and greater confidence in the raters' ability to discern movement differences. Although those working in nonverbal behavior more generally had described difficulties in teaching observers to differentiate and identify the "micro-units" of interaction, Lomax reported that his team "discovered that gross contrasts in movement style between culturally different groups were relatively easy to define and agree on. Not only that, but when we incorporated graded examples of the steps in these movement scales in teaching films, we found that students could learn to make these distinctions at the same high level of agreement which we had established among the original coders."[57]

Lomax was particularly pleased because he had, all along, been shaping the system with the aim of making it easy for nonexperts to follow. In fact, Lomax's ultimate plan was to extend the kind of perceptual training the raters received to a massive popular audience. Though Lomax, Bartenieff, and Paulay produced academic presentations and publications on Choreometrics from the project's beginning, much of the team's energy was directed toward the production of films intended for a wider viewership.[58] The first of these, *Dance and Human History*, was released in 1974 and eventually distributed by the Extension Media Center at the University of California-Berkeley.[59] Approximately forty minutes in length, the film presented clips of world movement and dance alongside explanations of the Choreometric system and the team's conclusions. Early screenings took place at the American Museum of Natural History, the Whitney Museum, and the Museum of Modern Art in New York, as well as on local public broadcast television stations. In subsequent years, Lomax received a steady stream of requests for copies destined for college classrooms in media studies, anthropology, and music; public high schools; and scientific laboratories in both the United States and Europe. Margaret Mead included a ten-minute excerpt from *Dance and Human History* at her presidential address to the American Association for the Advancement of Science in 1976.[60] Three similar films followed in subsequent years: *Palm Play* (1977), *Step Play* (1977), and *The Longest Trail* (1984).[61] For the Choreometrics team, these movies were not afterthoughts, but a core part of the project from its earliest days. As Lomax put it in a set of undated notes from the late 1960s, his central goal was to "depict . . . the stylistic resources of human movement patterns so that all of them would be understandable to any human being," thus avoiding "one of the traditional errors of social science which has always been to develop a system of observation and thought, useful to the specialist, but essentially out of the reach and thus meaningless to the masses of human beings who make up society."[62] Film seemed to be one of the most direct ways of achieving this aim, especially if viewers were trained in new forms of observation with each watching and rewatching.

The second major project on Lomax's agenda was the manuscript for *Dancing: A World Ethnography of Dance and Movement Styles*, a draft of which he and Paulay began in the early 1970s and completed around 1981. Envisioned as a chimera of scholarly tome and coffee-table book, Lomax pitched the work to a variety of presses, both trade and academic.[63] At 870 manuscript pages, plus front matter and extensive appendixes, the book was filled to bursting, containing chapters devoted to accessible summaries and illustrations (both verbal and visual) of Choreometric principles as well as detailed coding data for nearly 600 dances and 140 movement variables. Lomax's concerns about taxing the public's mind with too many variables seems to have evaporated by this point, replaced by a renewed drive toward totality. The manuscript also included a blank version of the Choreometric Coding Book and instructions for its use. In part, Lomax was making an effort to be methodologically transparent, an exercise perhaps particularly important for a liminal figure in a new field of endeavor. More importantly, however, he envisioned the blank pages as an invitation to a kind of do-it-yourself instruction in movement observation.[64] Progressing through each section of the book, casual readers would find themselves gradually inducted into a new society of movement spectators. By the text's conclusion, he hoped, they might already be amateur coders, able to intuitively make sense of even mountains of raw data.

For Lomax, the goal was to make Choreometrics "understandable to anyone who wanted to use it—whether sociologist, filmmaker, or schoolboy; whether American, African, or Polynesian."[65] Still, after a 1977 screening of *Dance and Human History* at the American Museum of Natural History, a Laban-trained analyst named Lynne Norris wrote a concerned note to Lomax. Though sympathetic to his overall program, she was skeptical of Lomax's optimism about the speed with which new patterns of observation could be internalized: "One cannot learn choreometrics in the offhand manner that you suggest," she remarked. Even with the existence of supplementary written materials this was not, she chastised, a technique that could be "learned in a weekend."[66]

Kinesthetic Community

Lomax pressed on nonetheless, convinced that popular retraining in movement observation was crucial for countering the dangerous dominance of American and European culture in a progressively globalized media environment. "All of us," he contended, "are being swept into an ocean of the visible media, whose tides and currents arise in the offices of the advertising and entertainment industries, rather than in the human community. With every passing month as passive watchers we are being moulded and remade by what we are allowed to see and hear."[67] Movement behavior, he held, was particularly vulnerable to this subconscious shaping.

Such changes contributed to what Lomax called the "greying of culture"—a progressive, worldwide homogenization he feared would lead to the wholesale elimination of certain forms of cultural life. Troublingly, Lomax believed the world had already begun to see evidence of the imposition of bodily techniques developed in one culture on people of another. The globalization of markets, for instance, had "imposed the confining and stiff-waisted European" postural style almost everywhere. "Held up for universal admiration as the only way for a real human being to carry himself," "school children and soldiers in every clime were drilled in this carriage, often with ridiculous and unfortunate consequences." For Lomax, the most painful and disorienting demand of such "cultural conquests" was that "the vanquished conform to the invader's standards of physical comportment."[68] The only remedy was to teach the world's inhabitants to become "expert viewers," to learn how to "spot the false, the fake and the oversold, to pick out the beautiful and to see where one's own non-verbal culture stands in relation to the others being presented." These skills, Lomax was certain, were latent in all human beings, but Choreometrics would unearth and refine them.[69]

As Lomax explained, once an individual was trained to analyze movement style in dance, their day-to-day experience of nondance movement would also change, with new scientific understandings demolishing old prejudices. No longer would "shuffling" connote laziness; instead, it would tell a story about climatic adaptation, agricultural technology, and persistence. In fact, as Choreometrics-trained observers moved through a city, they would encounter hard evidence about the long course of human history in the body of every person they passed. A trip to the grocery store might teach as much as an afternoon at a natural history museum. Experiencing daily life in this way was, Lomax promised, "like looking through a microscope or underwater for the first time."[70]

If successful, Lomax was certain that this kind of perceptual shift could make real change in American society. In a 1972 correspondence with a Michigan public school teacher Lomax contended that many of the problems plaguing the United States' educational system were rooted in its "refusal to match classroom procedures with the cultural patterns of the communities" it served, relying instead on "one standardized, basically Germanic approach." If teachers could instead be sensitized to the varying movement preferences of their students, they could shift their own style as they moved from pupil to pupil. Body language would then "melt . . . into body language, and no one body" would dominate.[71] Lomax further likened this vision to the then-ongoing efforts to recognize African American Vernacular English as a dialect in educational settings.[72] Similarly, in 1968 he ventured that a wider appreciation of Black cultural contributions could help salve some of the country's deep wounds, writing in a letter to his agent that "the guys in the integration movement tell me that a loss of identity, a need to have a clear-cut history, is the problem of the Negro and is the source of more anger than economic conditions."[73] He hoped to gain the support of the Southern Christian Leadership Conference and the Ford Foundation to develop a national television series on Black heritage, explaining that such a program could represent a "most direct way to go after [racial] tension problems."[74]

In his conviction that film and notation could alter both the self and the social, Lomax's work echoes that of other Cold War figures invested in the idea that exposure to the right kinds of media could foster the development of tolerant, freethinking "democratic personalities." Alongside Buckminster Fuller, Margaret Mead, Gregory Bateson, and John Cage, American social scientists and psychologists of the 1950s and '60s created museum exhibitions, art installations, and educational programs, all premised on the idea that multimedia environments that encouraged active involvement could orient American citizens to the "perception and appreciation of difference" and serve as a bulwark against totalitarianism.[75]

But Lomax's concerns were not solely educational or political. If, as he contended, movement style was in fact a functional element of culture, its loss would put a number of populations in real danger. What would a herding culture be without its "graduated and flowing" style of energy? How would a hunter move on rough terrain without a deeply ingrained sense of restrained, linear movement? Indeed, how would *any* culture perpetuate its movement styles—and thus its lifeways more generally—if its members were ceaselessly bombarded with the movements of the West?[76]

Like many of his contemporaries, Lomax held a loosely cybernetic view of culture.[77] The best way to combat cultural greying, therefore, was to alter the surrounding feedback system. As such, Lomax was particularly keen to promote film screenings—and accompanying Choreometrics training—in places where "traditional" cultures seemed threatened. He remarked often that the individuals and communities who appeared in the Choreometrics films should "see the footage in toto and as often as they like, both in local screenings and over local television." Most social scientists, Lomax contended, had forgotten that the primary social function of film was to reinforce the culture the culture that had produced it. The Choreometrics team would not make that mistake.

The Paradox of Preservation

"All authors," Lomax wrote, "have their dreams. Mine runs this way: a folk dancer, an aboriginal choreographer, a student from some place away from the overwhelming mainstream picks up this book, looks through this atlas for his or her culture area and finds a pattern that is quite familiar—coming from his home or at least from his home ground." Though, until this point, the reader may have been suffocated by the barrage of Western cultural media, "now he discovers that there are many other aesthetic alternatives created far away from the urban art and pop market-places, including one by his own ancestors. This style he can feel in his joints and muscles belongs to him or is akin to the one he knows." He begins to "look at himself and his people with renewed esteem" and moreover "has acquired a clear and adequate way of explaining to the administrators of his local television station why his own tradition should be aired and how it should be photographed."[78] This was Lomax's holy grail. Numbers and figures, maps and diagrams, would awaken in even the casual reader a new sense of culture, of history, and even of his or her own body. This sensibility could then be shared and replicated.

Written recording also served another, less-acknowledged purpose: it concretized and thus implicitly elevated the cultural materials of "nonliterate" peoples. It was no accident that Lomax emphasized that the readers of *Dancing* would have "evidence, *printed evidence*, a text to support the claim that his aesthetic traditions deserve care, study, a place to live, and the means to grow"[79] and addressed his work "to all people who lack written history and the sense of assurance it brings."[80] Though by the 1960s scholars like Walter Ong and Marshall McLuhan had begun to publicly question the superiority of writing as a mode of communication, the written word still carried weight that movement alone did not.[81] At the same time, by analyzing the dance patterns of grocery store clerks and white-collar executives, Lomax sought to demonstrate to those in the modern "West" that their cultural identities were no less rooted in their physical bodies than their allegedly "primitive" counterparts.

Choreometrics also drew on Americans' eagerness to embrace the role of statistical subject. As they flocked to participate in Gallup polling, the Middletown studies, and the Kinsey experiments, these surveys seemed to hold out the promise of working as a kind of national glue, revealing the country to itself while allowing individuals to conceptualize their place within it. Writing about one woman who sent in data to Kinsey unprompted, historian Sarah Igo notes that "Somehow, the merging of her life history with thousands of others through the 'fabulous' technology of the Hollerith card puncher linked her to something almost cosmic in significance."[82] Choreometrics promised a comparable recognition, holding out the prospect of visibility to those individuals and communities left out of more conventional representations. Still, as Igo reveals, these midcentury surveys were not without their critics, though "most seemed to take it as a given that what was needed from the pollsters was *more* respondents, *more* representation, *more* analysis to correct their instrument's most visible flaws."[83]

The effects of Choreometrics were similarly complicated. The project did attract the attention of major anthropologists, and the documentaries were widely viewed; the Lomax archives include letters from a variety of people writing in thanks. Shortly after Lomax's death in 2002, the well-known ethnographic and documentary filmmaker John Bishop credited Lomax and Choreometrics as key influences on his shooting style: "If there is a single thing that Cantometrics and Choreometrics has given me as an ethnographic filmmaker, it is the appreciation of why people move and interact differently from culture to culture, how to perceive and describe the difference, and how not to be alienated by body language that is radically different from my own."[84] It is difficult to imagine a compliment that would have pleased Lomax more.

Although Bartenieff formally left the Choreometrics team in 1970, student evaluations of a DNB course on "Ethnic Dance" she taught in 1976 tell a similar story. One student explained that "living in New Haven among many third world people, I've really enjoyed observing body use and attitude with a new fresh attitude of my own. My body has become much freer for me to explore and, in the workshops I do with N.H. teachers and parents, I've become more supportive of people taking it easy and exploring their own body movement styles and then trying new and different ones." Another noted that "Over and over again I have felt that the real importance of E/S is as a *tool* to help one see more deeply and clearly and not as an end in itself. The beauty of it seems to be in its use." A third praised the course as a "great resource for communal awareness," a fourth for helping her improve her observational skills. "You

FIGURE 6.5. Alan Lomax (right) and John Bishop run the Steenbeck film editing station during the Choreometrics project, 1970s. Alan Lomax Collection (AFC 2004/004), American Folklife Center, Library of Congress. Courtesy of the Association for Cultural Equity.

can feel," she noted, "what another people's life is like as opposed to just trying to understand them intellectually.... It has enlarged my world view and added color to my life."[85]

On the other hand, in a pair of reviews in the newsletter for the Congress of Research on Dance in 1974, the anthropologists Joann W. Kealiinohomoku and Drid Williams savaged Choreometrics, questioning both its methodology and its conclusions. Kealiinohomoku focused on flaws in Lomax's approach, pointing out the limitations in Choreometrics's film sample and offering a slew of counterexamples to Lomax's suggestive correlations. Williams attacked the project on an even more fundamental level, opening her review as follows: "Reading Choreometrics is like finding oneself caught in a chapter of a science fiction novel. One knows that the authors of the project are human; they have human names and they would doubtless appear very human if one were to meet them over a cup of coffee, but their ignorance of the nature of the dance—of syntactical, grammatical, spatio-linguistic and above all semantic features of dances—seems so profound as to be explainable only by assuming that their minds were taken over by members of an alien race, probably Kryptonians."[86] For Williams, the model of dance Choreometrics promoted was coldly mechanical, too far removed from the creative experience of movement to hold any significance: "According to Choreometric theory, dancing does not negotiate insights or mediate meanings. Rather it 'effectively organizes joint motor activity.'"[87] To reduce dance to its component parts, to the actions of limbs and muscles and nerves, was, in Williams's view, to rob dance of its soul and character, its essential humanity.

Moreover, both Williams and Kealiinohomoku questioned the static cultural groupings upon which Choreometrics's analysis rested. "Lomax's hypothesis," Kealiinohomoku wrote, "is useful only for areas that are culturally homogeneous, where there has been little acculturation and minimal technological change. Surely there are few such societies today."[88] How, for example, could Choreometrics's model account for the creolization occasioned not by media pollution, but by the arrival of West Africans in the Americas through the slave trade? Thus, while Choreometrics's data-obsessed accumulations seemed to be a quintessential modernist project, its fundamental architecture assumed an atomistic premodern world that surely no longer existed, if it ever had in the first place. Even some students were quick to pick up on these tensions. After a guest lecture on Cantometrics and Choreometrics in a course at the Pennsylvania State University in 1968, one attendee observed that the "Pigmes' [sic] songs that Mr. Lomax recorded are surely a 'modern' version of those that existed fifty years ago. At least he cannot prove that they are not as 'modern' to the Pigmies as Dylan's interpretation of American folk songs is to the Americans today."[89]

Lomax's work also put him in the middle of an ongoing debate about dance, race, and cultural exchange in America. To argue that dancers had inherited affinities for certain movements provoked particular unease in the 1960s and '70s, a moment in which Black dancers were trying to make inroads in the traditionally white art of classical ballet. To them, Lomax's assertions were far from new. Instead, they smacked uncomfortably of an era in which Black dancers who attempted ballet were met with notes that "Negroes cannot be expected to do dances designed for another race," that the "leaps and pas de deux . . . were singularly incongruous to these performers, honestly as they worked at them."[90] In fact, as late as 1963, *New York Times* critic John Martin argued that African Americans were simply unsuited for ballet, as its "wholly European outlook, history, and technical theory are alien to him culturally, temperamentally and anatomically."[91]

While Lomax decried the media systems that kept "traditional" African dance off the airwaves, Arthur Mitchell—the first Black male member of the New York City Ballet—fought his own battle for media visibility. In 1957, NYCB director George Balanchine premiered the ballet *Agon*, which featured a *pas de deux* made for Mitchell and Diana Adams, a white ballerina from the American South. Though the two performed to great acclaim in New York and on world tours, Mitchell was not permitted to perform the role on commercial television until 1968. And while much of the angriest opposition to Mitchell's performance stemmed from the piece's interracial pairing, the uproar surrounding *Agon* also speaks to an ingrained cultural skepticism about the suitability of Black bodies for a "white" art form. Partly in response to barriers like these, Mitchell would found the Dance Theatre of Harlem in 1969, widely regarded as the first American Black classical ballet company.

Furthermore, while many Black performers and choreographers were—particularly in the 1960s—creating and publicizing new forms of dance that drew on Africanist traditions, they were also pushing back against overly simplistic or deterministic accounts of what Black concert dance could be.[92] As dancer Percival Borde put it to scholar Lynne Fauley Emery in 1970, "there is a caricature-concept of this idea of [American] movement, and since this caricature is widely imitated by Americans of all ethnic backgrounds it would be impossible to detect recognizably ethnic movements if such a thing existed."[93] Modern dancer and choreographer Rod Rodgers similarly noted that "while traditional black art is playing a vital role in the awakening of a black cultural identity, now it is equally important for black artists to discourage the crystallization of new stereotype limitations by not confining themselves to oversimplified traditional images."[94] Black choreographers fought to define their works as conscious acts of artistic creation, rather than mere translations of ingrained cultural heritage.

To Lomax and his collaborators' dismay, *Dancing* was never published. In part, the problems were practical: crushed under the weight of its own ambition, the volume simply grew too large for any publisher to take on.[95] But as the objections of Choreometrics's critics suggest, the project was also plagued by troubles more philosophical in nature. For Lomax totality was the goal, and it ultimately required a distant, cartographic view. Unfortunately, this perspective was ill-suited to capture the variable and idiosyncratic details of embodied daily life that had first fascinated him.

As Lomax himself acknowledged, excluded from the study's purview were "the sequences of movements, the gestures, the costumes, the dramas, the themes, the functions, the contexts in which particular dance sequences acquire their meanings."[96] Lomax's decision to elucidate the basic movement patterns of large *groups*, rather than the particularities of individual styles, also meant that information about the dancers themselves is absent from the record. Though they are intensely present when watching one of the films—it is they who compel the viewer's gaze and their actions that are the object of study—the viewer learns little about their lives or motivations.[97] They are static symbols, not fully realized beings.

Indeed, Choreometrics exposed a conflict between two different notions of the nature of dance: one that held it up as a form of original, individual expression with an infinite degree of variation, and another that suggested that it was a collection of basic cultural vernaculars rearranged with varying degrees of skill. Perhaps due to his orientation as a folklorist, Lomax embraced the latter stance, arguing that dance was fundamentally a collective cultural product. He recalled that a musicologist had once claimed that only 5 to 10 percent of even Beethoven's contributions to European music were truly new and argued that "in dance, I fancy, the percent of individual innovation must be far smaller."[98] Tellingly, though Lomax thanked the "film-makers, producers, and distributors" who "lavished time and money, risked their fortunes and often their lives to document and make known some otherwise neglected aspect of human culture," he did not thank the dancers whose groaning muscles and calloused feet made those films possible.[99]

Such a tactic—highlighting one phenomenon at the expense of others—is certainly not an uncommon one in the production of knowledge. It left Lomax, however, trapped between competing visions of Choreometrics's ultimate purpose. On the one hand, Lomax wanted to make claims about population-level relationships between movement style, history, and culture. He was interested in structures of performance, not particular performers, and his categories of analysis were thus "intentionally coarse," ignoring the differences between individuals in favor of an attention to the "baseline movement signature" of a given area.[100] On the other, Lomax hoped that the resulting collection of film excerpts, images, and coding sheets could be a resource for individual self-discovery. In these imaginings, the archive—as mediated by Lomax—would act as a mirror in which a person could recognize themself in all their particularity, generating a "sense of history, accomplishment, and self-pride."[101] This altered perception would, in turn, create new kinds of collectivities and a more tolerant, humane, diverse world.

That these aims were in tension should not come entirely as a surprise. Whatever a creator's aims, the ways in which information is collected, structured, and stored shapes and limits its ultimate uses. Lomax envisioned Choreometrics as important tool both for the "greying" cultures at the world's peripheries and for Western multicultural democracies like the United States. But it was precisely these latter individuals—the citizens of an increasingly multiethnic America—who articulated discomfort with the flattened, categorical portraits Choreometrics produced. The image reflected simply did not seem to be their own. Unexpectedly, Choreometrics appears to have

FIGURE 6.6. African American children playing singing games, Eatonville, Florida, June 1935. Photo: Alan Lomax. Library of Congress Prints and Photographs Division. Courtesy of the Association for Cultural Equity.

succeeded in achieving its most audacious aim: helping to alter the "human perceptual apparatus" and generating a new public consciousness of the diversity and significance of human movement. The reactions to this new awareness, however, were not what Lomax anticipated: not an enthusiastic embrace of Choreometrics's vision of totality but a critique of its essential incompleteness.

EPILOGUE

Movement in the Digital Age

In a recent book, the philosopher of art Alva Noë remarked on the potentially revolutionary power of a fictional system for recording dance on paper. "Inventing a way of writing the moving body," he argued, would be "transformative and imaginative in precisely the way that the invention of the writing of speech was but is no longer." Writing provided the foundation for the construction of "civilization, government, law, not to mention science," and he held that the creation of a similar tool for dance might provoke equally wide-ranging and unanticipated changes.[1] What Noë did not realize is that such a system does exist—and its effects have, in fact, been profound.

Over the course of nearly a century, Laban-inspired techniques were used to synchronize movement across time and space, to dig up seemingly concealed truths, and to attempt to forge communities at times of great anxiety and dissolution. They shaped bodies and minds and politics—for better and for worse—and made the case for the ongoing significance of the body in a world that sometimes seemed to want to ignore it. Notation also helped to mold a particularly twentieth-century mode of movement—an unnoticed but ubiquitous style that reached from the gestures of dancers to the stride of corporate executives.

This book, though, is less about any particular movement recording tools than about the varied people who used them: what they hoped to gain from the promise of mastering bodily experience, and what these desires can tell us about the worlds in which they lived. In retrospect it becomes clear how—especially in twentieth-century Europe and the United States—the management of movement came to function within larger projects of social, cultural, and political governance. These efforts often echoed one another in darkly ironic ways. Movement recording was deployed in state-building in Nazi

Germany *and* to manage the psychological fallout of the Holocaust. It was used in a bid to gain broader respect for the art of dance but *also* to minimize the contributions of dancers and their bodies. It was applied to critique industrial capitalism as well as to ensure its steady, ongoing functioning. The slippery potential of mediated movement, moreover, still holds allure in the twenty-first century. Indeed, even as Noë was writing, notated movement was being strategically deployed in a new array of digital devices to uncannily familiar ends. This epilogue will explore what we want from movement notation and analysis today—and what that constellation of aspirations says about our own time.

Bits and Pieces

With the release of LabanWriter in the late 1980s, the idea that a computer might help produce, reproduce, and store notation scores was no longer particularly shocking. Across the United States and Europe, notators began requesting copies of the program (or one of its competitors), and the software gradually became a mundane, useful reality. Behind the scenes, however, other kinds of efforts to deploy Labanotation in the digital realm were quietly underway—and had been for some time. In fact, many of the engineers who worked on Labanotation writing software had seen their efforts as mere stepping-stones on the path to solving a more interesting problem: the general modeling of human movement via computer.

One computer scientist at the University of Iowa, for example, explained in 1980 that his work on the writing program "NOTATE" had little to do with an interest in dance. Instead, he clarified, "I am interested in computerized mathematical models of the body that might advance scientific knowledge in research and theory" and hoped his experience with Labanotation would serve those ends.[2] The team at the University of Pennsylvania—which, by the late 1970s, was led by assistant professor Norman Badler and included researchers Joseph O'Rourke, Maxine Brown, and Stephen Smoliar, as well as Labanotator Lynne Weber—had similar designs. In 1979, Badler and Smoliar published a paper in *ACM Computing Surveys* that considered a variety of ways in which "information concerning and related to the movement of the human body" might be represented within a digital computer.[3] They explained, for example, that filmed or videotaped movements could be digitized, but that this method produced an "explosive amount of data" and would require a great deal of computationally intensive deconstruction.[4] Another option was natural language: relying, for example, on terms derived from classical ballet or descriptions by scholars of nonverbal behavior. Badler and

Smoliar rejected this technique as well, noting that though verbal description might be "compact," it required users to have mastered a body of preexisting knowledge. Natural language descriptions, moreover, were "subject to ambiguity and unavoidable imprecisions."[5]

Labanotation, in contrast, was cast as a "system which describes movement *explicitly* and has a well-structured set of options to accommodate various refinements of a description."[6] It was based on the "abstraction of the structure of the human body" as a series of interconnected joints, which could be replicated digitally as "a network of special-purpose processors," each of which could be given instructions about position and trajectory in space.[7] Building on this foundation, Badler and Smoliar outlined their novel

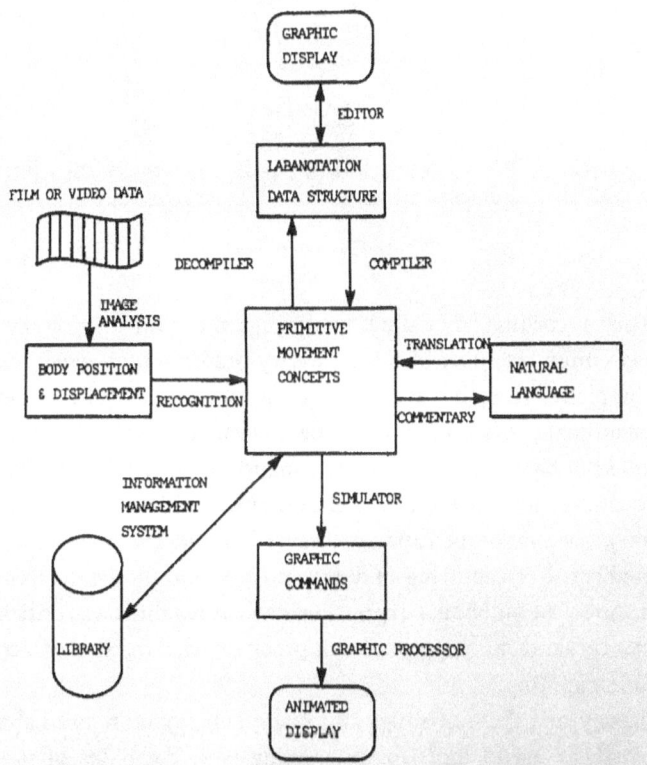

FIGURE 1. Comprehensive computer system for notating, modeling, analyzing, and describing human movement.

FIGURE E.1A-B. Diagrams from Norman I. Badler and Stephen W. Smoliar, "Digital Representations of Human Movement," *ACM Comput. Surv.* 11, 1 (March 1979), 19–38, https://doi.org/10.1145/356757.356760.

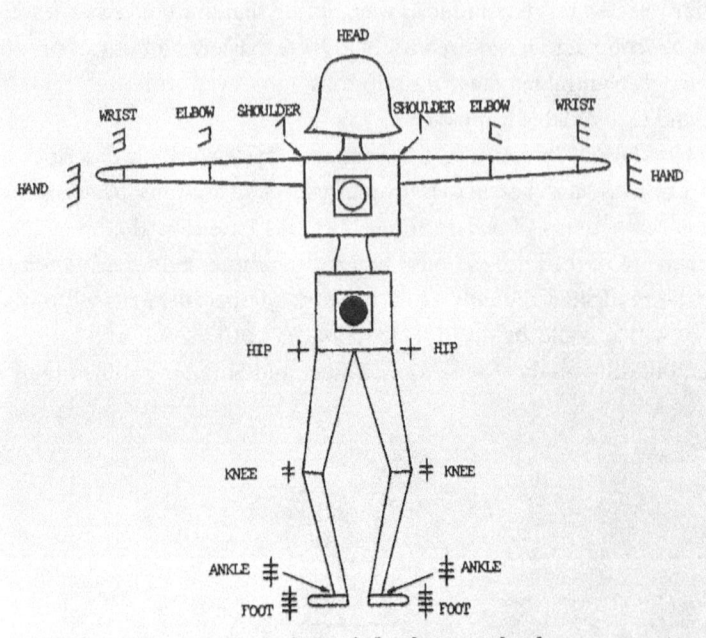

FIGURE 2   Abstraction of the human body into joints.

FIGURE E.1A-B. Continued

system for recording, modeling, analyzing, and simulating human movement via computer. Most of the necessary programming work had already been completed, though the configuration was not yet quite fast enough to produce animated graphics in real time. Instead, movement commands were interpreted in batches, and sequential snapshots were taken of the resulting graphic display to create a moving, permanent video or film record. As it happened, Labanotation's capacity to simplify and compress movement, its promoters' goal of capturing only the most important features and leaving out the "noise" of individual expression—a feature which was criticized by so many dancers and choreographers—is precisely what made it attractive in the world of computing.

In later years, these qualities also made Labanotation and Laban Movement Analysis useful tools for interacting with databases of prerecorded movements, including the significant collections assembled at Carnegie Mellon University and the University of Texas, Austin. (Laban Movement Analysis, or LMA, is now often used as an umbrella term encompassing various Laban-derived theories and notation systems.) Many researchers and animators were excited by the scope of these new collections, though their sheer size could make them a challenge to actually use. As one 2013 paper explained, "the space of human pose configurations is continuous, non-linear, and extremely high dimensional which makes searching in this space prohibitive."[8] In several cases, Laban-based systems were developed to provide solutions to this problem: one program, for example, allowed users to input a "motion key" that would search the database and return all movements exhibiting a given set of body positions or Effort parameters. Another produced Labanotation scores for each clip in a given library and then prompted searchers to input a new score and locate matching movement segments.[9]

The Badler team's work attracted interest and grant support from the National Science Foundation but also from NASA and the US military, who were interested in movement simulation as a tool for human factors research, particularly for mapping out how humans might navigate new machines, vehicles, and environments.[10] Relatively quickly, Penn became an important center for movement simulation research, and in 1995 Badler founded the Center for Human Modeling and Simulation, taking over the space that had previously housed the original ENIAC computer.[11] In the early 1980s the group would move on from an entirely Labanotation-based architecture, citing both the work required to teach the notation to would-be programmers and new developments in three-dimensional interactive computer graphics.[12] Even as they developed novel simulation systems, however, they maintained a number of conventions derived from Labanotation, including the set of body parts for which movement would be specified and the standard positions of those parts.[13] As Badler explained, Labanotation provided "a good set of default values" because it "incorporate[ed] so many assumptions about normal human movement."[14] The systems the Penn group produced, most notably the "Jack" movement simulator, were used by the federal government but also by private companies like Lockheed Martin, John Deere, Caterpillar, General Motors, and Ford; eventually, the work of running and licensing the software became so time-consuming that Jack was spun off into a private company, now owned by Siemens.[15]

The Badler group's move away from Labanotation proper, however, did not signal a total abandonment of Laban-based methods.[16] Though Badler and Smoliar had once found Effort/Shape notation too vague for their purposes, the lab returned to the technology in the late 1990s as they increasingly focused on computer animation.[17] At the 2000 SIGGRAPH International Conference on Computer Graphics and Interactive Techniques, the lab presented a method for using Effort/Shape to bring a missing sense of "naturalness" to computer-generated animated movements. Existing systems for producing gestures, they contended, had yet to produce convincing performances: "The

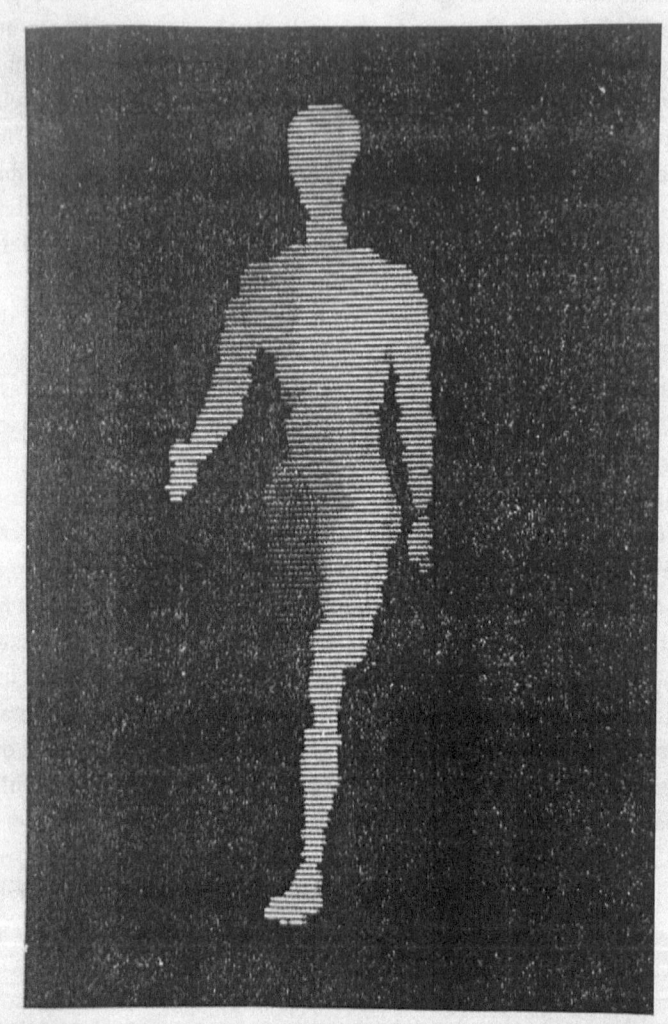

FIGURE 15 Body position at end of Labanotation segment of Figure 14

FIGURE E.2. Figure from Norman I. Badler and Stephen W. Smoliar, "Digital Representations of Human Movement," *ACM Comput. Surv.* 11, 1 (March 1979), 19–38, https://doi.org/10.1145/356757.356760. The animated body model was constituted by 300 overlapping spherical volumes.

characters seem to be doing the right things, but with a kind of robotic awkwardness that quickly marks the performance as synthetic."[18] Animated arms, for instance, could be made to reach out and grasp an object, but generally did so in an entirely goal-directed fashion, with little to no variation in speed or dynamics.

With the assistance of LMA consultant Janis Pforsich, the Penn group had created a program called EMOTE intended to remedy these deficiencies. Short for Expressive Motion Engine, EMOTE translated Effort and Shape qualities into mathematical formulas and employed them to modify the movements of animated characters. The Effort and Shape parameters could be simultaneously and continuously varied along a numerical scale and applied either to the body as a whole or to a subset of its parts.[19] A +1 Weight value, for example, corresponded to a very Strong movement, while a −1 value would produce a very Light movement. To arrive at the relevant algorithms, the Penn group combined principles derived from the existing body of Laban Movement Analysis literature—including texts by Laban, Bartenieff, and Carol-Lynne Moore—with traditional animation techniques and "much experimentation," guided by iterative feedback from Pforsich and three other Laban-trained Certified Movement Analysts.[20]

The creation of these movement descriptions was only one of the ways in which Laban-derived ideas shaped EMOTE. Its designers also drew heavily on Warren Lamb's legacy, particularly his claim that Posture-Gesture Mergers were an indication of personal authenticity. Directly citing Lamb's book, the Penn researchers explained that "a gesture localized in the limbs alone lacks impact, but when its Effort and Shape characteristics spread to the whole body, a person appears to project full involvement, conviction, and sincerity."[21] As such, EMOTE pushed users to apply Effort and Shape modifiers to all parts of the body, rather than only the one most obviously moved.

Perhaps even more significantly, Badler's team argued that LMA provided the "ground truth" that allowed animators to link movements with corresponding emotions.[22] EMOTE was expressly intended to create not only "believable" animations but characters whose movements were in some sense actually true to their inner psychological states.[23] EMOTE's engineers, in fact, hoped to create a future iteration of the program that would allow users to input an emotion in natural language and have it automatically translated into the relevant Effort and Shape parameters.[24] The adverb "carefully," for example, could be coded as a "Light and slightly Sustained Effort portrayed during arm movements and a little Retreating Shape displayed by the torso." "Proudly," on the other hand, would be rendered as a "Rising posture."[25]

EMOTE, therefore, quite literally encoded Laban's idiosyncratic claims about movement and its meaning into the computing architecture of the twenty-first century. Referring to the body of LMA literature simply as "empirical movement science," the program's developers saw it as a neutral tool for quickly solving a set of technical problems. As the history traced here has demonstrated, however, LMA was anything but. Instead, its assumptions about movement were rooted in a very particular conception of the body born out of the confluence of Weimar science, art, and politics and then redeployed in response to new social and political challenges across the twentieth century. Laban's work was indeed founded on the premise that the relationship between motion and emotion was governed by immutable, universal laws, operating without regard for time, place, or culture. But his own quest to deploy these laws was profoundly shaped by history, in particular by a hope that movement could create new forms of (ultimately racialized) community in a world that seemed out of control.

A number of other projects have either built directly on the Badler group's work with EMOTE or independently developed along similar lines. A later version of EMOTE called PERFORM, for example, correlated movement behavior not with transient emotions but with more permanent personality traits, relying on the judgments of two Certified Movement Analysts and the published work of Irmgard Bartenieff.[26] Initially conceived of as a tool for producing convincing background characters in animations, PERFORM has since been used in efforts to create virtual customer service agents endowed with believable personalities.[27] In Japan, a lab at the University of Tokyo deployed LMA in an attempt to construct robots that humans felt more comfortable interacting with. After developing a "mechanical realization" of each Effort and Shape element, they developed six "dances" intended to allow a wheeled, two-armed robot to communicate joy, anger, astonishment, and sadness to viewers, with the ultimate goal of "long-term human-robot symbiosis."[28] At the University of Manitoba engineers went even further, similarly attempting to use motion to communicate mental state, but via a flying quadrotor drone with no human features whatsoever.[29]

As such, whether or not you realize it, the movements of a favorite cartoon character or a virtual agent frustrating you at the airport may well have been designed in line with the precepts and presuppositions of LMA.[30] Even the drone buzzing around your head might be quietly bringing to life theories about movement rooted in the physiology, physics, aesthetics, and politics of more than a century ago. Some of the researchers behind these programs were aware that they were selecting one theory of embodied emotional expression among many, but they justified their decision because of LMA's practicality.[31]

As one engineer wrote, other approaches might have merit but were simply too "abstract to implement on the robot algorithm." Laban's theory of body movement, on the other hand, "describe[d] body motion quantitatively and covers a wide range of body expression," making it an appealingly easy tool on which to settle.[32]

The consequences of these choices were—and are—significant. Though LMA is by no means the only tool currently used in the simulation of human movement, it remains a significant—and growing—presence in computing and robotics. As a result, while Alan Lomax once dreamed of a media system characterized by a cacophonous variety of movement styles, the opposite seems to be occurring, particularly as LMA spreads across the globe. With each new program or device that bases its structure on LMA, a single, hegemonic view of human movement is being reinscribed. The message, however unintended, is clear: only one form of physical embodiment, one way of moving in the world, is normal. This state of affairs would be troubling in and of itself—one shudders to think of a world cemented in such unrelenting monotony—but it is particularly worrisome given how Laban movement theory has long treated issues of race, gender, class, and disability.

This issue becomes even more acute when one realizes that LMA is not only being deployed to simulate movement on screens or in machines but to assess the movements of living, breathing people. This goal, in fact, is an old one: as early as the 1960s, Warren Lamb had imagined how Aptitude Assessment might be improved with the aid of novel recording devices. As he put it in *Posture and Gesture*, "perhaps the weakest link in the study of physical behaviour is that observations are taken by the human eye. Although a convincing correlation has been achieved from a number of people observing the same subject, the recordings obtained will always be suspect."[33] In his more ambitious daydreams, Lamb sketched the image of a small control room, "in which we walk about, sit down, have a cup of tea, then eventually emerge with the essentials of our physical behaviour laid bare by a computer."[34] The human analyst could be removed from the equation entirely: a new kind of tool would take the observations and the computer would handle the analysis process via a preprogrammed algorithm. Lamb admitted that the precise technical details still needed some working out, but he was optimistic that they would resolve with time.

In the late 1970s, Lamb personally began experimenting with electromyography (EMG), a tool for recording electrical activity produced by muscle cells. He was delighted that it could record "muscle-movements in various parts of the body with a detail, continuity and accuracy impossible to the human eye."[35] In a collaboration with researchers at the University of California, twenty students were fitted with fourteen electrodes placed in strategic locations across their bodies and then asked to go about their normal activities for several hours as continuous EMG readings were taken.[36] Though the experiment does not seem to have gone much farther—or produced any publication—Lamb believed it represented a promising avenue for further development.[37] To Lamb it was always quite clear that the kind of data he was collecting was inherently amenable to both recording and evaluation by machine. He emphasized repeatedly that "there is nothing in the observations" which would prevent a machine from doing the work: "all that is required is recognition that this is a worth-while measurement to take."[38]

By the early decades of the twenty-first century, this recognition seemed to be gaining ground. The researchers in the lab at the University of Tokyo hoped, for example, to not only produce robots that appeared to possess recognizable human feelings but to endow those same machines with the capacity to automatically evaluate and respond to the emotional displays of the humans with whom they interacted. If humans could deduce emotional states from robot movements, they ventured, there was no reason that robots could not eventually analyze human movements in return.[39] In 2020 another team, this time at the Pennsylvania State University, reported on their work on a program to identify human emotions on film, correlating quantitative measures of LMA features with human-generated assessments of characters' feelings.[40] Though the researchers' ultimate goal was to recognize emotions in the so-called "wild," their initial set of movement samples were drawn from professionally produced full-length movies; these samples were then rated by anonymous workers recruited through the Amazon Mechanical Turk platform.[41] At the conclusion of the study, they contended that there was indeed strong evidence linking certain LMA features with emotional arousal.

A close analysis of their methodology, however, casts doubt on the validity of this claim. Not only were the Hollywood films that constituted their sample undoubtedly unrepresentative of the totality of human movement, but the researchers explicitly excluded data produced by two groups of raters from their final analysis. The first cohort was made up of "dishonest" contributors who the team determined were not sincerely rating the excerpts and were participating purely for monetary reasons. The second was constituted by "exotic" participants, a group the researchers defined solely by the fact that they had "given inconsistent annotations" compared either with other raters or with the "gold-standard" assessments that had been produced by a "small trusted group" prior to the study's start.[42] The experiment, in essence, operated from a tautological premise: only ratings that conformed with researchers' expectations could possibly be reliable.

The lab did acknowledge some of these issues, noting in passing, for example, that a singular truth about the relationship between movement and emotion might be difficult to establish and that "demographic factors," particularly ethnicity, did appear to influence how movements were perceived. "Our current model," they stated, "has largely ignored these potentially useful factors." These limitations, however, did not stop the team from pressing forward and suggesting ways in which their new system, ARBEE (Automated Recognition of Bodily Expression of Emotion), might eventually be used. As they explained, the aim was a system that could be deployed on live human movement in real time, continually monitoring emotional output via ever-present cameras. This kind of surveillance, they pointed out, could be useful in a variety of settings: "in public areas such as airports, metro or bus stations, or stadiums to help police identify potential threats," in inpatient psychiatric settings to "help assess and evaluate disorders . . . to predict danger to self and others from patients, or to track the progress of patients over time," and by police to "assess the identity of suspected criminals" or uncover "their emotions and deceptive motives during an interrogation."[43] It might even, they continued, be of use to advertisers, who could assess the emotional state of individuals via social media video posts and tailor their pitches in response.[44]

The ARBEE team was not alone in its ambitions. Among others, engineers at the University of Coimbra in Portugal had embarked on a similar project more than a decade before, and the Pentagon has made efforts to automate the movement assessments of both leaders and potential security threats as part of their ongoing ALEADMOVE project.[45] This work again mirrors Warren Lamb's vision for computerized movement analysis. In *Body Code*, Lamb expressed certainty that that the results of automated evaluations would soon be used as a new form of legal evidence, reporting that "an eminent barrister" had predicted that "in ten years' time, judges will be directing the attention of the jury to the postures and gestures of the accused or of witnesses rather than the spoken evidence. 'The jury may consider,' the judge may well say, 'that the arm and head gestures accompanying the accused's protestations of innocence denote insincerity—but you may read them as an expression of genuine conviction. It is for you to decide.'"[46] Lamb was similarly optimistic that individual movement signatures could come to serve as an alternative method for identification, as they were not only unique, but—unlike a characteristic scar or an unusually weak chin—could not be removed by plastic surgery.[47]

Laban's influence even lurked inside the 2020 viral video of four Boston Dynamics robots "dancing" to the Contours' 1962 hit "Do You Love Me." Over the course of approximately three minutes, the two humanoid Atlases, the dog-like Spot, and Handle, a seven-foot-tall wheeled robot designed for box-moving, engaged in a piece of complex choreography, their "fluid and expressive motions" engineered to impress.[48] These machines, the company seemed to say, were not the frightening, uptight robots of the past—who might have managed to eke out a few halting measures of "The Robot" but little else—but rather a new, friendlier breed imbued with an appealingly human sensibility. Tellingly, the piece's choreographer, Monica Thomas, explained that it was her Laban training that had provided her with the necessary analytical skills to design their movements; she recommended a similar course of study for all would-be robot choreographers.[49] The video, of course, had a commercial purpose: to showcase the capabilities of Boston Dynamics's robots to potential customers. A number of commentators suggested that all the cheerful cavorting was a ruse, distracting viewers from the potential lethality of technologies intended to serve the needs of DARPA, the NYPD, and the US Army.

In an evocative turn of events, nestled among the steps the robots performed was the twist, a dance that had been famously recorded in Labanotation in 1962 as part of the Dance Notation Bureau's own publicity stunt. In February of that year, Chubby Checker joined members of the DNB at New York's Hotel Empire for a special command performance and official recording session. As the iconic entertainer swiveled, three notators scribbled, throngs of excited teenagers squealed, and the *New York Herald Tribune* took pictures.[50] In a light human interest story, the paper later reported how "Mr. Checker's every twist, turn, grimace, and flutter of an eyelash" was captured

FIGURE E.3A. Still from Boston Dynamics, "Do You Love Me" (2020). Choreography by Monica Thomas.

for posterity in this "highly scientific dance script."[51] (Reportedly, the DNB overcame Checker's initial reticence by making clear that, if he declined, they would invite a rival twist performer—Joey Dee of New York's Peppermint Lounge—to take his spot, both at the Hotel Empire and in dance history.)

The Twist for Posterity?

Chubby Checker demonstrates a basic movement of the "Twist" which will be recorded for posterity at the Hotel Empire at 2 p. m. today. Experts from the Dance Notation Bureau will write in Labanotation—a highly scientific dance script—Mr. Checker's every twist, turn, grimace and flutter of an eyelash. The symbols (see inset), prepared by Billie Mahoney of the bureau, represent Mr. Checker's initial movements in the "Twist." The square with the black dot represents the pelvic twist, the black symbol in the center indicates feet and leg positions and the symbols at far right and far left show placings of the dancer's upper and lower arms at all points.

FIGURE E.3B. From "The Twist for Posterity?," *New York Herald Tribune*, February 11, 1962, C3.

A casual observer might see this juxtaposition—of art and military, control and expression, human and machine—as strange or at least unanticipated. By now, however, it should be clear that the story of notated movement is the history of the messy and productive entangling of these threads, a truth that at least some notators had themselves long recognized. During the development of the ill-fated Labanotation Selectric typewriter, Dance Notation Bureau member Jo Floyd became worried that ongoing production delays were not, as IBM claimed, the result of a tooling issue, suspecting instead that the company was stalling while it secretly worked to obtain an intellectual property claim on the notation itself. As she wrote in a memo to bureau leaders, such a move could lead to a cascade of dramatic consequences: "At least one researcher to my knowledge has been in touch with IBM recently, trying to arrange use of the L/N IBM Element as an input to a computer program. If this man (or some other of the several people at present attempting to set up a computer program that will process movement information) is successful, there will be a powerful tool available for research in any field concerned with movement," from anthropology to biology to psychology. "Unfortunately," she continued, "the military is likely to become interested."[52] In response, Floyd suggested that the bureau buy out the entire production run of the L/N Element, agreeing to lease the technology only to groups or individuals they trusted. Floyd turned out to be wrong about IBM's specific ambitions—the company had no immediate nefarious designs and ultimately made little money from the project—but her insight about the potential applications of digitized movement proved undeniably, presciently correct.

Meaning in Movement

Sometime in the late 1940s, an English engineer and businessman named William Carpenter quit his job. His thirty years as an industrial executive had left him depressed and adrift, despondent that the "ever rising standard of living" his salary provided produced no true contentment. In the first months after the change, he tried to build a new kind of life as a working farmer, hoping that land and labor would help him find the meaning that had eluded him. Alas, he recalled, "though the farm grew to be an island of green fertility, my problems remained."[53]

At the end of his tether, he checked himself in to the Withymead Center, a residential therapeutic institution founded during World War II by the psychotherapist Dr. Irene Champernowne and her husband Gilbert. Situated in the verdant fields of Devon, the center synthesized Irene's Jungian training with a variety of art therapeutic practices, prodding patients to take up pottery, painting, woodworking, and dance. Carpenter likely engaged in all of the above, but it was movement that spoke to him most. In therapeutic sessions led by a Laban-trained instructor—Dartington Hall was not far from Withymead and the Elmhirsts were among its most important backers—Carpenter began, in his words, to find himself. Later he took courses directly with Laban, and after one session in particular, the incipient change became complete. As he moved in new ways and with new awareness, "the revelation of the narrowness of my previous physical and emotional experiencing came with startling clarity. . . . [M]y life's course was transformed in that single session."[54]

So altered, in fact, was Carpenter that he went on to formally enroll in the Art of Movement Studio and eventually became deeply involved with the organization's administrative structure. Working closely with Laban, he began composing two book manuscripts intended to present Laban's theories to the general public and to reconcile Labanian thought with Jungian psychology.[55] After a sudden illness, Carpenter died in June, 1954, but his life had become so intertwined with the Labanian project that he spent his last months living in a trailer on the grounds of the Art of Movement Studio's new building site in Addlestone.

What is most poignant about Carpenter's account of this period—which can be pieced together from remaining fragments of his book manuscripts—is the clear sense of the pain he felt prior to encountering Laban's work. Older people, he explained, had suffered "all our lives" because of the ways in which their movements—and thus also their emotional selves—had been controlled and repressed. He alluded to a boyhood in which "huntin', shootin', and fishin'" were "the only acceptable forms of movement behavior" and outlined the "devastating" effects of a set of social conventions in which "expressive patterns of behaviour are forbidden." In the "race to the top of his vocation," he wrote, the young executive "trains himself to look businesslike; the student doctor limits his movements to an appropriate bedside manner; the articled solicitor adopts a deferential attitude to clients; the housewife obeys her suburban movement conventions lest she fail to keep up with her neighbors." The damage caused by this restriction, in Carpenter's opinion, was worse even than the notorious injuries produced by assembly line work: "While the factory worker runs a risk of physical mutilation; the narrow conventionalist mutilates the harmonious sequences, which are the foundations of inner contentment."[56] The result was a society of silent suffering.

For Carpenter, therefore, movement therapy was a liberation. He could at last perceive his own movements and as a result became able to alter them.

He did so, moreover, not alone but within a group, a setting he found particularly powerful. "The understandable shyness of joining such a Group for the first time," he wrote, "is dispelled when we realise that our companions are as uncritical of our movements as we are of theirs. . . . [I]n company with our fellow dancers we can find the courage to do what we would not dare to do in the matter of fact world," experiencing "emotions which we may have never before expressed nor even acknowledged as possible."[57] He no longer felt alone or without purpose, finding in movement the relief he had sought from "from the tensions of our 20th century."[58]

Not everyone, to be sure, shared Carpenter's transformational experience, but it was this promise—to bring meaning and depth to lives that seemed increasingly disconnected, empty, rigid, and rationalized—that drew many people to Laban's work and the notation systems that undergirded it. The irony is that on a broad scale, movement notation was as responsible for perpetuating the conventions that tormented Carpenter as it was for revealing and dispelling them. Warren Lamb may have pledged his clients freedom from "puppetry," but he in fact created new standards for the physical behavior of white-collar workers—standards that, decades later, would be used to program actual robots. Rudolf Laban offered his movement choir participants an all-encompassing spiritual awakening, but his notation simultaneously ensured that this new communal festival would be measured, controlled, and racialized, engineered to serve first the ends of the National Socialist state and later the British war industry.

Notators' efforts to argue for movement's importance were plagued by parallel tensions. The Dance Notation Bureau fought tirelessly to get their art form taken seriously but ended up obscuring the creativity and labor of the very dancers who produced it.[59] Alan Lomax's quest to capture the totality of the world's movement traditions similarly stumbled, its efforts to quantify and categorize at odds with the changing, ephemeral, and inherently contextual nature of the underlying phenomena. Movement therapists, too, struggled with the extent to which the traumas stored deep within the body could or even should be brought to light.

Despite these struggles, the idea that movement—properly managed—might have the capacity to transform not just the individual, but the world as a whole, still holds a seductive power. For many of those currently deploying movement notation or analysis in computing, robotics, policing, and national security, the attraction is transparency and control: the ability to puncture the body's opacity and see the mind within, the capacity to engineer exactly how a human will react to a drone or a robot, a consumer product or a film. In truth, these aspirations are phantasms. Movement is culturally, politically, and historically contingent, with both its substance and its meaning changing across time, place, and context. And though many of its features *can* be recorded by notation, film, or verbal description, to claim that any single method of doing so is complete or objective—or that it fully captures the experience of movement itself—is simply unrealistic.

Still, there is nothing inherently wrong about the desire to document movement, to accord it cultural importance, or to honor the people who practice it as an art form. Indeed, attempts to record movement that recognize their own subjectivity and mutability can provide enormous, transformative value. Leaders in the disability arts movement, for example, have begun thinking creatively about the multiple ways in which movement might be represented for blind or visually impaired audiences. While audio description for dance performances was traditionally provided in flat, neutral tones, with describers instructed to "report, rather than interpret," new experiments have encouraged describers to embrace narrative, emotion, and poetry and have created technologies that allow listeners to switch back and forth among multiple forms of description, text, and sound.[60] Laban-trained individuals have begun to participate in these efforts, considering how the various forms of notation and analysis in which they are expert might aid these undertakings.[61] Such projects not only provide disabled audiences with new access to dance, but serve as novel aesthetic creations in and of themselves, complicating and enriching the art form. Disability scholars and activists, whose work has long undermined claims about "natural" or universal forms of bodily experience, will be also crucial partners in broader, society-wide discussions about movement's analysis and meaning.[62]

Labanotation also still plays an important role in protecting the work of choreographers, a business that picked up in the 2010s and '20s as videocentric social media platforms became increasingly widespread, dances rapidly disseminated across the internet, and new issues of credit and compensation arose. As a result, new groups of choreographers—particularly Black choreographers working in commercial or popular genres, a group whose work has often been appropriated without their consent—have sought the legal protection Labanotation can provide. JaQuel Knight, the choreographer for Beyoncé's 2008 "Single Ladies" music video, for example, hired Dance Notation Bureau executive director Lynne Weber to notate and copyright the piece in 2020 and has since launched a company to assist other choreographers in doing the same. For Knight, Labanotation's symbolic power as a written text held particular appeal. As he recalled in an interview with *Billboard*, when he first saw the "Single Ladies" score his "jaw dropped. . . . It's out of this world to see my hard work and sweat put on paper."[63] There is an implicit assertion here that movement does and should matter and that notation is a way of making it visible in a world that sometimes tries to ignore it. At the same time, many notators now recognize the ways in which all forms of recording are products of their own contexts, critically engaging with movement scores as political, social, and cultural documents.[64]

A small number of roboticists are similarly trying to approach Laban-based notation and analysis in more nuanced ways. The collaborative engineer-dancer team of Amy LaViers and Catherine Maguire, for example, have pushed back on assertions that there are fixed relationships between motion and meaning and urge scientists and engineers to think about notation simply as one technique for partially representing the human body. Reminding designers that the context of movement always matters, they argue that "making a robot with motion that is meaningful to its users may have as much in common with the process of writing a best-selling novel as designing a structurally sound bridge."[65] Taking this kind of approach not only prevents the kind of homogenizing and discriminatory assumptions increasingly common in other parts of the industry, it opens up new possibilities for engaging with movement in its complex multiplicity, highlighting the way it changes over time and with the unique bodies that perform it.

The problem, again, is not the act of caring about movement. Though it has often been dismissed as ephemeral, feminized, racialized, and relegated to the fringes of academic study, movement does indeed deserve the same kind of dedicated attention as other areas of modern life. This legacy of condescension is one reason many of the individuals and groups in this story fought so doggedly to promote movement analysis and recording. We should not, however, allow the desire to highlight movement's importance to obscure the fact that notation systems are nothing more than human-made technologies, with all the possibilities and limitations that fact implies.

Taking movement seriously also means recognizing the ways in which it can be used for both good and ill—including how ecstatic forms of embodiment have also functioned as tools of social control. As the story told here illustrates, the widespread adoption of Laban-based tools owed less to their frictionless efficacy in any particular setting and more to their users' hope that managed movement could reconcile seemingly irreconcilable values: reason and feeling, individuality and cohesion, modernity and tradition. Today corporate exercise classes, power poses, and seemingly friendly robots similarly dangle the idea that expressive movement might sufficiently soften the hard edges of a world that seems ever more geared toward solitude, efficiency, and profit.[66] It is an understandable yearning, but also one that is both unrealistic and insufficient. Though movement can provide many things—beauty, physical release, solidarity, self-expression—it is too much to ask that individual forms of bodily practice remedy the larger structural problems that plague the world in which we live. As we look to the future, therefore, we should remain wary of those who refuse to recognize this nuance and complexity, and who promise that in the act of recording the moving body, everything will be gained and nothing will be lost.

Acknowledgments

In sitting down to write these acknowledgments, I have been struck—perhaps a bit like the historical figures that populate this book—by the impossibility of capturing the many kinds of debts I have incurred with just a handful of scribbles on a page. Also like them, I will proceed nonetheless, recognizing that I will inevitably come up short.

The seed of this book was likely planted early in my childhood, which I spent the better part of in a ballet studio. From the age of three, I was lucky to have a series of dance teachers who introduced me—both physically and intellectually—to movement in all its complexities. Most important were Kathy and Dennis Landsman, in whose Kansas studio I grew up in more ways than one. The memories are visceral: the smell of the dressing room, the record player's occasional skips, the wooden barre under my hands, and, of course, the feeling of movement.

I also recall a brochure for the school, regularly posted on the hallway bulletin board, which included a set of frequently asked questions from potential students. The one that comes back to me now read something like: "Why do I have to spend all this time learning technique? Isn't dance about freedom and individual creativity?" (The particular words are lost, as my own record-keeping was far from archival grade.) The answer provided was that technical skill was the necessary foundation for self-expression: it was only the properly trained body that could fully and authentically communicate.

As a historian of science, technology, and medicine, I did not anticipate that my research would take me back to the world of dance, but it has been an unexpected pleasure to explore these paradoxical pairs—freedom and control, joy and discipline—from a new angle. It has also only strengthened my gratitude to those who taught me to dance and whom I danced alongside. What I learned from them about strength, beauty, community, and dedication is incalculable.

I also, of course, have countless debts of more recent vintage. This project was initially conceived in the University of Pennsylvania's Department of History and Sociology of Science, where the sense of community and uniquely creative vision of the field produced a graduate career that I could not have envisioned elsewhere. Special thanks go to John Tresch, a model of intellectual curiosity and generosity; Beth Linker, whose astute insights about the history of medicine, the body, and disability are matched only by her humanity; and Susan Lindee, who provided many of my foundational intellectual experiences in the field and whose analytical sharpness continues to inspire. My particular gratitude also to Ruth Schwartz Cowan for both early encouragement and continued support. Thanks too to Rob Kohler, Nathan Ensmenger, Robby Aronowitz, David Barnes, and the late Riki Kuklick, whose cameo in chapter 3 was one of my happiest archival discoveries.

Further early support for the project was provided by a Social Science Research Council DPDF, codirected by Peter Galison and Caroline Jones, and a writing fellowship in the Max Planck Institute for the History of Science's Department II, led by the incomparable Lorraine Daston. The conversations and communities that flowed from these experiences still shape my work today. I am also grateful for an ACLS/Mellon Dissertation Completion Fellowship, which funded a final year of writing.

I had the immense good fortune to spend three years at the Columbia University Society of Fellows, where my fellow fellows created a true academic Eden. To Heidi Hausse, Chris Florio, María González Pendás, Max Mischler, Ben Breen, Arden Hegele, Carmel Raz, David Gutkin, Lauren Kopajtic, Ardeta Gjikola, Joelle Abi-Rached, and Rachel Nolan, my eternal thanks for brilliant questions, fascinating conversations, and lots of fun. Further gratitude to Eileen Gilooly, Reinhold Martin, Chris Brown, and Emily Bloom, for stalwart leadership and for making it all possible, as well as to Deborah Coen for her welcome and support.

A yearlong membership at the Institute for Advanced Study's School of Historical Studies provided more space for deep thinking, good company, and excellent food than anyone has the right to expect. My special thanks to Myles Jackson for helming the terrific group in the history of science with his signature verve and to Asheesh Siddique, Andrew Amstutz, Douglas Flowe, Peter Lake, Sandy Solomon, Rosanna Dent, Andrea Bohlman, Jonathan Bach, Yukiko Koga, Takashi Miura, Diana Kim, and Anna Wilson for wonderful conversations and lunches. I count myself especially lucky to have finally intersected with Ken Alder, whom I thank for his ongoing generosity, insights, and humor. Thanks to Josh Horowitz for making the days so enjoyable.

I am grateful to Jennifer Homans and the Center for Ballet and the Arts at New York University—where a fellowship facilitated new connections with scholars across the humanities, social sciences, and arts—as well as to funding from two Falk Research Grants and the Center for the Arts in Society at Carnegie Mellon University. I presented aspects of this project to a variety of groups and benefited from the feedback and questions from audiences at the History of Science Group at the Technische Universität Berlin; the MPIWG Research Group on Twentieth Century Histories of Knowledge about Human Variation; the New York History of Science Lecture Series; the Columbia University Seminar in Dance Studies; the New York Academy of Sciences's Lyceum Society; Princeton University's Colloquium in the History of Science; the STS Program at Harvard University's Kennedy School of Government; Johns Hopkins University's Colloquium in the History of Science, Medicine, and Technology; the National Institute for Art History in Paris; the University of Toronto's BMO Lab for Creative Research in the Arts, Performance, Emerging Technologies and AI; the University of Chicago's Committee on the Conceptual and Historical Studies of Science; the Richardson Seminar on the History of Psychiatry at Weill Cornell Medicine; the Yale Architecture Forum; Northwestern University's Science and Human Culture Program; and the History of Science Society and the Society for the History of Technology. Conversations with participants at the "Learning How" and "Invisible Labor" workshops at the Max Planck Institute for the History of Science and the "Movement Movement" conference at Philipps-Universität Marburg were also immensely constructive. Much work on the project was completed during my time as an assistant professor in the Department of History at Carnegie Mellon University, and I am further grateful to my colleagues at the Department of Social Sciences and Cultural Studies at the Pratt Institute for their imagination, intelligence, and good cheer.

Portions of this manuscript appeared previously in *History of the Human Sciences* ("The Living Record: Alan Lomax and the World Archive of Movement (1965–1985)," *History of the Human Sciences* 31 (5) (2018): 23–51, SAGE Publications Ltd, all rights reserved), *Information and Culture* ("Paper Dancers: Art as Information in Twentieth Century America," *Information & Culture* 52 (1), 2017, 1–30 © 2017 University of Texas Press, all rights reserved), *Historical Studies in the Natural Sciences*, *Grey Room*, and *Limn*, and I thank the editors and reviewers of those pieces for their insights and the journals for their permission to republish here. I am also enormously grateful to the many librarians and archivists who made this project possible, including the staff of the Brotherton Research Centre at the University of Leeds; the Bauhaus Archive; the University of Surrey's National Resource Center for Dance; the Trinity Laban Conservatoire; the Library of Congress and the American Folklife Center; Special Collections at the University of Maryland; the Dance Notation Bureau; the Dartington Hall Trust Records; and the New York Public Library for the

Performing Arts. Writing a book that repeatedly confronted the complicated and unending work of preservation made me appreciate their contributions all the more. Thanks too to Janka Kormos, Guilherme Hinz, and Sydney Skybetter for graciously sharing their own archival finds and contacts. Near the beginning of this project, I was fortunate to be able to attend one of the biennial conferences of the International Congress of Kinetography Laban, and I am still appreciative of the warm welcome I received as an interested interloper and for the opportunity to meet a number of the field's founding figures, including Ann Hutchinson Guest. In later years I was able to speak with Forrestine Paulay, Anna Lomax Wood, and Mark Sossin, and I thank them all for sharing their time and their valuable insights. I also appreciate the thoughtful work of the manuscript's anonymous reviewers, and Karen Darling and Fabiola Enríquez Flores at the University of Chicago Press have been a pleasure to work with. I thank Thomas Kiefer for his preparation of the index, and copyeditor Catherine Osborne's sharp eyes and insights were a special treat.

Though I have already mentioned her name once, María González Pendás deserves an extra avalanche of gratitude for making me smarter and bolder and for being the truest kind of friend. I trust the kind and brilliant Rachel Elder like no one else; Jenna Tonn and Isabel Gabel were lifelines in the pandemic and far beyond; Mary Mitchell is sharp and steadfast; and Elaine LaFay has been there since the days of our shared office. Intellectually and personally, these kinds of relationships are among the best parts of this profession. Special thanks too to Jenny Bangham, Stephanie Dick, Michal Friedman, Allegra Giovine, Emanuela Grama, Benjamin Gross, Matthew Hoffarth, Andy Hogan, Judy Kaplan, Christopher Phillips, Brittany Shields, Nellwyn Thomas, and Laura Quinton, all of whom have been important confidants and interlocutors at various stages of this book's life.

I am also profoundly grateful to friends, family, and other supporters who exist far from the confines of academic life, only a handful of whom I have the space to mention here. To Sarah Campana, who made Philadelphia a home and has provided sunlight and good sense ever since; to Cory Schneider, for his excitement and for telling me to just get it done; to Jessica Ballou, Suparna Salil, Lian Huang, and Kyrsten Skogstad, for fighting the good fight and for listening; to Douglas Goldmacher, for conversation and cocktails; to Jackie Masher, for joyful recalibration; to Elizabeth Parks, for seeing me and being there. I don't know how I was lucky enough to meet Alana Aylward, but her intelligence, humanity, and general moxie has inspired me since day one, and I look forward to continued journeys together. Thanks too to my mother, Leslie, an iconoclast who always encouraged me to go my own way; to my much-missed father, Carl; and to my sisters, Kiley and Taylor. Finally, my gladdest appreciation to Matthew Hersch, with whom I share far too many things to count, past, present, and future.

Notes

Introduction

1. "Dance Notation Bureau Bulletin," January 1943, Dance Notation Bureau; and John Martin, "Travel Notes: How the Country Feels About the Ballets and Vice Versa-Miscellany," *New York Times*, January 3, 1943.

2. For a discussion of some of the philosophical dilemmas inherent in Aristotle's ideas of self-motion, see Mary Louise Gill and James G. Lennox, eds., *Self-Motion: From Aristotle to Newton* (Princeton University Press, 1994). My own description of movement, moreover, is purposefully capacious, designed to include humans of all abilities and body types.

3. Jan Bremmer and Herman Roodenberg's 1991 edited collection *A Cultural History of Gesture* (Cornell University Press) brings together a particularly wide-ranging selection of this kind of scholarship. In particular, see Jan Bremmer, "Walking, Standing, and Sitting in Ancient Greek Culture," 15–35; and Jean-Claude Schmitt, "The Rationale of Gestures in the West: Third to Thirteenth Centuries," 59–70. See also Jean-Claude Schmitt, *La raison des gestes dans l'occident médiéval* (Gallimard, 1990). William McNeill's short and suggestive *Keeping Together in Time: Dance and Drill in Human History* (Harvard University Press, 1995) argues that dance and military drill played important and analogous roles in promoting group solidarity across history. More recently, see Beth Linker, *Slouch: Posture Panic in Modern America* (Princeton University Press, 2024).

4. Norbert Elias's famous thesis that the formation of modern European nation-states was entangled with increasing control over bodily manners is another influential—though not uncontroversial—model of this kind of work. See *The Civilizing Process* (Blackwell, 1969).

5. Originally published as Marcel Mauss, "Les Techniques Du Corps," *Journal de psychologie normal et pathologique* 32 (1935): 271–93. Reprinted as Marcel Mauss, "Techniques of the Body," *Economy and Society* 2 (1973): 70–88.

6. Mauss, "Techniques of the Body," 74.

7. Mauss, "Techniques of the Body," 70.

8. Marta Braun, *Picturing Time: The Work of Étienne-Jules Marey (1830–1904)* (University of Chicago Press, 1995); François Dagognet, *Étienne-Jules Marey: A Passion for the Trace*, trans. Robert Galeta (Zone Books, 1992); John W. Douard, "E.-J. Marey's Visual Rhetoric and

the Graphic Decomposition of the Body," *Studies in History and Philosophy of Science* 26, no. 2 (1995): 175–204; and Brian Price, "Frank and Lillian Gilbreth and the Motion Study Controversy, 1907–1930," in *A Mental Revolution: Scientific Management since Taylor*, ed. Daniel Nelson (Ohio State University Press, 1992), 58–76. See also Rebecca Solnit, *River of Shadows: Eadweard Muybridge and the Technological Wild West* (Penguin Books, 2003).

9. Anson Rabinbach, *The Human Motor: Energy, Fatigue, and the Origins of Modernity* (Basic Books, 1990); Jennifer Karns Alexander, "Efficiency and Pathology: Mechanical Discipline and Efficient Worker Seating in Germany, 1929–1932," *Technology and Culture* 47, no. 2 (2006): 286–310; Richard Gillespie, "Industrial Fatigue and the Discipline of Physiology," in *Physiology in the American Context, 1850–1940* (American Physiological Society, 1987); Charles S. Maier, "Between Taylorism and Technocracy: European Ideologies and the Vision of Industrial Productivity in the 1920s," *Journal of Contemporary History* 5, no. 2 (1970): 27–61; Daniel Nelson, *Frederick W. Taylor and the Rise of Scientific Management* (University of Wisconsin Press, 1980); and Steffan Blayney, *Health and Efficiency: Fatigue, the Science of Work, and the Making of the Working-Class Body* (University of Massachusetts Press, 2022).

10. "Mr. Rudolf Laban," *The Times*, July 3, 1958.

11. Hedwig Müller and Patricia Stöckemann, "... *Jeder Mensch ist ein Tänzer": Ausdruckstanz in Deutschland zwischen 1900 und 1945* (Anabas-Verlag, 1993); Ana Isabel Keilson, "Making Dance Modern: Knowledge, Politics, and German Modern Dance, 1890–1927" (PhD diss., Columbia University, 2017); Martin Green, *Mountain of Truth: The Counterculture Begins, Ascona, 1900–1920* (University Press of New England, 1986); and Isa Partsch-Bergsohn and Harold Bergsohn, *The Makers of Modern Dance in Germany: Rudolf Laban, Mary Wigman, Kurt Jooss* (Princeton Book Company, 2002).

12. Lilian Karina and Marion Kant, *Hitler's Dancers: German Modern Dance and the Third Reich*, trans. Jonathan Steinberg (Berghahn Books, 2003). See also Laure Guilbert, *Danser avec le IIIe Reich: Les danseurs modernes sous le Nazisme* (Éditions Complexe, 2000).

13. Carole Kew, "From Weimar Movement Choir to Nazi Community Dance: The Rise and Fall of Rudolf Laban's 'Festkultur,'" *Dance Research* 17, no. 2 (1999): 73–96; Arabella Stanger, "Dancing Nature, Dancing Artifice: Laban, Schlemmer, and Reactionary Living Diagrams," in *Dancing on Violent Ground: Utopia as Dispossession in Euro-American Theater Dance* (Northwestern University Press, 2021), 89–124; and Patricia Vertinsky, "Movement Practices and Fascist Infections: From Dance Under the Swastika to Movement Education in the British Primary School," in *Physical Culture, Power and the Body*, ed. Jennifer Hargreaves and Patricia Vertinsky (Routledge, 2007), 25–51.

14. See, for example, Evelyn Doerr, *Rudolf Laban: The Dancer of the Crystal* (Scarecrow Press, 2008); Karen Bradley, *Rudolf Laban* (Routledge, 2008); and Eden Davies, *Beyond Dance: Laban's Legacy of Movement Analysis* (Routledge, 2006). Overviews and annotated source compilations on Laban have also been enormously useful. See Dick McCaw, *The Laban Sourcebook* (Routledge, 2011); Valerie Preston-Dunlop and Susanne Lahusen, eds., *Schrifttanz: A View of German Dance in the Weimar Republic* (Dance Books, 1990); Dick McCaw and Pat Lehner, eds., *The Art of Movement: Rudolf Laban's Unpublished Writings* (Routledge, 2024); and John Hodgson, *Mastering Movement: The Life and Work of Rudolf Laban* (Theatre Arts Books, 2001). In addition, Hodgson's work assembling the Laban archive at the University of Leeds was invaluable to this project.

15. Key present-day organizations for research and scholarship on Laban-based systems include the International Council of Kinetography Laban/Labanotation (ICKL), the Laban/

Bartenieff Institute of Movement Studies (LIMS), and the Dance Notation Bureau and its Extension Center at The Ohio State University. For a sample of the academic literature emerging from scholars and contemporary practitioners, see, for example, Victoria Watts, "The Perpetual 'Present' of Dance Notation," *Ekphrasis* 2 (2014): 180–99; Mara Pegeen Frazier, "Labanotation Is Creative: How a Systems Perspective Reveals Generativity in Dance Notation and Its Archives," *Journal of Movement Arts Literacy* 7, no. 1 (2021): 105–31; and Crystal U. Davis, Selene Carter, and Susan R. Koff, "Troubling the Frame: Laban Movement Analysis as Critical Dialogue," *Journal of Dance Education* (October 18, 2021): 1–9.

16. See, e.g., John Martin, "New Method of Notation: Lieutenant Chiesa, an Italian Army Officer, Devises System Called 'Motography' for Recording Movement of Body," *New York Times*, July 15, 1934; John Martin, "A New Script: The Problem of Notation as Solved by Sol Babitz," *New York Times*, August 27, 1939; Rudolf Benesh and Joan Benesh, *An Introduction to Benesh Dance Notation* (Adam and Charles Black, 1956); John Martin, "Notation Campaign: New Benesh System Being Widely Promoted by Official British Agency," *New York Times*, May 12, 1957; and Noa Eshkol and Avraham Wachmann, *Movement Notation* (Weidenfeld & Nicholson, 1958). Dance Notation Bureau founder Ann Hutchinson Guest provides a useful, albeit interested, overview of these and other techniques, including older methods as well as twentieth-century systems originating in China, Korea, and Japan; see "Dance Notation," *Encyclopedia Britannica*, October 2, 2016, https://www.britannica.com/art/dance-notation.

17. Bruno Latour, "Drawing Things Together," *Knowledge and Society* 6 (1986): 1–40; Ursula Klein, *Experiments, Models, Paper Tools* (Stanford University Press, 2002); David Kaiser, *Drawing Theories Apart: The Dispersion of Feynman Diagrams in Postwar Physics* (University of Chicago Press, 2005); Lisa Gitelman, *Paper Knowledge: Toward a Media History of Documents* (Duke University Press, 2014); Carla Bittel, Elaine Leong, and Christine von Oertzen, eds., *Working with Paper: Gendered Practices in the History of Knowledge* (University of Pittsburgh Press, 2019); and Seth Rockman, "Forum: The Paper Technologies of Capitalism," *Technology and Culture* 58, no. 2 (2017): 487–505.

18. For a recent collection of essays dedicated to undermining such distinctions, see James Evans and Adrian Johns, eds., "Beyond Craft and Code: Human and Algorithmic Cultures, Past and Present," special issue, *Osiris* 38, no. 1 (2023).

19. Sarah E. Igo, *The Averaged American: Surveys, Citizens, and the Making of a Mass Public* (Harvard University Press, 2007); Dan Bouk, *How Our Days Became Numbered: Risk and the Rise of the Statistical Individual* (University of Chicago Press, 2015); Geoffrey C. Bowker and Susan Leigh Star, *Sorting Things Out: Classification and Its Consequences* (MIT Press, 2000); and Rebecca Lemov, *Database of Dreams: The Lost Quest to Catalog Humanity* (Yale University Press, 2015).

20. See, for example, Elena Aronova, Christine von Oertzen, and David Sepkoski, eds., "Data Histories," special issue, *Osiris* 32, no. 1 (2017); and Lorraine Daston, ed., *Science in the Archives: Pasts, Presents, Futures* (University of Chicago Press, 2017).

21. Michael Sawh, "What Is Google Going to Do with Your Fitbit Data? Anything It Likes," *Wired*, May 11, 2019; Dake Kang, "Chinese 'Gait Recognition' Tech IDs People by How They Walk," *AP News*, November 6, 2018; and Drew Harwell, "A Face-Scanning Algorithm Increasingly Decides Whether You Deserve the Job," *Washington Post*, November 6, 2019.

22. Roger Smith, *The Sense of Movement: An Intellectual History* (Process Press, 2019); Mark Paterson, *How We Became Sensorimotor: Movement, Measurement, Sensation* (University of Minnesota Press, 2021); Andreas Mayer, *The Science of Walking: Investigations into Locomotion*

in the Long Nineteenth Century (University of Chicago Press, 2020); and Irina Sirotkina and Roger Smith, *The Sixth Sense of the Avant-Garde: Dance, Kinaesthesia and the Arts in Revolutionary Russia* (Bloomsbury, 2017).

23. Robin Veder, *The Living Line: Modern Art and the Economy of Energy* (Dartmouth College Press, 2015); Ana Hedberg Olenina, *Psychomotor Aesthetics: Movement and Affect in Modern Literature and Film* (Oxford University Press, 2020); and Robert Brain, *The Pulse of Modernism: Physiological Aesthetics in Fin-de-Siecle Europe* (University of Washington Press, 2015).

24. Zeynep Çelik Alexander, *Kinaesthetic Knowing: Aesthetics, Epistemology, Modern Design* (University of Chicago Press, 2017).

25. Veder, *The Living Line*, 2.

26. J. Alexander, "Efficiency and Pathology"; Lisa Cartwright, *Screening the Body: Tracing Medicine's Visual Culture* (University of Minnesota Press, 1995); Hannah Landecker, *Culturing Life: How Cells Became Technologies* (Harvard University Press, 2007); and Lori Andrews and Dorothy Nelkin, "Whose Body Is It Anyway? Disputes over Body Tissues in a Biotechnology Age," *Lancet* 351 (January 3, 1998): 53–60.

27. Hillel Schwartz, "Torque: The New Kinaesthetic of the Twentieth Century," in *Incorporations*, ed. Jonathan Crary and Sanford Kwinter (Zone Books, 1992). See also, for example, Carrie Streeter, "Breathing Power and Poise: Black Women's Movements for Self-Expression and Health, 1880s-1900s," *Australasian Journal of American Studies* 39, no. 1 (2020): 5-46.

28. Karl Toepfer, *Empire of Ecstasy: Nudity and Movement in German Body Culture, 1910–1935* (University of California Press, 1997); Joseph S. Alter, *Yoga in Modern India: The Body Between Science and Philosophy* (Princeton University Press, 2005); and Mark Singleton, *Yoga Body: The Origins of Modern Posture Practice* (Oxford University Press, 2010).

29. Mauss, "Techniques of the Body," 87.

30. Schwartz, "Torque."

31. Jonathan Sterne, *The Audible Past: Cultural Origins of Sound Reproduction* (Duke University Press, 2003), 9.

32. See, for example, Sterne, *The Audible Past*; Emily Thompson, *The Soundscape of Modernity: Architectural Acoustics and the Culture of Listening in America, 1900–1933* (The MIT Press, 2004); and Alexandra Hui, Julia Kursell, and Myles W. Jackson, eds., "Music, Sound, and the Laboratory from 1750–1980," special issue, *Osiris* 28, no. 1 (2013).

33. See, for example: Norman Bryson, "Cultural Studies and Dance History," in *Meaning in Motion: New Cultural Studies of Dance*, ed. Jane C. Desmond (Duke University Press, 1997); and Susan Manning and Lucia Ruprecht, eds., *New German Dance Studies* (University of Illinois Press, 2012). See also Rebekah J. Kowal, Gerald Siegmund, and Randy Martin, *The Oxford Handbook of Dance and Politics* (Oxford University Press, 2017); and Lucia Ruprecht, *Gestural Imaginaries: Dance and Cultural Theory in the Early Twentieth Century* (Oxford University Press, 2019).

34. The most recent wave of humanistic interest in the body was born in the 1990s, though its influence continues into the present. For discussion of this historiography, see, for example, Caroline Bynum, "Why All the Fuss about the Body? A Medievalist's Perspective," *Critical Inquiry* 22, no. 1 (1995): 1–33; Roy Porter, "History of the Body Reconsidered," in *New Perspectives on Historical Writing*, ed. Peter Burke, 2nd ed. (Pennsylvania State University Press, 2001), 233–60; and Mark S. R. Jenner and Bertrand O. Taithe, "The Historiographical Body," in *Medicine in the Twentieth Century*, ed. Roger Cooter and John V. Pickstone (Routledge, 2003), 187–200.

35. Pamela H. Smith, *The Body of the Artisan: Art and Experience in the Scientific Revolution* (University of Chicago Press, 2004); Pamela H. Smith, *From Lived Experience to the Written Word: Reconstructing Practical Knowledge in the Early Modern World* (University of Chicago Press, 2022); Donald MacKenzie and Graham Spinardi, "Tacit Knowledge, Weapons Design, and the Uninvention of Nuclear Weapons," *American Journal of Sociology* 101, no. 1 (1995): 44–99; Michael Polanyi, *The Tacit Dimension* (University of Chicago Press, 2009); Harry Collins, *Tacit and Explicit Knowledge* (University of Chicago Press, 2010); and Christopher Lawrence and Steven Shapin, eds., *Science Incarnate: Historical Embodiments of Natural Knowledge* (University of Chicago Press, 1998).

36. In 2009, the anthropologist Richard Handler similarly urged scholars to "historicize the trend to see 'bodily practices' as an important topic for cultural and historical studies." See "Erving Goffman and the Gestural Dynamics of Modern Selfhood," *Past and Present* 203 (2009): 280.

37. Historians of who focus on dance explicitly tend to be more attentive to the complex political ends the art has served. See, among others, Kowal, Siegmund, and Martin, *The Oxford Handbook of Dance and Politics*; Adria L. Imada, *Aloha America: Hula Circuits Through the U.S. Empire* (Duke University Press, 2012); Joel Dinerstein, *Swinging the Machine: Modernity, Technology, and African American Culture Between the World Wars* (University of Massachusetts Press, 2003); Louis S. Warren, *God's Red Son: The Ghost Dance Religion and the Making of Modern American* (Basic Books, 2017); Elizabeth B. Schwall, *Dancing with the Revolution: Power, Politics, and Privilege in Cuba* (University of North Carolina Press, 2021); Julie Malnig, *Dancing Black, Dancing White: Rock 'n' Roll, Race, and Youth Culture of the 1950s and Early 1960s* (Oxford University Press, 2023); and Laura Quinton, "Britain's Royal Ballet in Apartheid South Africa," *The Historical Journal* 64, no. 3 (2021): 750–73.

38. R. Smith, *The Sense of Movement*, 343.

39. Ann Hutchinson Guest, *Labanotation: The System for Recording Movement* (New Directions, 1954), 7.

Chapter 1

1. Rudolf von Laban, *A Life for Dance: Reminiscences*, trans. Lisa Ullmann (New York: Theatre Arts Books, 1975), 184. Originally published as Rudolf von Laban, *Ein Leben für den Tanz: Erinnerungen* (Karl Reissner Verlag, 1935).

2. A. M., "Tänze die Man Lesen lernt," *Schweizer illustrierte Zeitung*, 1928, 1/1/266, John Hodgson Collection, Special Collections, University of Leeds, Leeds, UK.

3. Hans Brandenburg, "Laban als Stilreformer," *Singchor und Tanz*, December 15, 1929, 1/1/293, John Hodgson Collection, Special Collections, University of Leeds, Leeds, UK.

4. Werner Schuftan, "Labans Bewegungschrift," *Singchor und Tanz*, December 15, 1929, 1/1/293, John Hodgson Collection, Special Collections, University of Leeds, Leeds, UK.

5. Eventually, *Schrifttanz* widened its mission to include analysis of the modern dance scene more generally, but its theoretical core—and, indeed, its name—remained the recording of dance on paper. Almost every edition included the insertion of notated dance material, whether short excepts from well-known social dances, avant-garde expressionist works, or folk dances.

6. Herman Geo. Scheffauer, "German Composes Dance Alphabet," *New York Times*, March 27, 1927. See also, for example, "Dancing Notation: A New System by Rudolf Von Laban," *The Observer*, June 17, 1928.

7. M. M., "Von Laban Work in Print: German Choreographer Explains System of Dance Notation in First Translation Offered," *Los Angeles Times*, October 4, 1931.

8. John Martin, "The Dance: It Has a Language of Its Own," *New York Times*, October 21, 1928.

9. Schuftan, "Labans Bewegungschrift."

10. Alfred Schlee, "Editorial," *Schrifttanz*, July 1928, 1/3/19, John Hodgson Collection, Special Collections, University of Leeds, Leeds, UK. Translated in Preston-Dunlop and Lahusen, *Schrifttanz: A View of German Dance in the Weimar Republic*, 29.

11. Toepfer, *Empire of Ecstasy*, 4.

12. Felicia McCarren, *Dancing Machines: Choreographies of the Age of Mechanical Reproduction* (Stanford University Press, 2003); Kate Elswit, *Watching Weimar Dance* (Oxford University Press, 2014); Alys X. George, *The Naked Truth: Viennese Modernism and the Body* (University of Chicago Press, 2020); Susan Manning, *Ecstasy and the Demon: Feminism and Nationalism in the Dances of Mary Wigman* (University of California Press, 1993); and Michael Cowan, *Technology's Pulse: Essays on Rhythm in German Modernism* (IGRS Books, 2011).

13. The Swiss composer and educator Émile Jaques-Dalcroze was best known for his development of the *"rythmique"* (eurythmic) method of musical pedagogy. Published in a series of volumes beginning in 1906, eurhythmics was founded on the theory that musical principles were best taught through the medium of the body: Dalcroze's students engaged in physical exercises to develop an internal sense of rhythm, musical structure, and tone. Dalcroze's system quickly gained traction across the European continent; by 1926, over twenty-two thousand people had enrolled in eurythmics courses. Laban himself attended the Institut Jaques-Dalcroze at Hellerau in 1912, and several members of Laban's inner circle, including Suzanne Perrottet and Mary Wigman, were originally Dalcroze students. See Émile Jaques-Dalcroze, *Méthode Jaques-Dalcroze: Gymnastique rythmique* (Sandoz, Jobin & Cie., 1906); Laura Kuhn and Nicholas Slonimsky, "History of the Dalcroze Method of Eurythmics," in *Music Since 1900*, 6th ed., ed. Laura Kuhn and Nicholas Slonimsky (Schirmer Reference, 2001), 915–16; and James W. Lee, "Dalcroze by Any Other Name: Eurythmics in Early Modern Theatre and Dance" (Texas Tech University, 2003). See also Toepfer, *Empire of Ecstasy*; Peter Jelavich, *Munich and Theatrical Modernism: Politics, Playwriting, and Performance, 1880–1914* (Harvard University Press, 1985); Cowan, *Technology's Pulse*; and Julia A. Walker, "Eurhythmics and Bohemian Models of Affiliation," in *Performance and Modernity* (Cambridge: Cambridge University Press, 2021), 106–59.

14. Elswit, *Watching Weimar Dance*, xxx.

15. Tresa Randall, "Cultural Modernity, the Wigman School, and the Modern Girl," *Feminist Modernist Studies* 4, no. 3 (2021): 360–74.

16. The "blue flower" was a common Romantic symbol of metaphysical longing, first appearing in Novalis's 1802 novel *Heinrich von Ofterdingen*. See von Laban, *A Life for Dance*, 21.

17. Von Laban, *A Life for Dance*, 24.

18. Von Laban, *A Life for Dance*, 24.

19. Istvan Deak, *Beyond Nationalism: A Social and Political History of Habsburg Officer Corps, 1848–1918* (Oxford University Press, 1992), 88.

20. On the history and culture of the Wiener Neustadt academy, see Deak, *Beyond Nationalism*.

21. Green, *Mountain of Truth*, 87.

22. Von Laban, *A Life for Dance*, 37, 53.

23. On Monte Verità, see Green, *Mountain of Truth*.

24. On German modern dance see, for example, Manning, *Ecstasy and the Demon*; Keilson, "Making Dance Modern"; Elswit, *Watching Weimar Dance*; Ruprecht, *Gestural Imaginaries*; and Müller and Stöckemann, "*. . . Jeder Mensch ist ein Tänzer.*"

25. Solnit, *River of Shadows*.

26. Robert Brain, "Representation on the Line: Graphic Recording Instruments and Scientific Modernism," in *From Energy to Information: Representation in Science and Technology, Art, and Literature*, ed. Bruce Clarke and Linda Henderson (Stanford University Press, 2002); and Brain, *Pulse of Modernism*.

27. Elswit, *Watching Weimar Dance*, 26–59; and Noam Elcott, *Artificial Darkness: An Obscure History of Modern Art and Media* (University of Chicago Press, 2016).

28. In addition to Beauchamp-Feuillet and others, Laban was likely familiar with Stepanov notation, used at the Russian Imperial Ballet. See Sheila Marion, "Recording the Imperial Ballet: Anatomy and Ballet in Stepanov's Notation," in *Dance on Its Own Terms*, ed. Melanie Bates and Karen Eliot (Oxford University Press, 2013): 309–40.

29. Raoul-Auger Feuillet, *Chorégraphie, ou l'art d'écrire la danse* (Paris, 1700). See also Ken Pierce, "Dance Notation Systems in Late 17th-Century France," *Early Music* 26, no. 2 (May 1998): 286–99; Gabriella Karl-Johnson, "From the Page to the Floor: Baroque Dance Notation and Kellom Tomlinson's The Art of Dancing Explained," *Signs & Society* 5, no. 2 (2017): 269–92; and Wendy Hilton, *Dance and Music of Court and Theater: Selected Writings of Wendy Hilton* (Pendragon Press, 1997). For an analysis of the political context of Beauchamp-Feuillet notation as well as its conceptual and ideological similarities to Linnean nomenclature, see Susan Leigh Foster, "Choreographing Empathy," *Topoi* 24 (2005): 81–91.

30. Pierce, "Dance Notation Systems in Late 17th-Century France," 287.

31. Laban's 1920 *Die Welt des Tänzers* also includes references to Zorn notation (1887) and Arthur Saint-Léon's *Sténochorégraphie* (1852), as well as more to general histories of writing including Karl Faulmann's *Illustrierte Geschichte der Schrift* (1880). See Rudolf Laban, *Die Welt des Tänzers* (Walter Seifert, 1920), 262–63.

32. Karl-Johnson, "From the Page to the Floor," 289; Linda Tomko, "Dance Notation and Cultural Agency: A Meditation Spurred by 'Choreo-Graphics,'" *Dance Research Journal* 31, no. 1 (1999): 3.

33. Rudolf von Laban, *Choreographie* (Eugen Diederichs, 1926), 54–64. Laban's book's title, however, was a direct tribute to Feuillet's earlier work.

34. Von Laban, *A Life for Dance*, 3.

35. Rudolf Laban, *Principles of Dance and Movement Notation* (Macdonald & Evans, 1956), 7–8. Laban also credits the choreographer Kurt Jooss with encouraging him to replicate Feuillet's right-left division of the body.

36. Rudolf von Laban, "Grundprinzipien der Bewegungsschrift," *Schrifttanz*, July 1928, 1/3/19, John Hodgson Collection, Special Collections, University of Leeds, Leeds, UK; and Rudolf von Laban, *Schrifttanz: Methodik, Orthographie, Erläuterungen* (Universal Edition, 1928).

37. Laban had dabbled in a variety of mystic traditions since his days as a student in Paris and while at Monte Verità was inducted into the Ordo Templi Orientis by Theodor Reuss; see Green, *Mountain of Truth*, 104. On Laban's dance practice as a fundamentally religious endeavor, see Marion Kant, "Laban's Secret Religion," *Discourses in Dance* 1, no. 2 (2002): 43–62. He was also profoundly influenced by childhood experiences with Sufi dervishes.

38. For a broader account of the intersections between these fields, see Bruce Clarke and Linda Dalrymple Henderson, eds., *From Energy to Information: Representation in Science and Technology, Art, and Literature* (Stanford University Press, 2002).

39. See, for example, Albrecht Knust, "Die Bewegungsschrift (Kinetographie Laban) ist von Jedermann leicht erlernbar," *Der Tanz*, 1933, 1/1/258, John Hodgson Collection, Special Collections, University of Leeds, Leeds, UK.

40. Laban, *Die Welt des Tänzers*, 65; Celia Applegate and Pamela Potter, "Germans as the 'People of Music': Genealogy of an Identity," in *Music and German National Identity* (University of Chicago Press, 2002), 1–35.

41. Rudolf von Laban, "Das Tänzerische Kunstwerk," *Die Tat*, November 1927. Translated in Vera Maletic, *Body, Space, Expression: The Development of Rudolf Laban's Movement and Dance Concepts* (De Gruyter, 1987), 10.

42. Brandenburg, "Laban als Stilreformer."

43. Suzanne Perrottet also played an important role in notation's development, particularly through her expertise in Dalcroze eurythmics.

44. It is likely for this reason that then-common discussions of the concept of *Einfühlung*, or empathy, rarely occur in Laban's work. Laban's hope was for performer and audience to merge, rather than for one to respond empathically to the other. For a concise account of the history of the concept of *Einfühlung*, see Robin Curtis, "An Introduction to Einfühlung," trans. Richard George Elliott, *Art in Translation* 6, no. 4 (2014): 353–75.

45. Laban, *Die Welt des Tänzers*, 80.

46. Laban, *Die Welt des Tänzers*, 9.

47. Laban, *Die Welt des Tänzers*, 13.

48. Laban, *Die Welt des Tänzers*, 38.

49. Laban, *Die Welt des Tänzers*, 10. For a discussion of the development and impact of a new epistemology of embodied, nonrational, "kinaesthetic knowing" during this period in design, art, and architecture, see Z. Alexander, *Kinaesthetic Knowing*.

50. Soraya de Chadarevian, "Graphical Method and Discipline: Self-Recording Instruments in Nineteenth-Century Physiology," *Studies in History and Philosophy of Science* 24, no. 2 (1993): 267–91; Rabinbach, *The Human Motor*; and Brain, *Pulse of Modernism*, 5–36.

51. Brain, *Pulse of Modernism*. On physiological aesthetics in the United States, see Veder, *Living Line*. On "psychomotor aesthetics" in Russia and the Soviet Union, see Olenina, *Psychomotor Aesthetics*.

52. Brain, *Pulse of Modernism*, xxvi.

53. Brain, *Pulse of Modernism*, 227.

54. Laban, *Die Welt des Tänzers*, 64.

55. Laban, *Die Welt des Tänzers*, 30.

56. Laban's notes on Bell, Lange, and Langley are undated, though they were most likely composed in the late 1930s shortly after his move to England, when he spent much of his time attempting to reconstruct the unpublished notes and manuscripts he left behind in Germany. This context—as well as references to these figures and concepts in Laban's previously published works—suggest they can be read as reasonable reflections of his thinking throughout the 1920s and 30s.

57. See also Charles Bell, *The Anatomy and Philosophy of Expression as Connected with the Fine Arts*, 3rd ed. (London, 1865).

58. Charles Bell, *Idea of a New Anatomy of the Brain* (London, 1811). See also Carin Berkowitz, *Charles Bell and the Anatomy of Reform* (University of Chicago Press, 2015).

59. Carl Lange, *Om sindsbevägelser* (Copenhagen, 1885). See also Berkowitz, *Charles Bell and the Anatomy of Reform*; Leonard Carmichael, "Sir Charles Bell: A Contribution to the History of

Physiological Psychology," *Psychological Review* 33, no. 3 (1926): 188–217; Frederick Cummings, "Charles Bell and the Anatomy of Expression," *The Art Bulletin* 46, no. 2 (1964): 191–203; and Claudia Wassmann, "Reflections on the 'Body Loop': Carl Georg Lange's Theory of Emotion," *Cognition and Emotion* 24, no. 6 (2010): 974–90.

60. J. N. Langley, *The Autonomic Nervous System, Part I* (W. Heffer & Sons Ltd., 1921).

61. Rudolf Laban, "Charles Bell (1774–1842)," n.d., L/E/70/19, Rudolf Laban Archive, University of Surrey.

62. Rudolf Laban, "Nervous Organs and Functions," n.d., L/E/37/1, Rudolf Laban Archive, University of Surrey.

63. Rudolf Laban, "Structural and Functional Shapes," n.d., L/E/37/2, Rudolf Laban Archive, University of Surrey.

64. Rudolf Laban, "Nerve Irradiation," n.d., L/E/37/5, Rudolf Laban Archive, University of Surrey.

65. Anne Harrington, *Reenchanted Science: Holism in German Culture from Wilhelm II to Hitler* (Princeton University Press, 1999), xvii.

66. Laban, "Structural and Functional Shapes."

67. As quoted in Rudolf Bode, "Laban School, Man in Rhythm and Space (English Translation)," 1916, 1/1/255, John Hodgson Collection, Special Collections, University of Leeds, Leeds, UK.

68. Bode, "Laban School, Man in Rhythm and Space."

69. See, among others, Laura Otis, "The Metaphoric Circuit: Organic and Technological Communication in the Nineteenth Century," *Journal of the History of Ideas* 63, no. 1 (2002): 105–28; Stephen Kern, *The Culture of Time and Space, 1880–1918* (Harvard University Press, 1983); and Rabinbach, *The Human Motor*, 50. On atomic theory and occultism, see Mark Morrison, *Modern Alchemy: Occultism and the Emergence of Atomic Theory* (Oxford University Press, 2017).

70. T. H. Huxley, "On the Physical Basis of Life," *Fortnightly Review* 5 (1869): 129–45.

71. Ernst Haeckel, *Kristallseelen: Studien über das anorganische Leben* (Alfred Kröner, 1917).

72. Haeckel's conceptualization of protoplasm—as a substance with its own "memory," but also one susceptible to alteration by external forces—allowed him to explain both the individual inheritance of acquired characteristics and speciation. See Georgy S. Levit and Uwe Hossfeld, "Natural Selection in Ernst Haeckel's Legacy," in *Natural Selection: Revisiting Its Explanatory Role in Evolutionary Biology* (Springer, 2021), 105–36. See, more broadly, Brain, *Pulse of Modernism*, 37–63; Gerald L. Geison, "The Protoplasmic Theory of Life and the Vitalist-Mechanist Debate," *Isis* 60, no. 3 (1969): 272–92; and Daniel Liu, "The Cell and Protoplasm as Container, Object, and Substance, 1835–1861," *Journal of the History of Biology* 50, no. 4 (2017): 889–925.

73. Laban, *Die Welt des Tänzers*, 64, 102, 69.

74. In linking gestural and protoplasmic movement, Laban echoed the work of the French psychophysiologist Alfred Binet, whose study of the "psychology of movements" encompassed both the vibrations of protoplasm and the voluntary gestures of animals and humans. See Brain, *Pulse of Modernism*, 55.

75. Laban, *Die Welt des Tänzers*, 69.

76. Laban, *Die Welt des Tänzers*, 51.

77. Laban, "Nerve Irradiation."

78. Laban "Nerve Irradiation." See also Ruprecht, *Gestural Imaginaries*, 71–77.

79. Though the substance of Laban's work clearly resonated with the crowd and group psychology of the era, he generally did not engage explicitly with figures like Gustave Le Bon

or Sigmund Freud. It is difficult to imagine, however, that he was not at least aware of these discourses, given their shared political and psychological concerns. For an introduction to the group psychology of the period, see, for example, Robert A. Nye, *The Origins of Crowd Psychology: Gustave Le Bon and the Crisis of Mass Democracy in the Third Republic* (Sage, 1975); Daniel Pick, "Freud's Group Psychology and the History of the Crowd," *History Workshop Journal* 40 (Autumn 1995): 39–61; and Gillian Swanson, "Collectivity, Human Fulfillment and the 'Force of Life': Wilfred Trotter's Concept of the Herd Instinct in Early 20th-Century Britain," *History of the Human Sciences* 27, no. 1 (2014): 21–50.

80. In addition to Haeckel, Laban's interest in crystallography was shaped by Victor Goldschmidt, a mineralogist whose works included the three-volume *Index der Krystallformen*, a catalog of the known crystalline forms of all minerals, published between 1886 and 1891, and the nine-volume *Atlas der Krystallformen*, published between 1913 and 1923. Goldschmidt also proposed that the ratios governing crystalline forms were the same as those governing musical theory, a principle which Laban believed applied to dance as well. See Laban, *Die Welt des Tänzers*, 36; Victor Goldschmidt, *Über Harmonie und Complication* (Julius Springer, 1901).

81. Martin, "The Dance: It Has a Language of Its Own."

82. For further discussion of the politics embedded in Laban's icosahededra—politics that tellingly mirrored those of Kinetographie itself—see Stanger, "Dancing Nature, Dancing Artifice."

83. Laban, *Die Welt des Tänzers*, 8.

84. Fritz Klingenbeck, "Kleiner rückblick," *Singchor und Tanz* 46, no. 24 (December 15, 1929): 306.

85. Knust, "Die Bewegungsschrift (Kinetographie Laban) ist von Jedermann leicht erlernbar."

86. On the political currents in "Red Vienna," see Anson Rabinbach, *The Crisis of Austrian Socialism: From Red Vienna to Civil War, 1927–1934* (University of Chicago Press, 1983).

87. Bücher noted that there were likely both physiological and psychological reasons rhythmic movement produced these reactions, though he declined to address them in depth. Like Laban, however, his thinking seems to have been particularly influenced by Wundt, whom he cites in passing, and who was a colleague at Leipzig. See Karl Bücher, *Arbeit und Rhythmus* (Leipzig, 1899).

88. "Labans Wiener Tanzfestzug," *Singchor und Tanz* 46, no. 13 (July 1, 1929): 182–83. On the significance of dance and body culture more generally to Viennese modernism and politics, see George, *The Naked Truth*, especially 214–28.

89. Harry Prinz, "Laban's Festzug der Gewerbe in Wien," *Der Tanz*, 1929.

90. "Labans Wiener Tanzfestzug."

91. Prinz, "Laban's Festzug der Gewerbe in Wien."

92. Von Laban, *A Life for Dance*, 143.

93. Von Laban, *A Life for Dance*, 152. Laban can also be understood as a participant in the evolving relationship between festive culture and German nationalism. See George L. Mosse, *The Nationalization of the Masses: Political Symbolism and Mass Movements in Germany from the Napoleonic Wars Through the Third Reich* (H. Fertig, 1975). See also Mary Wigman, "Rudolf von Laban," *Singchor und Tanz* 46, no. 24 (December 15, 1929): 295.

94. Von Laban, *A Life for Dance*, 79, 96.

95. Helga Hain, "Gruppentanz," n.d., 1/1/355, John Hodgson Collection, Special Collections, University of Leeds, Leeds, UK; and Hans-Joachim Kurras, "Wie Der Zweitanz Aufmichwirkte," n.d., 1/1/355, John Hodgson Collection, Special Collections, University of Leeds, Leeds, UK.

96. Benedict Anderson, *Imagined Communities: Reflections on the Origin and Spread of Nationalism* (Verso, 1983).
97. Martin Gleisner, "Schrifttanz und Laientanz," *Schrifttanz* 1, no. 2 (October 1928): 20–22. Translated in Preston-Dunlop and Lahusen, *Schrifttanz*, 45.
98. Gleisner, "Schrifttanz und Laientanz."
99. Rudolf von Laban, "Tanzkomposition und Schrifttanz," *Schrifttanz* 1, no. 2 (October 1928): 19–20. Translated in Preston-Dunlop and Lahusen, *Schrifttanz*, 39.
100. "Titan, The Celebratory Chorus of the Hamburg Movement Choir," *Hamburg Echo*, July 30, 1928. English translation reproduced in McCaw, *The Laban Sourcebook*, 115.
101. Fritz Böhme to Joseph Goebbels, November 8, 1933, Bundesarchiv 50.01 237, reproduced in Karina and Kant, *Hitler's Dancers*, 197.
102. Rudolf Laban to Otto von Keudell, February 11, 1935, Bundesarchiv 50.01 237, reproduced in Karina and Kant, *Hitler's Dancers*, 217.
103. Karina and Kant, *Hitler's Dancers*; and Frank-Manuel Peter, "The German Dance Archive, Cologne," *Dance Chronicle* 32, no. 3 (2009): 476–89.
104. "Tanzformen—Tanzprache—Tanznotation: Eine Rundfrage," *Singchor und Tanz* 45, no. 12 (1928): 170–74.
105. Rudolf Laban, "Der laientanz in kultureller und und pädagogischer bedeutung," June 21, 1930, reprinted in Müller and Stöckemann, "... *Jeder Mensch ist ein Tänzer*," 96–99. See similar sentiments in von Laban, *A Life for Dance*, 133.
106. Rudolf Laban, "Radiation and Emanation, Part II," n.d., L/E/37/10, Rudolf Laban Archive, University of Surrey.
107. See Martin Gleisner, *Tanz für Alle: Von der Gymnastik zum Gemeinschaftstanz* (Hesse & Becker, 1928).
108. Marie Luise Lieschke, "Bericht über die Chorische Tagung in Essen vom 4–11 Juli 1934," n.d., John Hodgson Collection, Special Collections, University of Leeds, Leeds, UK.
109. Lieschke, "Bericht über die Chorische Tagung in Essen vom 4–11 Juli 1934."
110. Fritz Böhme, "Deutscher Tanz und Volkstanz," in *Deutsche Tanzfestspiele*, ed. Rudolf Laban (C. Reissner, 1934) 1/1/332, John Hodgson Collection, Special Collections, University of Leeds, Leeds, UK.
111. Hans Brandenburg, "Von Deutscher Tanzkunst: Rückblick und Ausblik," in *Deutsche Tanzfestspiele*, ed. Rudolf Laban (C. Reissner, 1934). 1/1/261, John Hodgson Collection, Special Collections, University of Leeds, Leeds, UK.
112. Laban also had a long relationship with the Wagner Festival at Bayreuth, including providing the choreography for the 1930 staging of *Tannhäuser*.
113. Herbert S. Levine, *Hitler's Free City: A History of the Nazi Party in Danzig, 1925–1939* (University of Chicago Press, 1973), 129.
114. Albrecht Knust and Rudolf Laban, June 23, 1935, 1/1/159, John Hodgson Collection, Special Collections, University of Leeds, Leeds, UK. Earlier movement choirs had not necessarily been separated by gender, but such practices became more common over time.
115. Marie Luise Lieschke to Otto von Keudell, October 25, 1935, Bundesarchiv 50.01 237, reproduced in Karina and Kant, *Hitler's Dancers*, 232.
116. Lieschke to von Keudell, October 25, 1935.
117. *Wir Tanzen* (Reichsbund für Gemeinschaftstanz in der Reichstheaterkammer, 1936).
118. *Wir Tanzen*.
119. *Wir Tanzen*.

120. Joseph Goebbels, "Die Tagebücher, Sämtliche Fragmente," June 21, 1936, quoted in Karina and Kant, *Hitler's Dancers*, 119.

121. Karina and Kant, *Hitler's Dancers*, 120. Laban's friend and collaborator, Felicia Sachs, however, notes that Laban himself greatly admired Riefenstahl's work. Felicia Sachs, interview by John Hodgson, December 30, 1974, 13n, 1/4/28, John Hodgson Collection, Special Collections, University of Leeds, Leeds, UK.

122. The following account of Laban's final months in Germany relies on the narrative provided in Karina and Kant, *Hitler's Dancers*, 124–35.

123. Rudolf von Laban to Marie Luise Lieschke, September 3, 1937, reproduced in Karina and Kant, *Hitler's Dancers*, 58.

124. Laban to Lieschke, September 3, 1937.

125. See, for example, Susan Au, "A Man of Movement: Rudolf Laban, 1879–1958," *Dance Magazine*, June 1979, 104.

126. Manning, *Ecstasy and the Demon*; Guilbert, *Danser avec le IIIe Reich*; Karina and Kant, *Hitler's Dancers*; Manning and Ruprecht, *New German Dance Studies*; and Susan Manning, "Modern Dance in the Third Reich, Redux," in *The Oxford Handbook of Dance and Politics*, ed. Rebekah J. Kowal, Gerald Siegmund, and Randy Martin (Oxford University Press, 2017), 395–416. See also Pamela M. Potter, *Art of Suppression: Confronting the Nazi Past in Histories of the Visual and Performing Arts* (University of California Press, 2016).

127. Felicia Sachs interview by John Hodgson.

128. Von Laban, *A Life for Dance*, 38.

129. Jeffrey Herf, *Reactionary Modernism: Technology, Culture, and Politics in Weimar and the Third Reich* (Cambridge University Press, 1984).

130. Rudolf Bach, "Vom Wesen der Gruppenregie," n.d., 1/1/357, John Hodgson Collection, Special Collections, University of Leeds, Leeds, UK.

131. Joseph Lewitan, "Der Tanz von Morgen," *Der Tanz* 3, no. 11 (November 1930): 2–3.

132. Rudolf von Laban, "Das Choreographische Institut Laban," 1930, 1/1/265, John Hodgson Collection, Special Collections, University of Leeds, Leeds, UK.

133. For the best-known critique of the Tiller Girls, see Siegfried Kracauer, *The Mass Ornament: Weimar Essays*, trans. Thomas Y. Levin (Harvard University Press, 1995).

134. For a fuller discussion of the potentially mystical references in the piece, see Julia A. Walker, "Performance and Modernity: Enacting Change on the Globalizing Stage," in *Performance and Modernity* (Cambridge University Press), 147–59. Carole Kew suggests that Goebbels may have also objected to the piece's depiction of defeated German soldiers; see "From Weimar Movement Choir to Nazi Community Dance," 82.

Chapter 2

1. Charles Chaplin, *My Autobiography* (New York: Simon and Schuster, 1964), 415.

2. *Modern Times* (United Artists, 1936).

3. Ellen Graff, *Stepping Left: Dance and Politics in New York City, 1928–1942* (Duke University Press, 1997); and Mark Franko, *The Work of Dance: Labor, Movement, and Identity in the 1930s* (Wesleyan University Press, 2002).

4. See Rabindranath Tagore and L. K. Elmhirst, *Rabindranath Tagore, Pioneer in Education: Essays and Exchanges between Rabindranath Tagore and L. K. Elmhirst* (John Murray, 1961); Leonard K. Elmhirst, *Poet and Plowman* (Visva-Bharati, 1975); Uma Das Gupta, "Tagore's Ideas

of Social Action and the Sriniketan Experiment of Rural Reconstruction, 1922–41," *University of Toronto Quarterly* 77, no. 4 (2008): 992–1004; and Krishna Dutta and Andrew Robinson, *Rabindranath Tagore: The Myriad-Minded Man* (Bloomsbury, 1995).

5. "Prospectus for Dartington Hall," 1925, as quoted in Victor Bonham-Carter, *Dartington Hall: The History of an Experiment* (Phoenix House, 1958), 47.

6. For a history of Dartington Hall's evolution from its founding through the late 1940s, see Anna Neima, "Dartington Hall and the Quest for 'Life in Its Completeness,' 1925–45," *History Workshop Journal* 88 (2019): 111–33.

7. Michael Young, *The Elmhirsts of Dartington* (Routledge & Kegan Paul, 1982), 226. See also Tagore and Elmhirst, *Rabindranath Tagore*, 101–11.

8. "Letter from Rudolf Laban to Felicia Sachs," May 12, 1941, L/E/52/58, Rudolf Laban Archive, University of Surrey. See also "The New Amongst the Old: Dartington Hall's Historic Setting for Experiments in Modern Living," *The Bystander*, July 24, 1940, 114.

9. Ann Hutchinson Guest, "The Jooss-Leeder School at Dartington Hall," *Dance Chronicle* 29 (2006): 168, 181. On the development of Laban's Choreutics and Eukinetics, see Maletic, *Body, Space, Expression*.

10. Rudolf Laban, "Dartington—How I See It," n.d., L/E/23/2 , Rudolf Laban Archive, University of Surrey.

11. Rudolf Laban, "Dartington Hall," n.d., L/E/20/31, Rudolf Laban Archive, University of Surrey.

12. Rudolf Laban to Leonard Elmhirst, March 10, 1939, L/E/45/16, Rudolf Laban Archive, University of Surrey.

13. Though in private Jooss never fully forgave Laban for his work with the Nazis, he remained a steadfast supporter in public.

14. "Claim Rhythm Helps Factory Workers," *Dundee Evening Telegraph*, October 3, 1942; and "Rhythmisation," *Birmingham Mail*, August 3, 1942. The press's tendency to cast Laban as a refugee continued for the remainder of his life and beyond. See, for example, Roland Hill, "Reconciliation after Three Generations," *The Times*, June 2, 1975.

15. See correspondence in L/E/50/50, Rudolf Laban Archive, University of Surrey.

16. "Letter from C. C. Martin to Rudolf Laban," August 12, 1941, L/E/50/50, Rudolf Laban Archive, University of Surrey.

17. "Letter from Rudolf Laban to C. C. Martin," September 20, 1941, L/E/50/50, Rudolf Laban Archive, University of Surrey.

18. As quoted in Trevor Boyns and John Richard Edwards, *A History of Management Accounting: The British Experience* (Routledge, 2013), 208.

19. In an effort to reflect this focus, the relevant professional organization eventually transitioned from referring to its members as "cost and works accountants" to "cost and management accountants."

20. "Business Efficiency Exhibition," n.d., Rudolf Laban Archive, University of Surrey.

21. Blayney, *Health and Efficiency*. For an overview of the literature on this subject, see Arthur J. McIvor, *A History of Work in Britain, 1880–1950* (Palgrave, 2001), 93–110. On the divergent reception of scientific management in the United States and the United Kingdom, see Craig R. Littler, *The Development of the Labour Process in Capitalist Societies* (Heinemann Educational Books, 1982). On scientific management in Europe and the United States more broadly, see, among others, Maier, "Between Taylorism and Technocracy"; Rabinbach, *The Human Motor*; Nikolas Rose, *Governing the Soul: The Shaping of the Private Self*, 2nd ed. (Free Association Books, 1999), 61–94.

22. Frank and Lillian Gilbreth were known primarily for their motion studies, in which they observed workers' movements using a variety of tools and prescribed the "one best way" of accomplishing any given task, usually by minimizing the total number of movements performed. Charles Bedaux and his followers claimed the ability to scientifically calculate the degree of fatigue caused by a given working action and, therefore, the time a worker would require for recovery. Bedaux used these calculations to determine how many such actions could expected of a laborer in a given length of time and to set pay rates and bonuses. The system was employed particularly widely in the United Kingdom. For a contemporary critique of the Bedaux system, see W. F. Watson, *The Work and Wage Incentives: The Bedaux and Other Systems* (Hogarth Press, 1934). On the relationship between the Gilbreths' movement study work and Taylor's time study program, see Price, "Frank and Lillian Gilbreth and the Motion Study Controversy."

23. McIvor, *A History of Work in Britain*, 166.

24. "Letter from F. C. Lawrence to Rudolf Laban," August 22, 1941, L/E/66/22, Rudolf Laban Archive, University of Surrey.

25. "Oral History with F. C. Lawrence, Conducted by John Hodgson," June 21, 1973, 1/4/10-/1/4/24, John Hodgson Collection, Special Collections, University of Leeds, Leeds, UK.

26. Sometimes the symbols on the second line were replaced by verbal descriptions; in later years, this line sometimes contained additional Effort notation. "Laban Lawrence Industrial Notation," January 1943, L/E/22/3, Rudolf Laban Archive, University of Surrey.

27. Rudolf Laban and F. C. Lawrence, "Laban Lawrence Industrial Rhythm and Lilt in Labour," 1942, 3, T/AD/3/D/6, The Dartington Hall Trust Records, Devon Heritage Center.

28. Laban and Lawrence, "Laban Lawrence Industrial Rhythm and Lilt in Labour," 4. See also Latour, "Drawing Things Together."

29. Laban and Lawrence, "Laban Lawrence Industrial Rhythm and Lilt in Labour," 5.

30. The particular terms used varied somewhat over time ("mass" sometimes appeared as "weight," for example), but the basic categories remained consistent.

31. Laban and Lawrence, "Laban Lawrence Industrial Rhythm and Lilt in Labour," 11.

32. "Laban Lawrence Industrial Notation."

33. Jean Newlove, "Log-Book, Visit to Dartington Estate," May 10, 1943, L/E/71/15, Rudolf Laban Archive, University of Surrey; and Jean Newlove, "Log-Book, Laban-Lawrence Industrial Rhythm (Mars)," April 1, 1943, L/E/71/15, Rudolf Laban Archive, University of Surrey.

34. Newlove, "Log-Book, Visit to Dartington Estate."

35. Laban and Lawrence, "Laban Lawrence Industrial Rhythm and Lilt in Labour," 14.

36. See, for example, F. C. Lawrence to R.L. Webster, "Letter to Hoover Limited," September 1, 1943, L/E/51/47, Rudolf Laban Archive, University of Surrey; and F. C. Lawrence to Ernest Bevin, "Letter to Minister of Labour and National Service," July 2, 1942, L/E/51/80, Rudolf Laban Archive, University of Surrey.

37. "Report upon Miss Newlove's Visit to Dartington Hall Ltd.," June 4, 1943, L/E/71/15, Rudolf Laban Archive, University of Surrey.

38. Rudolf Laban, "The Job Effort Graph and Its Application," n.d., L/E/62/27, Rudolf Laban Archive, University of Surrey; Rudolf Laban, "Some Simple Facts about Wage-Policies, Effort Measurement, Etc.," n.d., L/E/64/66, Rudolf Laban Archive, University of Surrey; and "Laban Lawrence Industrial Notation."

39. Indeed, in one letter to an official in the Ministry of Labour, Lawrence suggested that Industrial Notation might also be used in the "training of Time and Motion Study Men in more

intense observation and in better understanding of Movement and Work, which should increase the value of their capabilities; and in a new method of Payment by Results which we have evolved." The idea, presumably, was to draw on the existing pool of movement specialists while shifting their approach to the one Laban and Lawrence favored. See F. C. Lawrence to E. Watson Smyth, January 2, 1943, L/E/51/80, Rudolf Laban Archive, University of Surrey.

40. Laban and Lawrence, "Laban Lawrence Industrial Rhythm and Lilt in Labour," 16.

41. Rudolf Laban, "Rhythm of J. Lyons & Company., Greenford, Tea Factory," 1944, L/E/72/12, Rudolf Laban Archive, University of Surrey.

42. "Concerns Our Offer to Mars-Bar Limited," December 12, 1942, L/E/72/12, Rudolf Laban Archive, University of Surrey.

43. F. C. Lawrence, "Dance into Industry," n.d., L/E/ 65/13, Rudolf Laban Archive, University of Surrey.

44. Newlove, "Log-Book, Laban-Lawrence Industrial Rhythm (Mars)"; Newlove, "Log-Book, Visit to Dartington Estate"; "Tyresoles Ltd., Wembley. Log-Book.," April 13, 1942, 1/1/148, John Hodgson Collection, Special Collections, University of Leeds, Leeds, UK.

45. Rudolf Laban, "Industrial Kinetography," 1942, L/E/71/9, Rudolf Laban Archive, University of Surrey. On rhythm and industrial work, see also Blayney, *Health and Efficiency*, 22.

46. Laban, "Industrial Kinetography."

47. David Nye, *American Technological Sublime* (MIT Press, 1994).

48. Dinerstein, *Swinging the Machine*.

49. Danielle Robinson has traced an analogous process in the early twentieth-century United States, as modern dance professionals repackaged, codified, and commodified ragtime through dance manuals, diminishing the form's claims to "self-expression, spontaneity, and individuality," even as these qualities were used to market social dance to the public. Robinson further demonstrates how this "refinement" process was racialized, intended to turn a genre associated with Blackness into one suitable for White audiences and performers. See "The Ugly Duckling: The Refinement of Ragtime Dancing and the Mass Production and Marketing of Modern Social Dance," *Dance Research* 28, no. 2 (2010): 188. See also Danielle Robinson, *Modern Moves: Dancing Race During the Ragtime and Jazz Eras* (Oxford University Press, 2015).

50. Paton Lawrence & Company, "What We Would Like to Tell Your Workpeople About LILT IN LABOUR; Mars Confections Ltd.," January 25, 1943, L/E/72/14, Rudolf Laban Archive, University of Surrey.

51. "Rhythmisation."

52. Sidney E. Rolfe, "Manpower Allocation in Great Britain During World War II," *ILR Review* 5, no. 2 (1952): 173–94. At first, only women between the ages of twenty and thirty were eligible for conscription; this age limit was later increased. Limited exceptions were granted for mothers with particularly young children.

53. *New Statesman*, March 14, 1942, L/E/71/1, Rudolf Laban Archive, University of Surrey. In July 1942, Lawrence wrote a letter to Ernest Bevin himself arguing that LLIR could—among its other virtues—solve precisely this problem, allowing the "more rapid absorption of women into industry, and, in particular, the speedier employment of part time women." Lawrence to Bevin, "Letter to Minister of Labour and National Service," July 2, 1942.

54. Peggy's story, along with others, appeared in *War Factory*, a 1943 book authored by the Mass-Observation group under the directorship of Tom Harrison and intended to provide an on-the-ground sense of factory life during the period. See Tom Harrison, ed., *War Factory: A Report by Mass-Observation* (Victor Gollancz Ltd, 1943). On the mass observation movement, see

Boris Jardine, "Mass-Observation, Surrealist Sociology, and the Bathos of Paperwork," *History of the Human Sciences* 31, no. 5 (2018): 52–79.

55. Harrison, *War Factory*, 33.

56. Harrison, *War Factory*, 30.

57. "Industrial Welfare Work: New Profession Growing Up in War-Time Factories," *The Press and Journal/Aberdeen Journal*, October 6, 1943, 2.

58. Though many of the requests for personnel changes that were submitted to the National Service Offices were ultimately granted, the government made clear that companies that made too many such requests would not be looked upon kindly. As a result, requests for transfers, reassignments, or firings remained strikingly low. Nevertheless, as an earlier Mass Observation report noted, "many firms still almost ignore the rather elementary fact that their employees are human beings with human failings, and that the sanction which was so long the basis of industrial employment, namely the ever-pending moment of potential unemployment of the worker, has now disappeared." See *People in Production: An Enquiry into British War Production; A Report Prepared by Mass Observation for the Advertising Services Guild* (Penguin Books, 1942).

59. McIvor, *A History of Work in Britain*, 104.

60. See Blayney, *Health and Efficiency*; Rose, *Governing the Soul*, 61–94.

61. For another case linking the increasing employment of women to the intensification of scientific management see Margery Davies, *Woman's Place Is at the Typewriter* (Temple University Press, 1984).

62. "Laban Lawrence Industrial Notation."

63. Paton Lawrence & Company, "What We Would Like to Tell Your Workpeople About LILT IN LABOUR."

64. "Mars Confections, Ltd., General Notes," December 14, 1942, L/E/72/12, Rudolf Laban Archive, University of Surrey.

65. "Oral History with F. C. Lawrence, Conducted by John Hodgson."

66. "Mars Confections Limited, Record of Visit," November 12, 1942, L/E/72/12, Rudolf Laban Archive, University of Surrey.

67. Newlove, "Log-Book, Laban-Lawrence Industrial Rhythm (Mars)."

68. Paton Lawrence & Company, "What We Would Like to Tell Your Workpeople About LILT IN LABOUR."

69. Paton Lawrence & Company, "What We Would Like to Tell Your Workpeople About LILT IN LABOUR."

70. "Glaxo Laboratories Limited, Report of Observation of Operators Engaged on Hand Packing Units," January 1944, L/E/62/14, Rudolf Laban Archive, University of Surrey.

71. "St. Olave's Curing & Preserving Co. Ltd., Record of Visit," November 20, 1942, 1/1/149, John Hodgson Collection, Special Collections, University of Leeds, Leeds, UK.

72. "Report: J. Lyons," February 1944, L/E/55/01, Rudolf Laban Archive, University of Surrey.

73. The consultants were interviewed in Dick McCaw, *An Eye for Movement: Warren Lamb's Career in Movement Analysis* (Brechin Books Limited, 2006), 27.

74. Olive Moore, "Man of the Month: Rudolf Laban," *Scope: Magazine for Industry*, October 1954.

75. Rudolf Laban, "Training Schedule, The Laban-Lawrence Method of Industrial Rhythm, Tyresoles Ltd.," 1942, L/E/71/1, Rudolf Laban Archive, University of Surrey.

76. £450 in 1942 is equivalent to approximately £16,800 (or $20,500) in 2022.

NOTES TO PAGES 83-88

77. "Letter from F. C. Lawrence to P. G. Hamilton," March 7, 1942, 1/1/148, Special Collections, University of Leeds, Leeds, UK.

78. "Report Upon the Laban Lawrence Training Course at the Wembley Factory of Tyresoles Limited," May 15, 1942, 1/1/148, University of Leeds Special Collections.

79. "Report Upon the Laban Lawrence Training Course at the Wembley Factory of Tyresoles Limited."

80. "Letter from Lisa Ullmann to F. C. Lawrence," May 27, 1942, L/E/71/1, Rudolf Laban Archive, University of Surrey.

81. N. Hillson, "Wembley Weekly," April 22, 1942, L/E/71/1, Rudolf Laban Archive, University of Surrey.

82. "Letter from Lisa Ullmann to F. C. Lawrence."

83. "Letter from P. G. Hamilton to F. C. Lawrence," March 18, 1942, L/E/55/01, Rudolf Laban Archive, University of Surrey.

84. "Letter from F. C. Lawrence to Lisa Ullmann," March 20, 1942, 1/1/148, Special Collections, University of Leeds, Leeds, UK.

85. "Letter from F. C. Lawrence to P. G. Hamilton."

86. "Letter from F. C. Lawrence to Major Palmer," May 16, 1942, 1/1/148, Special Collections, University of Leeds, Leeds, UK; and "Letter from F. C. Lawrence to Lisa Ullmann."

87. Newlove, "Log-Book, Laban-Lawrence Industrial Rhythm (Mars)."

88. See McCaw, *An Eye for Movement*, 27.

89. Rudolf Laban and F. C. Lawrence, "Man and the Commonwealth," July 1945, L/E/46/37, Rudolf Laban Archive, University of Surrey.

90. Moore, "Man of the Month: Rudolf Laban."

91. Laban and Lawrence, "Man and the Commonwealth."

92. Rudolf Laban, "Letter to 'Friends' Re: Industrial Work," n.d., L/E/52/57, Rudolf Laban Archive, University of Surrey.

93. Rudolf Laban, "Lecture Notes, Manchester Association of Engineers," 1944, L/E/55/01, Rudolf Laban Archive, University of Surrey.

94. Rudolf Laban, "The Stage of Work," n.d., L/E/64/68, Rudolf Laban Archive, University of Surrey.

95. "Tyresoles: Report on Activities," March 1942, L/E/55/01, Rudolf Laban Archive, University of Surrey.

96. Rudolf Laban, *Gymnastik und Tanz* (Gerhard Stalling, 1926), 15, 128.

97. Laban and Lawrence, "Man and the Commonwealth."

98. Laban, *Die Welt des Tänzers*, 126. Translation drawn from "The World of the Dancer," John Hodgson Collection, University of Leeds Special Collections 1/1/36.

99. Laban, 122. Translation drawn from "The World of the Dancer," John Hodgson Collection, University of Leeds Special Collections 1/1/36.

100. This orientation toward embodied modes of experiencing divinity is consonant with Laban's long involvement in Rosicrucianism and Freemasonry. As Marion Kant discusses, Laban understood dance as central to establishing a new "religion of the act"; see "Laban's Secret Religion," 44.

101. Though Weber posited that, by the early twentieth century, the religious underpinnings of the Protestant Ethic had become largely vestigial, new research has brought fresh attention to the varied ways in which spiritual practice and industrial (and postindustrial) capitalism have continued to be codependent. See, for example: Bethany Moreton, *To Serve God and Walmart:*

The Making of Christian Free Enterprise (Harvard University Press, 2010); Richard Callahan, Jr., Kathryn Lofton, and Chad E. Seales, "Allegories of Progress: Industrial Religion in the United States," *Journal of the American Academy of Religion* 78, no. 1 (2010): 1–39; Kathryn Lofton, "The Spirit in the Cubicle: A Religious History of the American Office," in *Sensational Religion: Sensory Cultures in Material Practice*, ed. Sally M. Promey (Yale University Press, 2014), 135–58; and Fred Turner, "Millenarian Tinkering," *Technology and Culture* 59, no. 4 supplement (2018): S160–82. See also Max Weber, *The Protestant Ethic and the Spirit of Capitalism*, trans. Stephen Kalberg (Oxford University Press, 2010).

102. Laban, *Die Welt des Tänzers*, 122.

103. Rudolf Laban, "Working Rhythms and Their Observation and Notation," n.d., L/E/65/1, Rudolf Laban Archive, University of Surrey.

104. "Laban Lawrence Industrial Notation."

105. "Special Report on the Effort Situation in the Mill, J. Lyons and Company (Confidential)," March 27, 1944, L/E/72/7, Rudolf Laban Archive, University of Surrey.

106. Laban and Lawrence, "Man and the Commonwealth."

107. Rudolf Laban, "Human Movement in Work," 1944, L/E/55/01, Rudolf Laban Archive, University of Surrey.

108. For a history of the British labor movement written amid—and in response to—these conflicts, see E. J. Hobsbawm, "Trends in the British Labor Movement Since 1850," *Science and Society* 13, no. 4 (1949): 289–312.

109. Laban, "Industrial Kinetography."

110. "Derby and Joan," *Derby Daily Telegraph*, September 24, 1942.

111. Rudolf Laban, "Early Notes on Industry for Our Instructors!," n.d., L/E/64/52, Rudolf Laban Archive, University of Surrey.

112. Laban, "The Stage of Work."

113. "J. Lyons Tea: Notes on Incentives," 1943, L/E/55/01, Rudolf Laban Archive, University of Surrey.

114. Laban, "The Stage of Work."

115. "Special Report on the Effort Situation in the Mill, J. Lyons and Company (Confidential)."

116. "Special Report on the Effort Situation in the Mill, J. Lyons and Company (Confidential)."

117. "Special Report on the Effort Situation in the Mill, J. Lyons and Company (Confidential)."

118. For an account of similar strategic efforts to promote workers' sense of corporate belonging in the United States, see Ryan M. Acton, "The Search for Social Harmony at Harvard Business School, 1919–1942," *Modern Intellectual History* (2022): 1–27.

119. Leonard Elmhirst, "Faith and Works at Dartington," 1937, as quoted in Young, *The Elmhirsts of Dartington*, 216.

120. Dorothy was particularly taken by Aldous Huxley's thinking on the subject, frequently returning to a 1935 talk Huxley gave at one of Dartington's Sunday Evening Meetings on the subject of "Religion in the Modern World." See Young, *The Elmhirsts of Dartington*, 174.

121. Anson Rabinbach, "The Aesthetics of Production in the Third Reich," *Journal of Contemporary History* 11, no. 4 (1976): 43–74. Rabinbach also highlights the ways in which the Nazi program was itself influenced by the English garden city movement at the turn of the twentieth century and notes, in a further parallel to Industrial Rhythm, that "what had begun as a revolt against mechanization became, by 1936, an adornment of industrial production itself."

122. Lino Camprubí, *Engineers and the Making of the Francoist Regime* (MIT Press, 2014); and María González Pendás, "Modernity Consecrated: Architectural Discourse and the Catholic

Imagination in Franquista Spain," in *Modern Architecture and Religious Communities, 1850-1970*, ed. Kate Jordan and Ayla Lepine (Routledge, 2018), 30-48.

123. Rudolf Laban, *Modern Educational Dance*, ed. Lisa Ullmann, 2nd ed. (Frederick A. Praeger, 1968), 4.

124. Laban, *Modern Educational Dance*, 6. For further discussion of the ways in which Duncan herself grappled with the aesthetics and ethics of the machine age, as well an analysis of discourses surrounding technology, labor, and dance in the early twentieth century, see McCarren, *Dancing Machines*, 65-97, 129-58.

125. See, for example, "Notes from Central Europe," *The Dancing Times* (1928), 393-97; "Laban Appointed Ballet Master at Berlin," *The Dancing Times* (1930), 248-50.

126. Rosa Widman, "The Modern 'Absolute' Dance," *Journal of Physical Education and School Hygiene* 16, no. 48 (1924): 138-41. See also Vertinsky, "Movement Practices and Fascist Infections."

127. On the history of physical education, see John Welshman, "Physical Culture and Sport in Schools in England and Wales, 1900-1940," *The International Journal of the History of Sport* 15, no. 1 (1998): 54-75; Sheila Fletcher, *Women First: The Female Tradition in English Physical Education 1880-1980* (Athlone Press, 1984); David Kirk and Patricia Vertinsky, eds., *The Female Tradition in Physical Education: Women First Reconsidered* (Routledge, 2016); and Patricia Vertinsky and Sherry McCay, eds., *Disciplining Bodies in the Gymnasium: Memory, Monument and Modernism* (Routledge, 2004).

128. F. M. G. Willson, *In Just Order Move: The Progress of the Laban Centre for Movement and Dance, 1946-1996* (Athlone Press, 1997). Though Laban was always formally attached to these projects, Ullmann (along with a number of other female colleagues) was responsible for nearly all of the practical work required to create, teach, and popularize modern educational dance.

129. John Blackie, *Inside the Primary School* (Her Majesty's Stationery Office, 1967). See also Fletcher, *Women First*.

130. Alec Clegg, "Enormous Untapped Potential of the Open University," *The Times*, January 2, 1973.

131. Laban, *Modern Educational Dance*, 32.

132. Laban, *Modern Educational Dance*, 106.

133. Laban, *Modern Educational Dance*, 8.

134. Rudolf Laban, "Industrial Rhythms in Dance Education," n.d., L/E/31/39, Rudolf Laban Archive, University of Surrey.

135. B. N. Knapp, "Review of Effort: Economy of Human Movement (Second Edition)," *Ergonomics* 17, no. 4 (1974): 555-56.

136. Willson, *In Just Order Move*; Fletcher, *Women First*; and Vertinsky, "Movement Practices and Fascist Infections."

Chapter 3

1. John Martin, "Scriveners," *New York Times*, June 9, 1940.

2. Martin, "Scriveners."

3. Eve Gentry was professionally known as Henrietta Greenhood until 1945. Helen Priest Rogers and Ann Hutchinson Guest are also slightly later married names, though these names will be used throughout for the sake of clarity.

4. Martin, "Scriveners."

5. Martin, "Scriveners."

6. Martin, "Scriveners." In the 1940s and '50s, there were three major centers of Labanotation promotion and research, all led by powerful personalities: Albrecht Knust in Germany, Lisa Ullmann and Sigurd Leeder in Great Britain, and Ann Hutchinson Guest at the Dance Notation Bureau in New York City. They shared a basic approach but differed in certain matters of emphasis and orthography.

7. Martin, "Scriveners."

8. The Bureau's concern with standard-setting, first domestically and then internationally, is consistent with the increasing attention to standards displayed by many engineering bodies in the early twentieth century and post–World War II years. See JoAnne Yates and Craig N. Murphy, *Engineering Rules: Global Standard Setting Since 1880* (Johns Hopkins University Press, 2021).

9. John Martin, "Solid Progress of the Laban Method in Hands of Lively Local Bureau," *New York Times*, July 7, 1957.

10. "Memo to Members," n.d., Dance Notation Bureau. "Kinetographie" was also sometimes anglicized to "Kinetography" (as in the International Council of Kinetography Laban), though I have generally retained the original German form for the sake of clarity. This chapter often relies on documents I consulted at the Dance Notation Bureau's offices in New York City. The material is private and uncatalogued, and these sources are therefore cited by collection name only. My special thanks go to the Dance Notation Bureau for granting me access and to Mei-Chen Lu, Director of Library Services, for her assistance.

11. Jack Anderson, "Helen Priest Rogers, 85, Teacher and Dance Notation Authority," *New York Times*, March 9, 1999.

12. Notably, Martin's ideas about how dance worked on an audience were, much like Laban's, deeply rooted in the operations of the physical body. In his 1933 book, *The Modern Dance*, Martin argued that dance communicated through a kind of kinetic transfer between dancer and viewer: that as dancers moved on stage, spectators' muscles responded sympathetically, reactions which in turn produced psychological and emotional effects. John Martin, *The Modern Dance* (A. S. Barnes, 1933).

13. Martin, "Scriveners."

14. Neil Genzlinger, "Ann Hutchinson Guest, Who Fixed Dance on Paper, Dies at 103," *New York Times*, April 15, 2022.

15. Lynne Weber, "Ann Hutchinson Guest, Dancer, Notator, Founder, Educator, Innovator (1918–2022)," *Library News from the Dance Notation Bureau* 16, no. 3 (2022): 7.

16. "Letter from Rudolf Laban to Ann Hutchinson," December 18, 1945, Dance Notation Bureau.

17. "Letter from Rudolf Laban to Ann Hutchinson," May 19, 1951, Dance Notation Bureau.

18. "Memo to Members."

19. "Letter from Ann Hutchinson to Bobby," March 13, 1952, Dance Notation Bureau.

20. "Letter from Ann Hutchinson to Bobby."

21. Selma Jeanne Cohen, "Dance Notation Conversation: Facts on the Universal System of Recording Human Movement—Labanotation," 1955, Dance Notation Bureau.

22. Cohen, "Dance Notation Conversation."

23. "Symbols in the Orient," *Dance Notation Record* 9, no. 3/4 (1958): 29.

24. In essence, the DNB presented itself as a quintessentially Mertonian scientific institution, unsurprising at a moment when the sociologist's theory of scientific norms was at the

height of its public influence. See Robert K. Merton, "The Normative Structure of Science," in *The Sociology of Science: Theoretical and Empirical Investigations* (University of Chicago Press, 1973), 267–78.

25. "Letter from Ann Hutchinson to Bobby."

26. Cohen, "Dance Notation Conversation."

27. John Martin, "The Dance Is Attuned to the Machine," *New York Times*, February 24, 1929; and George Balanchine, "Recording the Ballet," *Dance Observer*, November 1950, 132.

28. Martin, "The Dance Is Attuned to the Machine."

29. Selma Jeanne Cohen, "Notation, Anyone? Dance Notation Bureau Celebrates its 20th Anniversary," *Dance Magazine*, July 1960, 45.

30. Martin, "Scriveners."

31. Lorraine Daston and Peter Galison, *Objectivity* (Zone Books, 2007).

32. McCaw, *The Laban Sourcebook*. See also chapter 1.

33. On the relationship between technology, gender, and professional identity formation, see, for example, Margarete Sandelowski, *Devices and Desires: Gender, Technology, and American Nursing* (University of North Carolina Press, 2000); and Ruth Oldenziel, *Making Technology Masculine: Men, Women, and Modern Machines in America, 1870–1945* (Amsterdam University Press, 2004). On female information workers more generally, see JoAnne Yates, *Control Through Communication: The Rise of System in American Management* (Johns Hopkins University Press, 1989); and Jennifer Light, "When Computers Were Women," *Technology and Culture* 40, no. 3 (1999): 455–583.

34. Earl Ubell, "Dance Notation Steps into a New Era," *New York Times*, October 24, 1976.

35. Some choreographers also developed idiosyncratic personal notation systems or recorded movement passages in words, though they were often indecipherable to other readers.

36. Guest, *Labanotation*, 7.

37. "Letter from Muriel Topaz to Charles B. Fahs, Rockefeller Foundation," March 25, 1958, Dance Notation Bureau.

38. Ann Hutchinson, "Letter to Members," c. 1950, Dance Notation Bureau.

39. Marshall McLuhan, *The Gutenberg Galaxy: The Making of Typographic Man* (University of Toronto Press, 1962); Walter Ong, *Orality and Literacy: The Technologizing of the Word*, 2nd ed. (Routledge, 2002); Jack Goody, *The Domestication of the Savage Mind* (Cambridge University Press, 1977); and Claude Levi-Strauss, *Tristes Tropiques* (Plon, 1955). To be fair, not all these thinkers saw this as an unqualified good.

40. Bernard Taper, "Choreographer—I," *The New Yorker*, April 16, 1960.

41. Hutchinson, "Letter to Members."

42. For more on the deep ties between modern concert dance and African American dance—and on the efforts to publicly distinguish the two—see Susan Manning, *Modern Dance, Negro Dance: Race in Motion* (University of Minnesota Press, 2006).

43. Susan Leigh Foster, "The Ballerina's Phallic Pointe," in *Corporealities: Dancing Knowledge, Culture and Power*, ed. Susan Leigh Foster (Routledge, 1996), 7.

44. Edward R. Murrow, *In Search of Light: The Broadcasts of Edward R. Murrow, 1938–1961*, ed. Edward Bliss (Da Capo Press, 1997), 102.

45. See Brett Spencer, "Rise of the Shadow Libraries: America's Quest to Save Its Information and Culture from Nuclear Destruction During the Cold War," *Information and Culture* 49, no. 2 (2014): 145–76; Fernando Vidal and Nélia Dias, *Endangerment, Biodiversity and Culture* (Routledge, 2016); Joanna Radin, *Life on Ice: A History of New Uses for Cold Blood* (University

of Chicago Press, 2017); and David Sepkoski, *Catastrophic Thinking: Extinction and the Value of Diversity from Darwin to the Anthropocene* (University of Chicago Press, 2020). The DNB would make a similar case for its relevance in the wake of the terrorist attacks on the United States on September 11, 2001. As that year's fall newsletter noted: "Our staff has responded with a renewed sense of mission. We have become more conscious than ever of the role we play in protecting this most fragile of arts. With your help we will carry forward this work, to keep the dance heritage intact for a future that now looks less secure." See "11 September 2001," *DNB Bulletin* 5, no. 1 (2001): 1, https://www.dancenotation.org/dnbulletin.

46. L.T., "Shorthand Is Used for Choreography," *The Evening Citizen*, March 20, 1948.

47. "Membership List, Dance Notation Bureau," 1967, Dance Notation Bureau. Hamilton was one of the commission members most distressed by plans to refuse medical treatment to Japanese survivors of the atomic bombing. See Susan Lindee, *Suffering Made Real: American Science and the Survivors at Hiroshima* (University of Chicago Press, 1997).

48. John Martin, "They Score a Dance as Others Do Music," *New York Times*, July 2, 1950.

49. John Martin, "Concerning Notation," *New York Times*, February 20, 1944.

50. Ann Hutchinson, "The Meaning of 'The Language of Dance,'" June 1958, Dance Notation Bureau.

51. In 1939, just before the bureau's work began, Kirstein wrote that while existing notation systems like Stepanov and Feuillet were "logically conceived and invitingly rendered, each equipped with provocative diagrams calculated to fascinate the speculative processes of a chess champion," they were also "all equally worthless" from a practical standpoint, "so difficult to decipher" on even a basic level that students were thrilled to be able to painstakingly make sense of even a short solo sequence. See Lincoln Kirstein, *Ballet Alphabet: A Primer for Laymen* (Kamin Publishers, 1939).

52. "Letter from Lincoln Kirstein to Lucy Venable," June 21, 1965, Dance Notation Bureau.

53. "Where Dance Notation Fails," *Dance News*, December 1943.

54. Cohen, "Dance Notation Conversation."

55. "Letter from Muriel Topaz to Charles B. Fahe, Rockefeller Foundation."

56. Taper, "Choreographer—I."

57. Anthea Kraut, *Choreographing Copyright: Race, Gender, and Intellectual Property Rights in American Dance* (Oxford University Press, 2015), 190–91; and Susan Leigh Foster, *Dances That Describe Themselves: The Improvised Choreography of Richard Bull* (Wesleyan University Press, 2002).

58. The DNB thus participated in a longer historical trend toward locating authorship in the single creative individual rather than in a more complex network of persons, materials, institutions, and cultural practices. See Martha Woodmansee, "The Genius and the Copyright: Economic and Legal Conditions of the Emergence of the 'Author,'" *Eighteenth-Century Studies* 17, no. 4 (1984): 425–48. In recent years, however, particularly in scientific fields, such assumptions are beginning to be questioned once again. See, for example, Mario Biagioli and Peter Galison, eds., *Scientific Authorship: Credit and Intellectual Property in Science* (New York: Routledge, 2003); Smith, *The Body of the Artisan*; and Blaise Cronin, "Collaboration in Art and in Science: Approaches to Attribution, Authorship, and Acknowledgment," *Information and Culture* 47, no. 1 (2012): 18–37.

59. "Toe Writing: Ballet Dancers Learn How to Put Muscles in Black and White," *Time Magazine*, April 11, 1941, Folder DNB_TM_28_894, NYPL for the Performing Arts, Dance Notation Bureau Collection.

60. Martin, "They Score a Dance as Others Do Music."
61. Martin, "Concerning Notation."
62. *Baker v. Selden*, 101 U.S. 99 (1879).
63. Tobi Tobias Interview with Hanya Holm, Audio, 1974, NYPL for the Performing Arts, Hanya Holm Papers.
64. Terese Sekora, "Dance Notation: A History of the Dance Notation Bureau, 1940–1952" (master's thesis, Texas Woman's University, 1979).
65. Nelson Landsdale, "Concerning the Copyrighting of Dances," *Dance Magazine*, June 1952.
66. Kraut suggests that an additional factor in the decision in favor of Holm was the distancing of musical theater dance from the racially marked genres of jazz and tap and its increasing association with the world of modern dance and ballet. At the same time, Broadway choreographers were eager to use copyright claims to gain status in an art form that seemed uneasily "middlebrow." See Kraut, *Choreographing Copyright*, especially the chapter "'High-Brow' Meets 'Low-Down.'"
67. John Martin, "Hanya Holm's Works Are First to Be Registered," *New York Times*, March 30, 1952.
68. One famous early attempt to copyright dance—Loie Fuller's 1892 application for protection for her signature "Serpentine Dance"—failed for this reason, though the case's outcome was also affected by racial and gendered dynamics. See Caroline Joan S. Picart, *Critical Race Theory and Copyright in American Dance: Whiteness as Status Property* (Palgrave Macmillan, 2013). Participatory social or "folk" dances also remained ineligible for copyright, a decision that drew in part on racialized stereotypes. Among others, see Kraut, *Choreographing Copyright*; Richard L. Schur, *Parodies of Ownership: Hip-Hop Aesthetics and Intellectual Property Law* (University of Michigan Press, 2009); Anthea Kraut, *Choreographing the Folk: The Dance Stagings of Zora Neale Hurston* (University of Minnesota Press, 2008); and Brenda Dixon Gottschild, *Digging the Africanist Presence in American Performance: Dance and Other Contexts* (Greenwood Press, 1996).
69. Borge Varmer, "Copyright Law Revision, Study No. 28: Copyright in Choreographic Works," Studies Prepared for the Subcommittee on Patents, Trademarks, and Copyrights (Washington, DC: Committee on the Judiciary, United States Senate, 1961), III.
70. Varmer, "Copyright Law Revision, Study No. 28: Copyright in Choreographic Works."
71. "Meeting of the Associate Members of the Dance Notation Bureau," October 10, 1948, DNB Series 9.2, Org. 2, Irmgard Bartenieff papers, Special Collections in Performing Arts, University of Maryland Libraries.
72. "Preserving the Art of Dance in Written Form for Posterity: The Dance Notation Bureau's Program and Future Plans," 1965, Dance Notation Bureau.
73. "Preserving the Art of Dance in Written Form for Posterity."
74. "LIMS Institute Programs/Policies," 1980, Org. 1, Series 9.3; Box 1, Folder 21, Irmgard Bartenieff papers, Special Collections in Performing Arts, University of Maryland Libraries.
75. Recent work on technology and art in the long 1960s suggests a larger context for why IBM might have been interested in working with the DNB. See, e.g., W. Patrick McCray, *Making Art Work: How Cold War Engineers and Artists Forged a New Creative Culture* (MIT Press, 2020); and Matthew Wisnioski, "Why MIT Institutionalized the Avant-Garde: Negotiating Aesthetic Virtue in the Postwar Defense Institute," *Configurations* 21, no. 1 (2013): 85–116.
76. "Lexington Engineers Relate Story Behind Unique Product," *IBM News*, 1973, Dance Notation Bureau.

77. Promotional Film, IBM Selectric Typewriter, 1961, https://www.youtube.com/watch?v=v NUEUth7qjc.

78. "Letter from Herbert Kummel to Dance Notation Bureau Members," October 26, 1971, Dance Notation Bureau.

79. George Gent, "Device Converts Typewriters for Notation of Choreography," *New York Times*, December 20, 1973.

80. Drid Williams, "A Note on Human Action and the Language Machine," *Dance Research Journal* 7, no. 1 (1974): 8–9. For an account of both the typewriter's symbolic weight as a modernizing, universalizing technology and the somewhat analogous challenges of adapting it to the Chinese language, see Thomas S. Mullaney, *The Chinese Typewriter: A History* (MIT Press, 2017).

81. Jo Floyd, *Manual for Use with the Labanotation-IBM Selectric Typewriter Element* (Dance Notation Bureau, Inc., 1974). A second, French-language manual was produced in Paris three years later; see Yvette Alagna, *Manuel d'Utilisation de La Sphere d'Impression IBM Labanotation (Cinétographie)* (Paris: Centre National d'Ecriture du Mouvement, 1977).

82. Floyd, *Manual for Use with the Labanotation-IBM Selectric Typewriter Element*, 11.

83. Floyd, *Manual for Use with the Labanotation-IBM Selectric Typewriter Element*, 29.

84. "Lexington Engineers Relate Story Behind Unique Product."

85. Floyd, *Manual for Use with the Labanotation-IBM Selectric Typewriter Element*, ii.

86. "Letter from Lucy Venable to Muriel Topaz," October 24, 1976, Dance Notation Bureau.

87. "Letter from Allan Miles to Muriel Topaz," October 26, 1976, Dance Notation Bureau.

88. In later years, however, some argued that computerizing Labanotation would provide an ideal opportunity to reevaluate its conventions, drawing on mathematicians and scientists' "unbiased impartial view" of the subject. See for example "Labanotation Computer Project," *Action! Recording! Newsletter from the Language of Dance Center*, January 1978, Dance Notation Bureau.

89. "Letter from Ann Hutchinson Guest to Muriel Topaz," October 19, 1976, Dance Notation Bureau.

90. On the development of computer graphics, see Jacob Gaboury, *Image Objects: An Archaeology of Computer Graphics* (MIT Press, 2021).

91. Maxine D. Brown and Stephen W. Smoliar, "A Graphics Editor for Labanotation," in *SIGGRAPH* (Philadelphia, 1976), 60–65; and Lynne Weber and Stephen W. Smoliar, "The Computer as a Tool for Labanotation," *Action! Recording! Newsletter from the Language of Dance Center*, September 1977, Dance Notation Bureau.

92. Brown and Smoliar, "A Graphics Editor for Labanotation," 63; Ubell, "Dance Notation Steps into a New Era"; and "Preface to the DNB Computer Development Program," n.d., Org 6, Series 9.3, Irmgard Bartenieff papers, Special Collections in Performing Arts, University of Maryland Libraries.

93. Ubell, "Dance Notation Steps into a New Era." These predictions were themselves rooted in faulty notions about the short-term effects of the invention of the printing press. See Scott D. N. Cook, "Technological Revolutions and the Gutenberg Myth," in *Internet Dreams: Archetypes, Myths, and Metaphors*, ed. Mark Stefik (MIT Press, 1997), 67–82.

94. "Preface to the DNB Computer Development Program."

95. See Ann Hutchinson Guest, "Captivating Computer," *Action! Recording! Newsletter from the Language of Dance Center*, September 1977, https://dnbtheorybb.blogspot.com/2016/01/action-recording.html.

96. Datapro Research Corporation, "UNIVAC Series 70," November 1972, http://www.bitsavers.org/pdf/univac/series_70/datapro/70C-877-21_7211_UNIVAC_Series_70.pdf.

97. "Letter from Herbert Kummel to Carol Kaplinski," October 12, 1977, Dance Notation Bureau. Less powerful than a mainframe, the most basic PDP-11 cost the equivalent of $80,000 when released in 1970.

98. Lucy Venable, "Labanwriter: There Had to Be a Better Way," *Dance Research* 9, no. 2 (1991): 78.

99. Venable, "Labanwriter"; and M. Howlett, R. Howlett, D. Miller, and W. Buckley, "Computerised Movement Notation—A Disc Programme for the BBC Model B Microcomputer and Dot Matrix Printer Using Kinetography Laban," *Action! Recording! Newsletter from the Language of Dance Center*, April 1986, 5.

100. Venable, "Labanwriter."

101. Elise Ivancich Dunin, "A Guide on How to Write Labanotation with the Apple Macintosh Computer Using LCs LN" (University of California, Los Angeles, 1987). See also D. Sealey, "NOTATE: Computerized Programs for Labanotation," *Journal for the Anthropological Study of Human Movement* 1, no. 2 (1980): 70–74; and Judith Allen, "Recording Movement in Labanotation on Computer: User's Manual" (University of Iowa, with the cooperation of the Computer Assisted Instructional Laboratory of the Weeg Computing Centre, October 1979). For discussion of additional programming developments during this period, see János Fügedi, "Dance Notation and Computers," *Yearbook for Traditional Music* 23 (1991): 101–11; Venable, "Labanwriter."

102. Lucy Venable and George Karl, *Manual for the LabanWriter Program* (Department of Dance, College of the Arts, The Ohio State University, 1987), Dance Notation Bureau.

103. "Dance on a Macintosh," *Arts Advocate: The Alumni and Friends Newsletter of the Ohio State University College of the Arts*, Spring 1990, Dance Notation Bureau.

104. Billie Mahoney, "How I Got Hooked on Labanotation (Part Two)," *Library News from the Dance Notation Bureau*, Summer 2016, 4, https://www.dancenotation.org/dnb-library-news.

105. Ubell, "Dance Notation Steps into a New Era."

106. Stephen Strauss, "Computer to Capture Dance Before Choreography Is Lost," *The Globe and Mail*, November 26, 1982.

107. Joseph Menosky, "Video Graphics & Grand Jetés," *Science 82*, May 1982.

108. Strauss, "Computer to Capture Dance Before Choreography Is Lost."

109. Susan Leigh Foster, *Choreographing Empathy: Kinesthesia in Performance* (Routledge, 2011), 44. For further discussion of the historiography of dance authorship and its entanglement with issues of race, gender, commercialization, and changing aesthetic values see also notes 57, 58, 66, and 68 above.

110. See, for example, Hilary Rose, "Hand, Brain, and Heart: A Feminist Epistemology for the Natural Sciences," *Signs* 9, no. 1 (1983): 73–90.

111. See John Cage and Alison Knowles, *Notations* (Something Else Press, 1969). This is not to say that Cage's work was apolitical or lacked its own methods of control. On this subject see Fred Turner, *The Democratic Surround: Multimedia and American Liberalism from World War II to the Psychedelic Sixties* (University of Chicago Press, 2013); Alexandre Popoff, "John Cage's Number Pieces: The Meta-Structure of Time-Brackets and the Notion of Time," *Perspectives of New Music* 48, no. 1 (2010): 65–82; and James Pritchett, *The Music of John Cage* (Cambridge University Press, 1996).

112. Claude Shannon, "Communication in the Presence of Noise," *Proceedings of the IRE* 37, no. 1 (1949): 10–21.

113. Hutchinson Guest, "The Jooss-Leeder School at Dartington Hall."

114. Dance Notation Bureau, "Guidelines for Checkers," 1976, Dance Notation Bureau.

115. See, for example, Frazier, "Labanotation Is Creative"; Davis, Carter, and Koff, "Troubling the Frame"; and Valarie Williams, "Writing Dance: Reflexive Processes-at-Work Notating New Choreography," *Journal of Movement Arts Literacy* 4, no. 1 (2018): Article 7. See also Daston and Galison, *Objectivity*.

116. "Letter from Lucy Venable to Lincoln Kirstein," June 17, 1965, Dance Notation Bureau.

117. "Letter from Ann Hutchinson Guest to Irmgard Bartenieff," April 22, 1976, Personal Correspondence, Series 2.1, Corr 3, Irmgard Bartenieff papers, Special Collections in Performing Arts, University of Maryland Libraries.

118. Latour, "Drawing Things Together," 15.

Chapter 4

1. Sloan Wilson, *The Man in the Gray Flannel Suit* (Simon and Schuster, 1955), 12.

2. Barbara Merlin, "Man in the Gray Flannel Suit Is Uncannily Familiar," *Los Angeles Times*, August 7, 1955.

3. Orville Prescott, "Books of The Times," *New York Times*, July 18, 1955.

4. William H. Whyte, Jr., *The Organization Man* (Simon and Schuster, 1956), 173.

5. Whyte, *The Organization Man*, 174.

6. For the purposes of clarity, this chapter will generally refer to the technique as "Aptitude Assessment," as this was the name used for the majority of the period covered here.

7. McCaw, *An Eye for Movement*, 21.

8. See Robert Leach, *Theatre Workshop: Joan Littlewood and the Making of Modern British Theatre* (University of Exeter Press, 2006).

9. Leach, *Theatre Workshop*, 79, 86.

10. Joan Littlewood, *Joan's Book: Joan Littlewood's Peculiar History as She Tells It* (Methuen, 1994), 773.

11. McCaw, *An Eye for Movement*, 18.

12. Laban, "The Job Effort Graph and Its Application."

13. "The Laban Lawrence Test for Selection and Placing," n.d., L/E/64/89, Rudolf Laban Archive, University of Surrey.

14. "Laban Lawrence Youth Advice Bureau," n.d., L/E/46/9, Rudolf Laban Archive, University of Surrey. The program ultimately failed because its initial clients were not impressed with the team's assessments; Laban worried that the predictive power for children was not as robust as it was for adults.

15. See, for example, Merve Emre, *The Personality Brokers: The Strange History of Myers-Briggs and the Birth of Personality Testing* (Random House, 2018); Matthew Hoffarth, "Building the Hive: Corporate Personality Testing, Self-Development, and Humanistic Management in Postwar America, 1945–2000" (PhD diss., University of Pennsylvania, 2018); Nadine Weidman, "Between the Counterculture and the Corporation: Abraham Maslow and Humanistic Psychology in the 1960s," in *Groovy Science: Knowledge, Innovation, and American Counterculture*, ed. David Kaiser and W. Patrick McCray (University of Chicago Press, 2016); and Kira Lussier, "Temperamental Workers: Psychology, Business, and the Humm-Wadsworth Temperament Scale in Interwar America," *History of Psychology* 21, no. 2 (2018). For earlier predecessors, see Jeremy Blatter, "Screening the Psychological Laboratory: Hugo Münsterberg, Psychotechnics, and the Cinema, 1892–1916," *Science in Context* 28, no. 1 (2015): 53–76.

16. Though the use of the lie detector only sometimes drew on the same rhetoric of self-development as the growing phalanx of personality and aptitude tests—it was more often

deployed by American corporations and government agencies to ferret out everything from petty theft to communist sympathies and homosexuality—it too was used as a mass screening device that promised to reveal the an individual's true thoughts through telltale bodily clues. See Ken Alder, *The Lie Detectors: The History of an American Obsession* (Free Press, 2007); and Michael Pettit, *The Science of Deception: Psychology and Commerce in America* (University of Chicago Press, 2012), 194–227.

17. Nick Tiratsoo and Jim Tomlinson, "Exporting the 'Gospel of Productivity': United States Technical Assistance and British Industry 1945–1960," *The Business History Review* 71, no. 1 (1997): 41–81; Nick Tiratsoo, "The 'Americanization' of Management Education in Britain," *Journal of Management Inquiry* 13, no. 2 (2004): 118–26; and Matthias Kipping and Ove Bjarnar, eds., *The Americanisation of European Business: The Marshall Plan and the Transfer of U.S. Management Models to Europe* (Routledge, 1998).

18. On the new scholarly interest in nonverbal behavior see, for example, Seth Barry Watter, "Scrutinizing: Film and the Microanalysis of Behavior," *Grey Room* 66 (2017): 32–69; Wendy Leeds-Hurwitz, "Notes in the History of Intercultural Communication: The Foreign Service Institute and the Mandate for Intercultural Training," *Quarterly Journal of Speech* 76 (1990): 262–81; James McElvenny and Andrea Ploder, eds., *Holisms of Communication: The Early History of Audio-Visual Sequence Analysis* (Language Science Press, 2021); Steve J. Heims, *The Cybernetics Group* (MIT Press, 1991); Valerie Manusov, "A History of Research on Nonverbal Communication: Our Divergent Pasts and Their Contemporary Legacies," in *APA Handbook of Nonverbal Communication*, ed. David Matsumoto, Hyisung C. Hwang, and Mark G. Frank (American Psychological Association, 2016), 3–15; and Martha Davis, *Understanding Body Movement: An Annotated Bibliography* (Indiana University Press, 1972).

19. Lamb did acknowledge that some movements—specific hand gestures, for example— were culturally specific, but argued that they were confined to the superficial realm of "gesture" and did not affect the underlying grammar of "posture."

20. "Letter from Ann Hutchinson to Bobby."

21. McCaw, *An Eye for Movement*, 137.

22. Warren Lamb, *Posture and Gesture: An Introduction to the Study of Physical Behaviour* (Gerald Duckworth & Co., 1965); Warren Lamb and David Turner, *Management Behaviour* (Gerald Duckworth & Co., 1969); and Warren Lamb and Elizabeth Watson, *Body Code: The Meaning in Movement* (Routledge, 1979). As Lamb described it, the analytical categories described in the following paragraphs evolved over time, fully coalescing in the early 1960s. See McCaw, *An Eye for Movement*, 131.

23. Lamb and Turner, *Management Behaviour*, 59.

24. Lamb and Turner, *Management Behaviour*, 59.

25. Lamb, *Posture and Gesture*, 63.

26. Lamb, *Posture and Gesture*, 15.

27. Warren Lamb, "Points Affecting Association of P.L.C. and CO. with W.L.," December 1, 1962, Warren Lamb Archive, University of Surrey. The Warren Lamb Archive was not processed or catalogued during my visits to the collection, so these sources are generally cited by archive name only. The cataloging process remains incomplete, but current box and folder numbers have been provided when available.

28. Warren Lamb, "Proposal for One-Day Appreciation Course in Aptitude Assessment for Personnel Administration Ltd.," October 1, 1963, Warren Lamb Archive, University of Surrey.

29. "Movement Betrays You," *Mainly for Women* (BBC Television, October 13, 1959).

30. Warren Lamb, "The Development of Action Profiling," n.d., Warren Lamb Archive, University of Surrey.

31. "Letter from Warren Lamb to Judith Kestenberg," June 1, 1966, Warren Lamb Archive, University of Surrey.

32. "Inhibiting," *Financial Times*, January 27, 1966; Harry Weaver, "Do You Give Yourself Away as You Wave Goodbye?," *Daily Mail*, June 29, 1966; Stephen Aris, "Mr. Lamb's Body Semantics for the Ideal Executive," *Sunday Times*, n.d.; "Appearances Are Not Deceptive," *Business*, n.d.; "Appearance Counts," *The Director*, n.d.; Special Correspondent, "Executive Counselling for Top Company Men," January 15, 1965; Tony Clifton, "Right Arm for the Job," *The Sunday Times*, November 16, 1969. These articles are available in L/E/64/61, Rudolf Laban Papers, University of Surrey.

33. "The Work of Warren Lamb-B.B.C. TV's 'Tomorrow's World,'" *The Laban Art of Movement Guild Magazine*, no. 39 (November 1967): 48-50, https://labanguildinternational.org.uk/wp-content/uploads/2021/09/LabanMagNo39Nov1967.pdf. Archival copies of the *Laban Art of Movement Guild Magazine* are also available in the Special Collections of the New York Public Library for the Performing Arts.

34. "The Work of Warren Lamb-B.B.C. TV's 'Tomorrow's World,'" and "Warren Lamb Associates," n.d., Org 6, Series 9.2, Irmgard Bartenieff papers, Special Collections in Performing Arts, University of Maryland Libraries.

35. A similar technique was employed in the earlier marketing of the Laban-Lawrence Test; see "The Laban Lawrence Test for Selection and Placing."

36. As Merve Emre has discussed in her study of the Meyers-Briggs Type Indicator, this lack of validation was very much the norm for personality and aptitude testing of the era; Nadine Weidman similarly chronicles Abraham Maslow's argument that humanistic psychology might require a new kind of scientific method. See Emre, *The Personality Brokers*; Weidman, "Between the Counterculture and the Corporation."

37. "Appearances Are Not Deceptive." See also McCaw, *An Eye for Movement*, 195.

38. See, for example, Deborah Du Nann Winter and Ellen Goldman, "Molecular Study of Action Profiling," Action Profilers International Conference, Antwerp, Belgium, 1987, Warren Lamb Archive, University of Surrey. See also Timothy J. Colton, "Nonverbal Behavior and Political Leadership: Methodological Issues and Research Frontiers," October 2010, Warren Lamb Archive, University of Surrey. As Colton discusses, studies of inter-rater reliability produced somewhat more promising results than the more minimal research on the system's validity.

39. As Michael Gordin has discussed, "counter-establishment sciences" often seek to replicate mainstream science's formal structures—its institutes, journals, conferences, credentialing procedures, and modes of presentation—without embracing its underlying methodologies. See *On the Fringe: Where Science Meets Pseudoscience* (Oxford University Press, 2021), 43.

40. See, for example, Aris, "Mr. Lamb's Body Semantics for the Ideal Executive."

41. "Letter from C. D. Ellis to G. H. Ladhams," December 20, 1950, L/E/62/1, Rudolf Laban Archive, University of Surrey; Iain Murray, "If It Wiggles, Give It a Job," *The Business Observer*, May 9, 1971; and "Appearances Are Not Deceptive."

42. Lamb, *Posture and Gesture*, 11.

43. "Letter from Warren Lamb to Judith Kestenberg," March 22, 1967, Warren Lamb Archive, University of Surrey.

44. Lamb and Turner, *Management Behaviour*, 1969, 43; and "Warren Lamb Associates."

45. Murray, "If It Wiggles, Give It a Job."

46. See, for example, Executive Search Ltd, "Aptitude Assessment for Group Marketing Executive," March 16, 1966, Warren Lamb Archive, University of Surrey; and Special Correspondent, "Executive Counselling for Top Company Men."

47. Warren Lamb, "Aptitude Assessment for Guidance on Placing and Development," December 4, 1963, Warren Lamb Archive, University of Surrey.

48. In the early 1960s, Lamb noted that a number of firms were interested in adding movement analysis to their existing contracts with the "Big Four" consulting companies. See Lamb, "Points Affecting Association of P.L.C. and CO. with W.L." On the context, see Michael Ferguson, *The Rise of Management Consulting in Britain* (Ashgate, 2002).

49. Moore, "Man of the Month: Rudolf Laban."

50. Weaver, "Do You Give Yourself Away as You Wave Goodbye?"

51. Warren Lamb, "Aptitude Assessment for General Manager (Air Freight)," October 8, 1963, Warren Lamb Archive, University of Surrey.

52. Warren Lamb, "Your Movements Are Revealing," n.d., Warren Lamb Archive, University of Surrey.

53. On the history of deception in the design of personality and character assessments, see Pettit, *The Science of Deception*, 194–227.

54. Davies, *Beyond Dance*.

55. Michael P. Hornsby-Smith, "Student Experiences in Industry," *The Vocational Aspect of Education* 23, no. 55 (1971): 81.

56. Committee on Manpower Resources for Science and Technology, "The Brain Drain: Report of the Working Group on Migration (Jones Report)," 1967, quoted in M. P. Hornsby-Smith, "Our Industrial Image," *Management Decision* 4, no. 2 (1970): 15.

57. Cambridge University Management Group, *Attitudes to Industry: A Survey of Undergraduate Attitudes to Industry by the Cambridge University Management Group*, BIM Occasional Paper (BKT City Print Ltd., 1969), 10; and Hornsby-Smith, "Student Experiences in Industry."

58. Cyril Sofer, *Students and Industry: Essays by Members of the Cambridge Balance Group* (W. Heffer & Sons Limited, 1966), 7.

59. Sofer, *Students and Industry*, 30.

60. Sofer, *Students and Industry*, 31.

61. Tiratsoo, "The 'Americanization' of Management Education in Britain," 121.

62. Tiratsoo, "The 'Americanization' of Management Education in Britain," 121.

63. Lamb and Turner, *Management Behaviour*, 130.

64. Lamb and Turner, *Management Behaviour*, 133.

65. Lamb and Turner, *Management Behaviour*, 82.

66. Warren Lamb, "The Tension of Failure: An Approach through the Analysis of Individual Movement" (Association of Industrial Medical Officers (London Group), November 18, 1963), WL/3/80, Warren Lamb Archive, University of Surrey.

67. Lamb and Turner, *Management Behaviour*, 134.

68. Lamb and Turner, *Management Behaviour*, 144.

69. Lamb and Turner, *Management Behaviour*, vii.

70. Lamb and Turner, *Management Behaviour*, 138.

71. Lamb, *Posture and Gesture*, 71.

72. Lamb, *Posture and Gesture*, 79.

73. Lamb, *Posture and Gesture*, 79. See also "The Puppets Have Taken Over" (BBC Home Service, April 13, 1967).

74. "Letter from Warren Lamb to Irmgard Bartenieff," December 5, 1963, Warren Lamb Archive, University of Surrey.

75. Lamb, "The Development of Action Profiling."

76. "Letter from Warren Lamb to Irmgard Bartenieff," December 5, 1963.

77. Warren Lamb, "Inspired Managers," *The Financial Times*, April 12, 1966.

78. Lamb and Turner, *Management Behaviour*, 155.

79. Peter Chambers, "Matching Personalities to Create Effective Teams," *International Management*, November 1973.

80. Executive Search Ltd, "Aptitude Assessment for Group Marketing Executive."

81. Warren Lamb Archive, University of Surrey.

82. Lamb and Turner, *Management Behaviour*, 141.

83. Lamb and Turner, *Management Behaviour*, 142.

84. For a broader history of the rise of temporary and contingent forms of employment, see Louis Hyman, *Temp: How American Work, American Business, and the American Dream Became Temporary* (Viking, 2018).

85. "Letter from Jeremy Hemming to Warren Lamb," October 17, 1966. Warren Lamb Archive, University of Surrey.

86. "Letter from Warren Lamb to Jeremy Hemming," October 24, 1966, Warren Lamb Archive, University of Surrey.

87. "Letter from Jeremy Hemming to Warren Lamb," October 17, 1966.

88. "Inhibiting."

89. Tony Clifton, "Right Arm for the Job," *Sunday Times*, November 16, 1969, L/E/64/61, Rudolf Laban Archive, University of Surrey.

90. Clifton, "Right Arm for the Job."

91. Weaver, "Do You Give Yourself Away as You Wave Goodbye?"

92. Fernau Hall, "Review of Posture and Gesture," *Ballet Today*, 1965, Warren Lamb Archive, University of Surrey.

93. "Letter from Warren Lamb to Miss Herf," *Ballet Today*, 1965, Warren Lamb Archive, University of Surrey.

94. Lamb, *Posture and Gesture*, 115.

95. Lamb and Turner, *Management Behaviour*, 33; "Appearance Counts."

96. Lamb, "Aptitude Assessment for Guidance on Placing and Development."

97. Lamb, *Posture and Gesture*, 177.

98. Lamb, *Posture and Gesture*, 183.

99. He also suggested turning one's newfound powers of observation on others. For academics, for example, he suggested that conferences were a prime opportunity to practice movement observation, "especially when a lecture is boring." See Lamb, *Posture and Gesture*, 26.

100. Lamb, *Posture and Gesture*, 128.

101. Lamb, *Posture and Gesture*, 34.

102. Lamb, *Posture and Gesture*, 144.

103. Lamb and Watson, *Body Code*.

104. Watson, the co-director of an antiquarian book dealership and a former teacher and social worker, met Lamb through her husband, one of Lamb's consulting clients. See "'Body Code': Lymington Author Writes on Human Behaviour," *Lymington Times*, September 22, 1979; and McCaw, *An Eye for Movement*, 10.

105. "Forthcoming from Routledge & Kegan Paul: Body Code," 1979, Warren Lamb Archive, University of Surrey.

106. Lamb and Watson, *Body Code*, 153.

107. Sara Barrett, "This Body Language Goes a Touch Too Far," *Daily Mail*, April 1985, Warren Lamb Archive, University of Surrey. An early review of *Posture and Gesture* did indeed suggest that Aptitude Assessment could be used by undecided voters to evaluate candidates for local political office: see "Judge Your Candidate by His A.Q.," *Evening Standard*, 1965.

108. Barrett, "This Body Language Goes a Touch Too Far."

109. Pamela Ramsden, *Top Team Planning: A Study of the Power of Individual Motivation in Management* (Wiley, 1973); Pamela Ramsden and Jody Zacharias, *Action Profiling: Generating Competitive Edge through Realizing Management Potential* (Gower Press, 1993); Carol-Lynne Moore, *Executives in Action: A Guide to Balanced Decision-Making in Management* (Macdonald and Evans, 1982); and Carol-Lynne Moore, *Movement and Making Decisions: The Body-Mind Connection in the Workplace* (Dance & Movement Press, 1995).

110. In addition to sponsoring annual conferences and workshops, one of Action Profilers International's main tasks was to accredit analysts, set minimum consulting rates, and ward off imitators who might dilute or damage the brand. Exactly how and why movement analysis took hold in South Africa is unclear, though its most visible promoter was Erik Schmikl, a consultant and faculty member at the University of South Africa's Graduate School of Business Leadership.

111. The James Thornhill Consultancy to S. P. Hayklan, February 2, 1987, Warren Lamb Archive, University of Surrey.

112. "News from North America," *Action News: Action Profilers International*, May 1990, 3, Warren Lamb Archive, University of Surrey.

113. Davies, *Beyond Dance*, 121. Davies quotes Lamb at length, drawing both from his previous publications and her own interviews with him.

114. Lamb provided little evidence for these assertions beyond his own observations. Even a sympathetic 2018 blog post from the MoveScape Center acknowledged that "Lamb claimed to have hard evidence for the flow patterns he observed, but sadly, his notations have been lost." See Carol-Lynne Moore, "Holiday Challenge for Movement Analysts," *MoveScape Center*, November 20, 2018, https://movescapecenter.com/holiday-challenge-for-movement-analysts.

115. Davies, *Beyond Dance*, 125.

116. Davies, *Beyond Dance*, 117.

117. Davies, *Beyond Dance*, 130.

118. Lamb, *Posture and Gesture*, 41, 28.

119. Lamb and Watson, *Body Code*, 57.

120. Davies, *Beyond Dance*, 117, 125.

121. Judith Butler, *Gender Trouble: Feminism and the Subversion of Identity* (Routledge, 1990).

122. Warren Lamb, "To Make a Gesture," *Twentieth Century* CLXXVII, no. 1037 (1968): 30.

123. The influence of Lamb's personal biases is particularly evident here, as a strict reading of his own analytic framework would have required an analyst to disregard the "sitting" aspect of the protest as merely a culturally conditioned gesture rather than a more meaningful posture.

124. AT&T had been interested in the system since at least 1978, when they proposed to hire two analysts to offer six months of weekly movement mini-courses during the company's management retreats in Buck Hill Falls, Pennsylvania. See "New Program Proposal: AT&T with LIMS," 1978, Org. 1, Series 9.1, Irmgard Bartenieff papers, Special Collections in Performing Arts, University of Maryland Libraries.

125. Jane Maloney, "Further Tributes to Warren Lamb from the Action Profile Community," *Moving On* 13, no. 1 & 2 (2016): 8.

126. Charlotte Honda, "Standards and Qualification Material, Action Profilers International," May 1, 1990, Warren Lamb Archive, University of Surrey.

127. There was some discussion of how the field itself might be diversified in the early 1990s—for example, by offering a training scholarship to a person of color—but little seems to have come of it. See "Letter from Ellen Goldman to Erik Schmikl," July 31, 1990, Warren Lamb Archive, University of Surrey.

128. "API European Executive Hearts Meeting," February 25, 1987, Warren Lamb Archive, University of Surrey.

129. "Points of Interest," *Action News: Action Profilers International*, November 1983, Warren Lamb Archive, University of Surrey.

130. C. Wright Mills, *White Collar: The American Middle Classes*, 50th anniversary ed. (Oxford University Press, 2002 [1956]), xvii.

131. More recently, Arlie Hochschild has remarked upon similar phenomena; see *The Managed Heart: The Commercialization of Human Feeling* (University of California Press, 1979).

132. Luc Boltanski and Ève Chiapello, *The New Spirit of Capitalism*, trans. Gregory Elliott (Verso, 2005). See also Hoffarth, "Building the Hive"; Rose, *Governing the Soul*, 103–19.

133. Rakesh Khurana, *Searching for a Corporate Savior: The Irrational Quest for Charismatic CEOs* (Princeton University Press, 2002), 69; and Rakesh Khurana, *From Higher Aims to Hired Hands: The Social Transformation of American Business Schools and the Unfulfilled Promise of Management as a Profession* (Princeton University Press, 2007).

134. Warren Lamb Associates, "Proposal: Incorporating Non-Verbal Behavioural Factors into Recruitment and Training of Patrolmen," February 1974, Warren Lamb Archive, University of Surrey; "Letter from Warren Lamb to Officer James Head," February 12, 1974, Warren Lamb Archive, University of Surrey.

135. Warren Lamb Associates, "Proposal: Incorporating Non-Verbal Behavioural Factors into Recruitment and Training of Patrolmen."

136. Ralph A. Vignola and Warren Lamb, January 7, 1974, Warren Lamb Archive, University of Surrey. At least a few members of the South African Defense Force also received Action Profile training in the mid-1980s; see "API European Executive Hearts Meeting."

137. See, for example, Martha Davis et al., "Defensive Demeanor Profiles," *American Journal of Dance Therapy* 22, no. 2 (2000): 103–21; and Sandra Adiarte, "Movement Analysis in Forensics–An Interdisciplinary Approach" (conference paper, Academy of Criminalistic and Police Studies, Belgrade, October 2-3, 2018), https://jakov.kpu.edu.rs/bitstream/id/2232/Major. The Davis study was conducted with the support of the John Jay College of Criminal Justice, several New York district attorneys' offices, and members of the New York City Police Department, including a former head of the Criminal Assessment and Profiling Unit. Stan Walters, one of the Davis study's coauthors, has also had a continuing career as a "kinesic" consultant; his client lists include the NSA, the FBI, the US Army, Navy, and Air Force, ICE, CPB, and local police departments across the United States, as well as government and military agencies in Singapore, Canada, and the UAE.

138. See, for example, ongoing criticisms as in D. J. A. Edwards, "Book Reviews: Body Code," *South African Journal of Psychology* 11, no. 4 (1981): 158–59.

139. See, for example, Brenda Connors, "A Technical Report on the Nature of Movement Patterning, The Brain, and Decision Making, the President of Russia, Vladimir Putin" (Office of Net Assessment, Office of the Secretary of Defense, January 2008); Brenda Connors, "The Russian Leadership Tandem in Interaction: Insights from Movement Analysis" (Office of Net Assessment, Office of the Secretary of Defense, January 2011); Brenda Connors, "See a Reflection

of Saddam's Biography," *Naval War College Review* 57, no. 1 (2004): 113–16; Brenda Connors, Richard Rende, and Timothy Colton, "Decision-Making Style in Leaders: Uncovering Cognitive Motivation Using Signature Movement Patterns," *International Journal of Psychological Studies* 7, no. 2 (2015): 105–12; and Brenda L. Connors, Richard Rende, and Timothy J. Colton, "Predicting Individual Differences in Decision-Making Process from Signature Movement Styles: An Illustrative Study of Leaders," *Frontiers in Psychology* 4, Article 658 (2013).

140. Connors, "The Russian Leadership Tandem in Interaction," 2.

141. Brenda Connors, "No Leader Is Ever Off Stage: Behavioral Analysis of Leadership," *Joint Force Quarterly* 43 (2006): 85.

142. Ray Locker, "Pentagon Studies Putin Body Language for Hint of Intent," *USA Today*, March 6, 2014; Ray Locker, "Pentagon 2008 Study Claims Putin Has Asperger's Syndrome," *USA Today*, February 4, 2015.

143. Elizabeth F. Ralph, "The Pentagon's Secret Putin Diagnosis: What the World's Most Powerful Military Learned from Watching TV," *Politico*, February 5, 2015. Interest in Body Leads was fueled in part by the fact that Putin's Russia was in the middle of its invasion of Ukraine and annexation of Crimea.

144. Julian E. Barnes, "Pentagon Shrugs at Its Own Study of Putin's Body Language," *Wall Street Journal*, March 7, 2014.

145. Robert E. Alcock III and Naval War College, "The Contractor Shall Provide MPA Assessments of Selected Individuals from Live One-on-One Interviews (N0012419Q0140)," March 14, 2019, Sam.gov. See also Moore, *Executives in Action*; and Moore, *Movement and Making Decisions*.

146. Darren McClurg and Naval War College, "Justification for Sole Source/Brand Name Specific (N0012419Q0140)," February 2, 2019, Sam.gov. See also Dara Baker, "Leadership, Creativity, Military Innovation, and Future Warfighting: An Oral History of Admiral James R. Hogg, USN (Ret.), While Serving as Director of the CNO Strategic Studies Group from 1995 to 2013," ed. William G. Glenney, IV (Naval History and Heritage Command, July 2021), 54.

147. Connors, "No Leader Is Ever Off Stage."

Chapter 5

1. Elizabeth Rosen, *Dance in Psychotherapy* (Teachers College, Columbia University, 1957), 69.

2. Rosen, *Dance in Psychotherapy*, 69.

3. On postwar American anxieties see, for example, Elaine Tyler May, *Homeward Bound: American Families in the Cold War Era* (Basic Books, 1998).

4. Wilson, *The Man in the Gray Flannel Suit*.

5. Ellen Herman, *The Romance of American Psychology: Political Culture in the Age of Experts, 1940–1970* (University of California Press, 1995), 82–123; and Christina Jarvis, "'If He Comes Home Nervous': U.S. World War II Neuropsychiatric Casualties and Postwar Masculinities," *The Journal of Men's Studies* 17, no. 2 (2010): 97–115.

6. For an overview of the early years of dance therapy from a practitioner's perspective, see Fran J. Levy, *Dance/Movement Therapy: A Healing Art* (American Alliance for Health, Physical Education, Recreation and Dance, 1988).

7. Marian Chace, the founding dance therapist at St. Elizabeth's Hospital in Washington, DC, for example, had trained as a dancer with Mary Wigman and subsequently developed a therapeutic approach derived largely from her own intuition.

8. Dorothy Zegart, "Dance Groups for Psychotic Patients: A Survey Study" (Smith College School of Social Work, 1956), iii.

9. Dorothy Zegart, "Dance Groups for Psychotic Patients."

10. Gerald N. Grob, *From Asylum to Community: Mental Health Policy in Modern America* (Princeton University Press, 1991).

11. See, for example, Irmgard Bartenieff, "A Post-Conference Letter to the American Dance Therapy Association," 1972, Series 9.2, Org. Box 6, Irmgard Bartenieff papers, Special Collections in Performing Arts, University of Maryland Libraries.

12. On the postwar prevalence of the idea that the most effective path to a healthy society was the modification of individual psychology, see, among others, Herman, *The Romance of American Psychology*; Rose, *Governing the Soul*; Grob, *From Asylum to Community*; Jonathan M. Metzl, *The Protest Psychosis: How Schizophrenia Became a Black Disease* (Beacon Press, 2009); and Deborah Weinstein, *The Pathological Family: Postwar America and the Rise of Family Therapy* (Cornell University Press, 2013). On the postwar modification of the individual *body* as a political act, see David Serlin, *Replaceable You: Engineering the Body in Postwar America* (University of Chicago Press, 2004).

13. "Irmgard Bartenieff Interview Transcript, Carol Boggs Interviewing," 1975, R/S Box 21, Irmgard Bartenieff Papers, University of Maryland.

14. "Letter from Irmgard Bartenieff to Marya Mannes," April 6, 1936, Corr. Box 4, Folder 17, Irmgard Bartenieff papers, Special Collections in Performing Arts, University of Maryland. Bartenieff also spent some of this time translating Feuillet's *L'art des decrire la danse* into Kinetographie.

15. Both Irmgard and Michal Bartenieff were considered stateless persons at the time of their immigration, as Michal had never acquired German citizenship, and Irmgard had forfeited hers upon their marriage. See Mary Jo Shelly, "Letter from Bennington College," July 27, 1936, Corr. Box 4, Folder 55, Irmgard Bartenieff papers, Special Collections in Performing Arts, University of Maryland Libraries; Mary P. O'Donnell, "Letter from Teachers College," September 23, 1936, Corr. Box 4, Folder 55, Irmgard Bartenieff papers, Special Collections in Performing Arts, University of Maryland Libraries; and John Martin to Irmgard Bartenieff, February 26, 1937, Corr. Box 4, Folder 55, Irmgard Bartenieff papers, Special Collections in Performing Arts, University of Maryland Libraries.

16. "Irmgard Bartenieff Interview Transcript, Carol Boggs Interviewing," 19.

17. "Telegram from Alice Dombois," July 26, 1939, Corr. Box 3, Series 2.1, Irmgard Bartenieff papers, Special Collections in Performing Arts, University of Maryland Libraries.

18. Sister Kenny's work mobilizing polio patients was a significant influence on Bartenieff, as was a broader emphasis on holistic healing and the involvement of patients in their own recovery and rehabilitation. See Irmgard Bartenieff, "Transcript of Seminar on Physical Therapy at the Institute for Movement Exploration," December 4, 1975, R/S Box 23, Folder 9, Irmgard Bartenieff papers, Special Collections in Performing Arts, University of Maryland Libraries. See also Naomi Rogers, *Polio Wars: Sister Kenny and the Golden Age of American Medicine* (Oxford University Press, 2013); and Steven R. Flanagan and Leonard Diller, "Dr. George Deaver: The Grandfather of Rehabilitation Medicine," *Physical Medicine and Rehabilitation* 5, no. 5 (2013): 355–59.

19. "Letter from Irmgard Bartenieff to Rudolf Laban," October 12, 1947, Corr. Box 4, Folder 2, Irmgard Bartenieff papers, Special Collections in Performing Arts, University of Maryland Libraries.

20. "Letter from Irmgard Bartenieff to Rudolf Laban."

21. "Letter from Irmgard Bartenieff to Rudolf Laban."

22. See, for example, "Letter from A. David Gurewitsch to Irmgard Bartenieff," February 6, 1953, Corr. Box 4, Folder 5, Irmgard Bartenieff papers, Special Collections in Performing Arts, University of Maryland Libraries.

23. "Letter from Irmgard Bartenieff to Rudolf Laban."

24. Claire Schmais and Elissa Q. White, "An Interview with Irmgard Bartenieff," *American Journal of Dance Therapy* 4, no. 1 (1981): 18.

25. Bartenieff, *Body Movement*, 9.

26. "Letter from Irmgard Bartenieff to Rudolf Laban."

27. "Letter from Irmgard Bartenieff to Dance Notation Bureau," c. 1950, Org Box 2, Folder 1, Irmgard Bartenieff papers, Special Collections in Performing Arts, University of Maryland Libraries; and Chloe Gardner, "The Part of Movement in Occupational Therapy for the Mentally Ill," *The Laban Art of Movement Guild News Sheet*, no. 6 (January 1951): 21–25, https://labanguildinternational.org.uk/wp-content/uploads/2021/08/LabanMagNo6Jan1951.pdf.

28. "Letter from Irmgard Bartenieff to Dance Notation Bureau."

29. See mention in Rosen, *Dance in Psychotherapy*, 57.

30. Warren Lamb, "Report of Lecture-Demonstration Given in London by Miss Joan Carrington and Mr. Warren Lamb to a Group of Psychiatrists," *The Laban Art of Movement Guild News Sheet*, no. 9 (October 1952): 13, https://labanguildinternational.org.uk/wp-content/uploads/2021/09/LabanMagNo9Oct1952.pdf.

31. Lamb, "Report of Lecture-Demonstration," 12.

32. "Letter from Warren Lamb to Irmgard Bartenieff," February 6, 1958, Corr. Box 4, Folder 15, Irmgard Bartenieff papers, Special Collections in Performing Arts, University of Maryland Libraries.

33. "Letter from Irmgard Bartenieff to Warren Lamb," October 12, 1959, Corr. Box 4, Folder 15, Irmgard Bartenieff papers, Special Collections in Performing Arts, University of Maryland Libraries.

34. Irmgard Bartenieff, "Dance Therapy in Jacobi Hospital—Day Care Unit," n.d., Series 3.5, R/S Box 8, Irmgard Bartenieff papers, Special Collections in Performing Arts, University of Maryland Libraries.

35. Bartenieff and Martha Davis were in fact regular participants in a long-running academic seminar on family therapy whose other attendees included Albert Scheflen, Andrew Ferber, and C. C. Beels. See, for example, "Meeting Minutes, Family Research Seminar, Thirteenth Meeting," December 15, 1967, Series 3.4, R/S 5, Irmgard Bartenieff papers, Special Collections in Performing Arts, University of Maryland Libraries; and "Project Proposal, Families With and Without Identified Patients: Similarities and Differences (Sub-Project: Effort-Shape of Movement in Interaction)," 1967, Series 3.4, R/S 5, Irmgard Bartenieff papers, Special Collections in Performing Arts, University of Maryland Libraries. For the history of family and systems therapy—with a particular attention to the role of new recording and observational technologies in its practice— see Weinstein, *The Pathological Family*. See also Peter Sachs Collopy, "The Revolution Will Be Videotaped: Making a Technology of Consciousness in the Long 1960s" (PhD diss., University of Pennsylvania, 2015), 170–80.

36. "Letter from Irmgard Bartenieff to Warren Lamb," June 14, 1961, Corr. Box 4, Folder 15, Irmgard Bartenieff papers, Special Collections in Performing Arts, University of Maryland Libraries.

37. Irmgard Bartenieff with Dori Lewis, *Body Movement: Coping with the Environment* (Gordon and Breach, 1980), 154.

38. "Letter from Irmgard Bartenieff to Warren Lamb," May 21, 1962, Corr. Box 4, Folder 15, Irmgard Bartenieff papers, Special Collections in Performing Arts, University of Maryland Libraries; and "Letter from Irmgard Bartenieff to Warren Lamb," January 16, 1963, Corr. Box 4, Folder 15, Irmgard Bartenieff papers, Special Collections in Performing Arts, University of Maryland Libraries.

39. "Letter from Irmgard Bartenieff to Warren Lamb," January 16, 1963. See also Irmgard Bartenieff, "Explanation of Posture-Gesture-Rest Ratios," 1963, Series 3.5, R/S Box 8, Irmgard Bartenieff papers, Special Collections in Performing Arts, University of Maryland Libraries.

40. Martha Davis, "Rating Scale for Hospitalized Mental Patients," 1968, Series 3.5, R/S Box 8, Irmgard Bartenieff papers, Special Collections in Performing Arts, University of Maryland Libraries.

41. See, for example, I. Zwerling and J. F. Wilder, "An Evaluation of the Applicability of the Day Hospital in Treatment of Acutely Disturbed Patients," *The Israel Annals of Psychiatry and Related Disciplines* 2 (1964): 162–85. See also Grob, *From Asylum to Community*.

42. Israel Zwerling, "The Creative Arts Therapies as 'Real Therapies,'" *American Journal of Dance Therapy* 11, no. 1 (1989): 23. Reprinted from *Hospital and Community Psychiatry*, 1979.

43. "Specific Aims, Cognitive Development Service," 1970, Series 9.2, Org 6, Irmgard Bartenieff papers, Special Collections in Performing Arts, University of Maryland Libraries. Overcrowding in state schools and other institutions for the developmentally disabled also meant that many of these patients remained in the hospital long after any real mental health crisis had resolved.

44. "Specific Aims, Cognitive Development Service."

45. "Specific Aims, Cognitive Development Service."

46. Yasuhiko Taketomo, "Cognitive Adaptation: A Psychodynamic Significance of a Mental Retardation Project," Conference Paper, American Academy of Psychoanalysis, 3-5 December 1971. Series 9.2, Org Box 6, Irmgard Bartenieff papers, Special Collections in Performing Arts, University of Maryland Libraries.

47. Taketomo, "Cognitive Adaptation."

48. "Dance Therapy Clinical Research Project, Bronx State Hospital," Early 1970s, Series 3.5, R/S Box 6, Irmgard Bartenieff papers, Special Collections in Performing Arts, University of Maryland Libraries.

49. Irmgard Bartenieff, "First 'In-House' Report on Institute for Dance Therapists," June 1972, Pedagogy Series 5, Box 1, Irmgard Bartenieff papers, Special Collections in Performing Arts, University of Maryland Libraries.

50. "DNB Effort/Shape Dance Therapy Training Program Proposal," n.d., Pedagogy Series 5, Box 1, Irmgard Bartenieff papers, Special Collections in Performing Arts, University of Maryland Libraries.

51. "DNB Effort/Shape Dance Therapy Training Program Proposal."

52. Irmgard Bartenieff, "The Vocabulary of Effort/Shape in Relation to Dance Therapy," n.d., Pedagogy Series 5, Box 1, Irmgard Bartenieff papers, Special Collections in Performing Arts, University of Maryland Libraries.

53. "LIMS Questionnaire Responses," 1978, Series 9.1, Org. Box 1, Irmgard Bartenieff papers, Special Collections in Performing Arts, University of Maryland Libraries.

54. Bartenieff, *Body Movement*, 9.

55. Bartenieff, *Body Movement*, 9.

56. "LIMS Questionnaire Responses."

57. Irmgard Bartenieff, "Effort/Shape: A Tool in Dance Therapy," March 3, 1976, Series 4, Publications Box 10, Irmgard Bartenieff papers, Special Collections in Performing Arts, University of Maryland Libraries.

58. "ADTA General Governance," 1973, Series 9.2, Org. Box 6, Irmgard Bartenieff papers, Special Collections in Performing Arts, University of Maryland Libraries. For a discussion of humanistic psychology's similarly anxious relationship with these practices, as well as with body-based therapies themselves, see Jessica Grogan, *Encountering America: Humanistic Psychology, Sixties Culture, and the Shaping of the Modern Self* (Harper Perennial, 2013).

59. "Letter from Marian Chace to Irmgard Bartenieff," September 28, 1964, Series 2.2, Corr. Professional Box 4, Irmgard Bartenieff papers, Special Collections in Performing Arts, University of Maryland Libraries.

60. Jody Zacharias, "Executive Director's Confidential Report, Laban Institute of Movement Studies," February 1980, Series 9.1, Org. Box 1, Irmgard Bartenieff papers, Special Collections in Performing Arts, University of Maryland Libraries.

61. Bartenieff, *Body Movement*.

62. Miriam Roskin Berger, "Dedication to Irmgard Bartenieff, Second Annual Spring Dance Therapy Conference, New York Chapter of the American Dance Therapy Association and Bronx Psychiatric Center," June 1980, Series 9.2, Org. Box 6, Irmgard Bartenieff papers, Special Collections in Performing Arts, University of Maryland Libraries.

63. Bartenieff, *Body Movement*, viii.

64. Elissa Queyquep White, "Effort-Shape: Its Importance to Dance Therapy and Movement Research," in *Dance Therapy: Focus on Dance VII*, ed. Kathleen Criddle Mason (American Association for Health, Physical Education, and Recreation, 1974), 37.

65. Paul Schilder, *The Image and Appearance of the Human Body* (Kegan Paul, 1935); and Francine Hanley, "The Dynamic Body Image and the Moving Body: A Theoretical and Empirical Investigation" (PhD diss., Victoria University, 2004).

66. On Schidler, specifically, as well as on the complicated history of the relationship between psychoanalysis and neurology, see Katja Guenther, *Localization and Its Discontents: A Genealogy of Psychoanalysis and the Neuro-Disciplines* (University of Chicago Press, 2016). As Gerald Grob notes, psychiatry and clinical psychology were also, in some ways, drawing closer in the postwar years, as the "simultaneous development of psychotropic drugs and milieu therapy—in addition to electroshock, lobotomy, and psychotherapy—blurred the conventional distinction between psychological and somatic approaches." See *From Asylum to Community*, 124. See also a discussion of Schilder's interest in the lie detector as tool for accessing the body's inadvertent "confessions"—and the connections between lie detection and psychotherapy more broadly—in Pettit, *The Science of Deception*, 191.

67. Ellen Goldman, *As Others See Us: Body Movement and the Art of Successful Communication* (Routledge, 2003), 184.

68. Judith Kestenberg, "The Role of Movement Patterns in Development," *Psychoanalytic Quarterly* 34 (1965): 4. On the growth of ideas about kinesthetic mirroring in the twentieth century, see Foster, *Choreographing Empathy*. For a further discussion of the intricacies of the KMP, see K. Mark Sossin, "Reliability of the Kestenberg Movement Profile," *Movement Studies* 2 (January 1987): 23–28.

69. Kestenberg was especially influenced by Anna Freud's work on child development. See Judith Kestenberg, "The Role of Movement Patterns in Development II: Flow of Tension and Effort," *Psychoanalytic Quarterly* 34 (1965): 517–63.

70. Kestenberg, "The Role of Movement Patterns in Development," 26.

71. Kestenberg, "The Role of Movement Patterns in Development II."

72. Michal Shapira, *The War Inside: Psychoanalysis, Total War, and the Making of the Democratic Self in Postwar Britain* (Cambridge University Press, 2013); and Deborah Blythe Doroshow, *Emotionally Disturbed: A History of Caring for America's Troubled Children* (University of Chicago Press, 2019).

73. Judith Kestenberg, "A Proposal for a Social Center for Parents and Children," 1972, Series 2.2, Corr. Professional, Irmgard Bartenieff papers, Special Collections in Performing Arts, University of Maryland Libraries. On primary prevention, see John Spaulding and Philip Balch, "A Brief History of Primary Prevention in the Twentieth Century: 1908 to 1980," *American Journal of Community Psychology* 11, no. 1 (1983): 59–80. On twentieth century parenting advice literature, see Ann Hulbert, *Raising America: Experts, Parents, and a Century of Advice About Children* (Vintage, 2004).

74. In the 1970s, Kestenberg even developed a form of fetal movement notation, intended to make pregnant people and obstetric nurses aware of the "preferred movements of the fetus and newborn" to facilitate early mother-child attachment. See Judith S. Kestenberg, *Children and Parents: Psychoanalytic Studies in Development* (Jason Aronson, 1973), 227.

75. Indeed, Kestenberg's concerns about poorly managed childrearing crossed social and class lines; she seemed to worry equally about the poorly staffed daycare centers where working families placed their children and the "unsupervised" playgroups her more well-off clients arranged. She advocated for increased parental leave (though she considered maternal leave most important) and for the creation of professionally run childcare centers modeled after the Israeli kibbutz. See Kestenberg, "A Proposal for a Social Center for Parents and Children"; Judith Kestenberg, "Editorial," *Child Development Research News* 2, no. 1 (1979): 4.

76. Bruce Lambert, "Obituary: Milton Kestenberg, 79, Lawyer Who Won Holocaust Reparations," *New York Times*, November 21, 1991.

77. "Dr. Judith S. Kestenberg Talks to Kristina Stanton," *Free Associations* 2 (1991): 163.

78. Yolanda Gampel, "Judith S. Kestenberg," *Echoes of the Holocaust*, April 2000, http://www.holocaustechoes.com/obituaries.html.

79. For example, see Judith Kestenberg, "Psychoanalytic Contributions to the Problem of Children of Survivors from Nazi Persecution," *Israel Annals of Psychiatry and Related Disciplines* 10 (1972): 311–25. A summary of Kestenberg's work on this subject as well as its influence on the larger Group for the Psychoanalytic Study of the Effect of the Holocaust on the Second Generation can be found in Martin S. Bergmann and Milton E. Jucovy, eds., *Generations of the Holocaust* (Basic Books, 1982).

80. "Dr. Judith S. Kestenberg Talks to Kristina Stanton," 161.

81. Allan Young, *The Harmony of Illusions: Inventing Post-Traumatic Stress Disorder* (Princeton University Press, 1995); Ben Shephard, *A War of Nerves: Soldiers and Psychiatrists in the Twentieth Century* (Harvard University Press, 2003); and Leena Akhtar, "From Masochists to Traumatized Victims: Psychiatry, Law, and the Feminist Anti-Rape Movement of the 1970s" (PhD diss., Harvard University, 2017).

82. Young, *The Harmony of Illusions*, 4.

83. Deborah Lipstadt, *Denying the Holocaust: The Growing Assault on Truth and Memory* (Free Press, 1993); Robert A. Kahn, *Holocaust Denial and the Law: A Comparative Study* (Palgrave Macmillan, 2004).

84. Christian Pross, *Paying for the Past: The Struggle Over Reparations for Surviving Victims of the Nazi Terror*, trans. Belinda Cooper (Johns Hopkins University Press, 1998); Ariel Colonomos and Andrea Armstrong, "German Reparations to the Jews after World War II: A Turning Point in the History of Reparations," in *The Handbook of Reparations*, ed. Pablo de Grieff (Oxford University Press, 2006); and Dagmar Herzog, *Cold War Freud: Psychoanalysis in an Age of Catastrophes* (Cambridge University Press, 2017), 89–122.

85. Herzog, *Cold War Freud*, 94.

86. Kurt R. Eissler, "Die ermordung von wie vielen seiner Kinder muss ein Mensch symptomfrei ertragen können, um eine normale Konstitution zu haben?," *Psyche* 17, no. 5 (1963): 241–91.

87. Herzog, *Cold War Freud*, 94–103.

88. Milton Kestenberg, "Discriminatory Aspects of the German Indemnification Policy: A Continuation of Persecution," in Bergmann and Jucovy, *Generations of the Holocaust*, 66.

89. M. Kestenberg, "Discriminatory Aspects of the German Indemnification Policy," 69.

90. Judith Kestenberg and Eva Fogelman, eds., *Children During the Nazi Reign: Psychological Perspectives on the Holocaust* (Praeger, 1994), x.

91. Herzog, *Cold War Freud*, 105; Miriam Rieck and Gali Eshet, "Die Bürden der Experten: Gespräche mit Deutschen und Israelischen Psychiatern über ihre Rolle as Gutachter in Entschädigungsverfahren," in *Die Praxis der Wiedergutmachung*, ed. Norbert Frei, José Brunner, and Constantin Goschler (Göttingen: Wallstein Verlag, 2009).

92. Susan Birnbaum, "Memories in a Holocaust Hourglass," *Jewish Telegraphic Agency Daily News Bulletin*, March 18, 1987.

93. Child Development Research, "The International Study of Organized Persecution of Children," accessed January 24, 2016, http://holocaustchildren.org. Many of Kestenberg's former students and collaborators also continue to work on questions related to trauma and the body, including Dr. Mark Sossin, who generously agreed to be interviewed for this chapter. See Judith Kestenberg and Mark K. Sossin, *The Role of Movement Patterns in Development* (Dance Notation Bureau, 1977); Beatrice Beebe et al., eds., *Mothers, Infants and Young Children Of September 11, 2001: A Primary Prevention Project* (Routledge, 2012); and Phyllis Cohen, Mark K. Sossin, and Richard Ruth, eds., *Healing After Parent Loss in Childhood and Adolescence: Therapeutic Interventions and Theoretical Considerations* (Rowman & Littlefield, 2014).

94. Judith Kestenberg and Ira Brenner, *The Last Witness: The Child Survivor of the Holocaust* (American Psychiatric Press, 1996), xii.

95. Paul Valent, "A Child Survivor's Appraisal of His Own Interview," in Kestenberg and Fogelman, *Children During the Nazi Reign*, 134.

96. Bartenieff, *Body Movement*, 151.

97. Theodor Adorno et al., *The Authoritarian Personality* (Harper & Row, 1950).

98. Jamie Cohen-Cole, *The Open Mind: Cold War Politics and the Sciences of Human Nature* (University of Chicago Press, 2014); and Jamie Cohen-Cole, "The Creative American: Cold War Salons, Social Science, and the Cure for Modern Society," *Isis* 100, no. 2 (2009): 219–62.

99. See, for example, Mark Sossin, "Development of Aggressive Patterns in Childhood," *Child Development Research News* 1, no. 1 (1978).

100. Kestenberg and Kahn, *Children Surviving Persecution*, x.

101. Judith S. Kestenberg, "A Multicreative Art Approach to Prevention with Infants, Toddlers, and Their Parents" (New England Council of Creative Therapies Conference, Brown University, 1979). Series 8.2, Other Publications Box 4, Irmgard Bartenieff papers, Special Collections in Performing Arts, University of Maryland Libraries. On theories of aggression in the postwar period more generally, see also Erika Milam, *Creatures of Cain: The Hunt for Human Nature in Cold War America* (Princeton University Press, 2019); Herzog, *Cold War Freud*, 123–50.

102. Warren Lamb, "Notes on the Meeting of the 'Research Board for the Correlation of Medical Science and Physical Education,'" November 29, 1944, E(L) 46:5, Warren Lamb Archive, University of Surrey.

103. Bartenieff, "A Post-Conference Letter to the American Dance Therapy Association." See also Irmgard Bartenieff, "Dance Therapy: A New Profession or a Rediscovery of an Ancient Role of the Dance?," *Dance Scope* (Fall–Winter 1972–73).

104. Bartenieff, "Dance Therapy in Jacobi Hospital—Day Care Unit."

105. "What Is Dance Therapy, Really? Program for 7th Annual Conference of the American Dance Therapy Association," October 20, 1972, Series 9.2, Org. Box 6, Irmgard Bartenieff papers, Special Collections in Performing Arts, University of Maryland Libraries. It is important to note that the 1969 ADTA code of ethics prohibited movement therapists working independently from referring to their clients as "patients," as they were not medical practitioners licensed by the state. In hospital settings, movement therapists were understood to be practicing under the supervision of physicians or other licensed professionals. See "ADTA Code of Ethics," October 1970, Series 4, Publications Box 10, Irmgard Bartenieff papers, Special Collections in Performing Arts, University of Maryland Libraries.

106. Bartenieff, *Body Movement*, 151.

107. Lauretta Bender and Franziska Boas, "Creative Dance in Therapy," *American Journal of Orthopsychiatry* 11, no. 2 (1941): 235–44. Franziska Boas was a dancer and dance therapist (and daughter of anthropologist Franz Boas); Lauretta Bender was a child neuropsychiatrist.

108. "Turtle Bay Music School Orientation Training in Arts in Therapy 1967–1968," 1967, Series 9.2, Org. Box 6, Irmgard Bartenieff papers, Special Collections in Performing Arts, University of Maryland Libraries.

109. Audrey Wethered, "Movement," L/E/39/38, Rudolf Laban Archive, University of Surrey.

110. Bartenieff, *Body Movement*, 144.

111. Bartenieff, *Body Movement*, 151.

112. Bartenieff, *Body Movement*, 13.

113. Bartenieff, *Body Movement*, 13.

114. Audrey Wethered, "Notes on Emotional and Psychological Conditions," n.d., AW/E/1/1, Audrey Wethered and Chloë Gardner Collection, University of Surrey.

115. Audrey Wethered, "Some Comments on Working with Groups," 1968, AW/E/1/2, Audrey Wethered and Chloë Gardner Collection, University of Surrey.

116. Penny Bernstein, "Editorial and Comments on Conference Demonstration Workshops," *ADTA Newsletter* 6, no. 3 (December 1972): 18. Series 8, Subseries 8.4, Other Publications, Irmgard Bartenieff papers, Special Collections in Performing Arts, University of Maryland Libraries.

117. "Sesame/Laban Centenary Festival," July 28, 1979, AW/E/1/10, Audrey Wethered and Chloë Gardner Collection, University of Surrey.

118. Joyhope Taylor, "The Dance Therapist's Function in Public Schools," *ADTA Newsletter* 8, no. 2 (September 1974): 4, Series 8, Subseries 8.4, Other Publications, Irmgard Bartenieff papers, Special Collections in Performing Arts, University of Maryland Libraries.

119. Bartenieff, *Body Movement*, 11.

120. Rosen, *Dance in Psychotherapy*, 81.

121. Rosen, *Dance in Psychotherapy*, 81.

122. Bartenieff, "Dance Therapy," 14.

123. Irmgard Bartenieff, "Renewed Humanity Promised in Choric Dance: Excerpts from Gymnastik Und Tanz," *Effort/Shaper* 1, no. 3 (June 1, 1977): 4.

124. "Material for Discussion of the Institute Name," March 8, 1978, Series 9.1, Org. Box 1, Irmgard Bartenieff papers, Special Collections in Performing Arts, University of Maryland Libraries.

125. Irmgard Bartenieff, "Effort Observation and Effort Assessment in Rehabilitation" (National Notation Conference, New York City, June 11, 1962). R/S Box 3, Irmgard Bartenieff papers, Special Collections in Performing Arts, University of Maryland Libraries.

126. Irmgard Bartenieff, "Exploring Interaction. Address at American Dance Therapy Association Conference," October 27, 1974, Series 3.4, R/S 4, Irmgard Bartenieff Papers, University of Maryland.

127. One significant exception can be found in William Carpenter's writings, which the epilogue will discuss in greater detail. See for example W. M. Carpenter, "Book Manuscript, Conflict and Harmony Between Man and Woman: A Study in Movement Expression," 1954, L/E/37/19, Rudolf Laban Archive, University of Surrey.

128. Evy Westbrook and Doreen Court, "Work with Subnormal Patients," n.d., AW/E/1/10, Audrey Wethered and Chloë Gardner Collection, University of Surrey.

129. "Letter from Joan Smallwood to Irmgard Bartenieff," April 30, 1973, n.d., Series 3.4, R/S Box 4, Irmgard Bartenieff papers, Special Collections in Performing Arts, University of Maryland Libraries.

130. Bartenieff, "Effort Observation and Effort Assessment in Rehabilitation." Bartenieff's views on the sources of movement diversity across racial and ethnic groups were profoundly shaped by the folklorist Alan Lomax and will be explored in greater depth in the following chapter.

131. Alan Lomax, unpublished manuscript for *Dancing: A World Ethnography of Dance and Movement Styles*, 1981, Atlas 6, MS 39.02.03, Alan Lomax Collection (AFC 2004/004), American Folklife Center, Library of Congress.

132. "Review of Seminar, Symposium 1972 Transcripts," June 9, 1972, Series 3.10, R/S Box 21, Irmgard Bartenieff papers, Special Collections in Performing Arts, University of Maryland Libraries.

133. Penny Bernstein, "Case Studies, Pittsburgh Child Guidance Center," 1971, Series 2.2, Corr. Professional Box 4, Irmgard Bartenieff papers, Special Collections in Performing Arts, University of Maryland Libraries; and Penny Bernstein, "Evaluations for Activity Movement Group," October 1971, Series 3.4, R/S Box 4, Irmgard Bartenieff papers, Special Collections in Performing Arts, University of Maryland Libraries.

134. "Letter from Judith Kestenberg to Warren Lamb," n.d., Warren Lamb Archive, University of Surrey.

135. For example, "Dance Therapy Positions," *ADTA Newsletter* 8, no. 1 (June 1974): 4. Series 8, Subseries 8.4, Other Publications, Irmgard Bartenieff papers, Special Collections in Performing Arts, University of Maryland Libraries.

136. Recently, racial bias in dance and movement therapy has come in for renewed scrutiny and is being interrogated by a new generation of practitioners and scholars. See, for example, Susan D. Imus, et al., "More Than One Story, More Than One Man: Laban Movement Analysis

Re-Examined," *American Journal of Dance Therapy* 44 (2022): 168–85; Meg H. Chang, "Dance/ Movement Therapists of Color in the ADTA: The First 50 Years," *American Journal of Dance Therapy* 38 (2016): 268–78; Meg H. Chang, "Cultural Consciousness and the Global Context of Dance/ Movement Therapy," in *The Art and Science of Dance/Movement Therapy*, ed. Sharon Chaiklin and Hilda Wengrower, 2nd ed. (Routledge, 2016); and Christine Caldwell, "Diversity Issues in Movement Observation and Assessment," *American Journal of Dance Therapy* 35 (2013): 183–200.

137. Chang, "Dance/Movement Therapists of Color in the ADTA," 271.

138. C. C. Beels, "Israel Zwerling: Lessons from a Career," *New Directions for Mental Health Services* 42 (1989): 8.

139. Grob, *From Asylum to Community*.

140. See also Anne Harrington, *Mind Fixers: Psychiatry's Troubled Search for the Biology of Mental Illness* (W.W. Norton, 2019).

141. See discussions in, for example, Joan Smallwood, "Statement on Dance/Movement Therapy," 1977, Series 8.2, Other Pubs Box 4, Irmgard Bartenieff papers, Special Collections in Performing Arts, University of Maryland Libraries; and Barbara F. Govine, "Politics and Dance Therapy," *ADTA Newsletter* 8, no. 2 (September 1974): 8. Series 8, Subseries 8.4, Other Pubs, Irmgard Bartenieff papers, Special Collections in Performing Arts, University of Maryland Libraries.

142. "American Dance Therapy Association," accessed May 21, 2023, https://www.adta.org.

143. See, discussions in, for example, Hans Falck, "Review of the Proceedings, Third Annual Conference of the American Dance Therapy Association," *National Association of Social Work Newsletter* IX, no. 3 (1968); Bartenieff, "A Post-Conference Letter to the American Dance Therapy Association"; and Martha Nastich, "Letter to Members of the ADTA," January 8, 1976, Series 9.2, Org. Box 6, Irmgard Bartenieff papers, Special Collections in Performing Arts, University of Maryland Libraries.

144. "Letter from Steven W. Malkus to Irmgard Bartenieff," June 20, 1975, Series 2.2, Corr. Professional, Irmgard Bartenieff papers, Special Collections in Performing Arts, University of Maryland Libraries; Dance Notation Bureau, "NEH Grant Proposal for Development of Notation Tools for Ethnic Dance Research," 1976, Series 9.2, Org. Box 3, Irmgard Bartenieff papers, Special Collections in Performing Arts, University of Maryland Libraries; and Zacharias, "Executive Director's Confidential Report, Laban Institute of Movement Studies."

145. In the end, the debate was moot, as the projects did not attract sufficient funding to move forward. See Zacharias, "Executive Director's Confidential Report, Laban Institute of Movement Studies," 3.

146. Sam Thornton, "Rudolf Laban—An Appreciation," Sesame-Laban Centenary Festival— 28–29 July 1979, AW/E/1/10, Audrey Wethered and Chloë Gardner Collection, University of Surrey.

Chapter 6

1. Alan Lomax, unpublished manuscript for *Dancing: A World Ethnography of Dance and Movement Styles, 1981*, I/38, MS 39.04.56, Alan Lomax Collection (AFC 2004/004), American Folklife Center, Library of Congress.

2. John Szwed, *Alan Lomax: The Man Who Recorded the World* (Penguin Books, 2011).

3. See Scott L. Matthews, "The Alan Lomax Archive," *American Quarterly* 68, no. 2 (June 2016): 429–37; Gage Averill, "Ballad Hunting in the Black Republic: Alan Lomax in Haiti,

1936-37," *Caribbean Studies* 36, no. 2 (2008): 3-22; Robert Baron, "'All Power to the Periphery': The Public Folklore Thought of Alan Lomax," *Journal of Folklore Research* 49, no. 3 (2012): 275-317; Rachel C. Donaldson, "Broadcasting Diversity: Alan Lomax and Multiculturalism," *The Journal of Popular Culture* 46, no. 1 (2013): 59-78; E. David Gregory, "Lomax in London: Alan Lomax, the BBC, and the Folk-Song Revival in London," *Folk Music Journal* 8, no. 2 (2002): 136-69; Aaron N. Oforlea, "[Un]Veiling the White Gaze: Revealing Self and Other in the Land Where the Blues Began," *Western Journal of Black Studies* 36, no. 4 (2012): 289-300; and Robert Gordon and Bruce Nemerov, eds., *Lost Delta Found: Rediscovering the Fisk University-Library of Congress Coahoma County Study, 1941-1942* (Vanderbilt University Press, 2005). As elements of the Lomax musical archive have begun to move online, scholarly discussions of the ongoing ethical ramifications of Lomax's work have only intensified. See, for example, Tanya Clement, "The Problem of Alan Lomax, or The Necessity of Talking Politics During the Lomax Year," *Sounding Out!* (blog), April 9, 2015, https://soundstudiesblog.com/2015/04/09/the-problem-of-alan-lomax-or-the-necessity-of-talking-politics-during-the-lomax-year; and Henry Adam Svec, "Folk Media: Alan Lomax's Deep Digitality," *Canadian Journal of Communication* 38, no. 2 (2013): 227-44.

4. On salvage anthropology, see Samuel J. Redman, *Prophets and Ghosts: The Story of Salvage Anthropology* (Harvard University Press, 2021). For an account of how "salvage" anthropology morphed into "urgent" anthropology in the 1960s—a project which resembled Lomax's own orientation somewhat more closely—see Adrianna H. Link, "Salvaging a Record for Humankind: Urgent Anthropology at the Smithsonian Institution, 1964-1984" (PhD diss., Johns Hopkins University, 2016). For an introduction to a series of distinct but related efforts to preserve purportedly vanishing cultures or populations, see, for example, Jenny Reardon, *Race to the Finish: Identity and Governance in an Age of Genomics* (Princeton University Press, 2004); Lemov, *Database of Dreams*; Vidal and Dias, *Endangerment, Biodiversity and Culture*; and Sepkoski, *Catastrophic Thinking*.

5. Alan Lomax FBI files, July 23, 1943, as quoted in Szwed, *Alan Lomax*, 203; and Ted Gioia, "The Red-Rumor Blues," *Los Angeles Times*, April 23, 2006.

6. For a fuller account of Lomax's biography, see Szwed, *Alan Lomax*.

7. John A. Lomax and Alan Lomax, *American Ballads & Folk Songs* (Macmillan, 1934); and John A. Lomax and Alan Lomax, *Negro Folk Songs as Sung by Lead Belly* (Macmillan, 1936).

8. Szwed, *Alan Lomax*, 104.

9. Alan Lomax, *Mister Jelly Roll: The Fortunes of Jelly Roll Morton, New Orleans Creole and "Inventory of Jazz"* (Duell, Sloan and Pearce, 1950).

10. Szwed, *Alan Lomax*, 235-40.

11. At the invitation of anthropologist Walter Goldschmidt, a summary of this material was published in Alan Lomax, "Folk Song Style: Musical Style and Social Context," *American Anthropologist* 61, no. 6 (1959): 927-54.

12. Lomax, "Folk Song Style," 927.

13. Anna Grimshaw, "Visual Anthropology," in *A New History of Anthropology*, ed. Henrika Kuklick (Oxford: Blackwell Publishing, 2008).

14. Ray L. Birdwhistell, *Introduction to Kinesics: An Annotation System for Analysis of Body Motion and Gesture* (Department of State, Foreign Service Institute, 1952); and Ray L. Birdwhistell, *Kinesics and Context: Essays on Body Motion Communication* (University of Pennsylvania Press, 1970). For more on Birdwhistell, see Watter, "Scrutinizing."

15. Szwed, *Alan Lomax*, 351.

16. Lomax's interest in the body as instrument may have also been influenced by his affiliation with Paris's Musée de l'homme during his time in Europe, where anthropologist André Leroi-Gourhan was in the middle of his own attempt to chart a kind of gestural history of humanity for the Hall of Arts and Techniques. See John Tresch, "Leroi-Gourhan's Hall of Gestures," in *Energies in the Arts*, ed. Douglas Kahn (MIT Press, 2019), 193–238.

17. Paulay remained part of Choreometrics throughout its life, while Bartenieff moved on to other projects in the late 1960s.

18. Murdock began work on the project that would become the Human Relations Area Files (HRAF) in the late 1920s in an effort to systematically gather, catalogue, index, and make accessible detailed information about the world's cultures. It was a massive collecting effort; Murdock and his collaborators pulled data from myriad, disparate sources, and detailed categories of recording included everything from Kinship to Magic to The Reproductive Cycle to Material Culture and Technology. On the creation of the HRAF, see Rebecca Lemov, "Filing the Total Human: Anthropological Archives from 1928 to 1963," in *Social Knowledge in the Making*, ed. Charles Camic, Neil Gross, and Michele Lamont (University of Chicago Press, 2011), 119–50; and Lemov, *Database of Dreams*.

19. Lomax, *Dancing*, 1981, I/31, MS 39.04.56, Alan Lomax Collection (AFC 2004/004), American Folklife Center, Library of Congress.

20. Lomax, *Dancing*, 1981, I/31, MS 39.04.56, Alan Lomax Collection (AFC 2004/004), American Folklife Center, Library of Congress.

21. See, for example: Margaret Bach, "Letter to Alan Lomax," September 4, 1972, MS 48.03.01, Alan Lomax Collection (AFC 2004/004), American Folklife Center, Library of Congress. On Gadjusek's own complicated relationships with the people he studied, see Warwick Anderson, *The Collectors of Lost Souls: Turning Kuru Scientists into Whitemen*, updated ed. (Johns Hopkins University Press, 2019).

22. Margaret Bach, "Letter to Alan Lomax," May 15, 1972, MS 48.03.01, Alan Lomax Collection, Library of Congress.

23. Lomax also contended that the sheer quantity of accumulated data would render the effects of the inclusion of errors merely incidental; see Lomax, *Dancing*, 1981, I/33, MS 39.04.56, Alan Lomax Collection (AFC 2004/004), American Folklife Center, Library of Congress. Anna Grimshaw has commented on a certain methodological naiveté that plagued early many ethnographic filmmakers; see "Visual Anthropology." Similar discussions of the film camera's status as an "objective recording" device—some of which Lomax participated in—also attended the creation of the National Anthropological Film Center in Washington, DC. See Adrianna Link, "Documenting Human Nature: E. Richard Sorenson and the National Anthropological Film Center, 1965–1980," *Journal of the History of the Behavioral Sciences* 52, no. 4 (2016): 371–91.

24. Alan Lomax, "Choreometrics and Ethnographic Filmmaking," *Filmmakers Newsletter*, February 1971, 27, MS 39.02.03, Alan Lomax Collection (AFC 2004/004), American Folklife Center, Library of Congress.

25. Sterne, *The Audible Past*, 332. Similar ironies and reclamation attempts are addressed in Gregg Mitman and Kelley Wilder, *Documenting the World: Film, Photography, and the Scientific Record* (University of Chicago Press, 2016).

26. Alan Lomax, Irmgard Bartenieff, and Forrestine Paulay, "The Choreometric Coding Book," in *Folk Song Style and Culture*, ed. Alan Lomax (American Association for the Advancement of Science, 1968), 263.

27. Lomax, *Dancing*, I/32, MS 39.04.56, Alan Lomax Collection (AFC 2004/004), American Folklife Center, Library of Congress.

28. See also Alan Lomax, "Cinema, Science, and Culture Renewal," *Current Anthropology* 13 (October 1973): 474–80. On future-oriented archival work in the sciences, see Lorraine Daston, "Sciences of the Archive," *Osiris* 27, no. 1 (2012): 156–87; and Daston, *Science in the Archives*.

29. Alan Lomax, *Dancing*, I/31, MS 39.04.56, Alan Lomax Collection (AFC 2004/004), American Folklife Center, Library of Congress.

30. Alan Lomax, "Brief Progress Report: Cantometrics-Choreometrics Projects," *Yearbook of the International Folk Music Council* 4, 25th Anniversary Issue (1972): 142–45.

31. Lomax, "Brief Progress Report."

32. Lomax, "Brief Progress Report."

33. Alan Lomax, "Body Vertical," n.d., 9.1.1.08, Alan Lomax Collection (AFC 2004/004), American Folklife Center, Library of Congress.

34. Lomax, "Body Vertical."

35. Alan Lomax and Irmgard Bartenieff, "Pilot Study Discussion," n.d., 9.1.1.07, Alan Lomax Collection (AFC 2004/004), American Folklife Center, Library of Congress.

36. Lomax, *Dancing*, I/8, MS 39.04.56, Alan Lomax Collection (AFC 2004/004), American Folklife Center, Library of Congress.

37. Lomax, *Dancing*, I/17-I/18, MS 39.04.56, Alan Lomax Collection (AFC 2004/004), American Folklife Center, Library of Congress.

38. Lomax, *Dancing*,1981, I/19, MS 39.04.56, Alan Lomax Collection (AFC 2004/004), American Folklife Center, Library of Congress.

39. Lomax, *Dancing*, II/12, MS 39.04.56, Alan Lomax Collection (AFC 2004/004), American Folklife Center, Library of Congress.

40. Alan Lomax, Irmgard Bartenieff, and Forrestine Paulay, "Dance Style and Culture," in *Folk Song Style and Culture*, ed. Alan Lomax (American Association for the Advancement of Science, 1968), 224.

41. See George W. Stocking Jr., *Race, Culture, and Evolution: Essays in the History of Anthropology*, new ed. (University of Chicago Press, 1982); and George W. Stocking, *The Ethnographer's Magic and Other Essays in the History of Anthropology* (University of Wisconsin Press, 1992).

42. Howard Brick, "Neo-Evolutionist Anthropology, the Cold War, and the Beginnings of the World Turn in U.S. Scholarship," in *Cold War Social Science: Knowledge Production, Liberal Democracy, and Human Nature*, ed. Mark Solovey and Hamilton Cravens (Palgrave Macmillan, 2012), 157.

43. Brick, "Neo-Evolutionist Anthropology," 159. See also Tresch, "Leroi-Gourhan's Hall of Gestures."

44. Lomax, *Dancing*, VI/I, MS 39.04.52, Alan Lomax Collection (AFC 2004/004), American Folklife Center, Library of Congress.

45. Lomax, *Dancing*, VI/I, MS 39.04.52, Alan Lomax Collection (AFC 2004/004), American Folklife Center, Library of Congress.

46. See Lynn Brooks Matluck, *Women's Work: Making Dance in Europe Before 1800* (University of Wisconsin Press, 2007).

47. Lomax's aims, therefore, might be compared productively to those of the feminist anthropologists chronicled by Alison Wylie; see "Doing Science as a Feminist: The Engendering of Archeology," in *Feminism in Twentieth-Century Science, Technology, and Medicine*,

ed. Angela N. H. Creager, Elizabeth Lunbeck, and Londa L. Schiebinger (University of Chicago Press, 2001), 23–45.

48. Lomax, *Dancing*, V2, MS 39.04.51, Alan Lomax Collection (AFC 2004/004), American Folklife Center, Library of Congress.

49. Lomax, *Dancing*, V3, MS 39.04.51, Alan Lomax Collection (AFC 2004/004), American Folklife Center, Library of Congress.

50. "Interview Transcript. Lomax and David Mayer," May 13, 1987, MS 47.01.05, Alan Lomax Collection (AFC 2004/004), American Folklife Center, Library of Congress.

51. "Choreometrics—Groundwork. Progress Report—Undated," n.d., MS 47.01.02, Alan Lomax Collection (AFC 2004/004), American Folklife Center, Library of Congress. Lomax also noted that Choreometrics's findings accorded well with contemporary research on the distribution of genetic traits.

52. Reardon, *Race to the Finish*.

53. Alan Lomax, Irmgard Bartenieff, and Forrestine Paulay, "A Handbook for the Analysis of Dance," 1970, MS 49.03.02, Alan Lomax Collection (AFC 2004/004), American Folklife Center, Library of Congress.

54. This focus on the acquisition and arrangement of value-neutral data and on quantitative tools was not unique to Lomax. See Theodore M. Porter, "Forward: Positioning Social Science in Cold War America," in Solovey and Cravens, *Cold War Social Science*; Mary Morgan, "Economics," in *The Cambridge History of Science: Modern Social Sciences*, ed. Theodore M. Porter and Dorothy Ross (Cambridge University Press, 2003), 7:275–305; and Joel Isaac, "Epistemic Design: Theory and Data in Harvard's Department of Social Relations," in Solovey and Cravens, *Cold War Social Science*.

55. Lomax, Bartenieff, and Paulay, "A Handbook for the Analysis of Dance."

56. On a similar scheme to alter perception on a mass scale, see Jardine, "Mass-Observation, Surrealist Sociology, and the Bathos of Paperwork."

57. Lomax, *Dancing*, II/7, MS 39.04.56, Alan Lomax Collection (AFC 2004/004), American Folklife Center, Library of Congress. For an account of difficulties training raters in this kind of observation, see Leeds-Hurwitz, "Notes in the History of Intercultural Communication."

58. See, e.g., Lomax, Bartenieff, and Paulay, "'Dance Style and Culture' and 'The Choreometric Coding Book'"; Irmgard Bartenieff and Forrestine Paulay, "Choreometric Profiles," in Lomax, *Folk Song Style and Culture*.

59. *Dance and Human History* (University of California Extension Media Center, 1974). A DVD titled *Rhythms of the Earth*, containing *Dance and Human History* as well as three other Choreometrics films and additional material, was released in 2008. See John Bishop, producer, *Rhythms of Earth: The Choreometrics Films of Alan Lomax and Forrestine Paulay* (Media-Generation, 2008).

60. "Letter from Bernice J. Gottschalk to Dr. Gregory Edwards," November 19, 1974, 13.01.09, Alan Lomax Collection (AFC 2004/004), American Folklife Center, Library of Congress; "Letter from Rolf Kuschel to Alan Lomax," 14 March 1979, 13.01.09, Alan Lomax Collection (AFC 2004/004), American Folklife Center, Library of Congress.

61. *Rhythms of Earth* (Media Generation, Association for Cultural Equity, 1974).

62. Alan Lomax, notes, late 1960s, 4.9.1.16, Alan Lomax Collection (AFC 2004/004), American Folklife Center, Library of Congress.

63. See: "Facts on DANCING: A WORLD ETHNOGRAPHY OF DANCE AND MOVEMENT STYLE," c. 1980, 39.04.01, Alan Lomax Collection (AFC 2004/004), American Folklife Center, Library of

Congress. The collection's official record lists only Lomax as the author of the text, though Paulay was also a contributor.

64. In broad strokes, this popular instruction in reading bodies bears a resemblance to the case of nineteenth-century physiognomy. See Sharrona Pearl, *About Faces: Physiognomy in Nineteenth Century Britain* (Harvard University Press, 2010).

65. Lomax, "Choreometrics and Ethnographic Filmmaking," 24.

66. "Letter from Lynne Norris to Alan Lomax," October 30, 1977, 13.01.09, Alan Lomax Collection (AFC 2004/004), American Folklife Center, Library of Congress.

67. Lomax, *Dancing*, 1981, I/36, MS 39.04.56, Alan Lomax Collection (AFC 2004/004), American Folklife Center, Library of Congress.

68. Lomax, *Dancing*, 1981, I/40, MS 39.04.56, Alan Lomax Collection (AFC 2004/004), American Folklife Center, Library of Congress.

69. Lomax, *Dancing*, 1981, I/40, MS 39.04.56, Alan Lomax Collection (AFC 2004/004), American Folklife Center, Library of Congress.

70. Lomax, *Dancing*, 1981, I/35, MS 39.04.56, Alan Lomax Collection (AFC 2004/004), American Folklife Center, Library of Congress.

71. "Letter from Alan Lomax to Roger Muldavin," August 16, 1972, 13.01.01, Alan Lomax Collection (AFC 2004/004), American Folklife Center, Library of Congress.

72. See John R. Rickford, "Labov's Contributions to the Study of African American Vernacular English: Pursuing Linguistic and Social Equity," *Journal of Sociolinguistics* 20, no. 4 (2016): 561–80.

73. Szwed, *Alan Lomax*, 360.

74. Szwed, *Alan Lomax*, 360.

75. Fred Turner, *The Democratic Surround: Multimedia and American Liberalism from World War II to the Psychedelic Sixties* (University of Chicago Press, 2013), 58; see also Cohen-Cole, *The Open Mind*.

76. On the broader twentieth-century discourse about the relationships between cultural diversity, biological diversity, and human survival, see Sepkoski, *Catastrophic Thinking*; Vidal and Dias, *Endangerment, Biodiversity and Culture*.

77. On cybernetics in the human, social, and behavioral sciences, see, among others, Heims, *The Cybernetics Group*; Andrew Pickering, *The Cybernetic Brain: Sketches of Another Future* (University of Chicago Press, 2010); Hunter Heyck, *Age of System: Understanding the Development of Modern Social Science* (Johns Hopkins University Press, 2015); Paul Erickson et al., *How Reason Almost Lost Its Mind: The Strange Career of Cold War Rationality* (University of Chicago Press, 2015); Bernard Dionysius Geoghegan, *Code: From Information Theory to French Theory* (Duke University Press, 2023).

78. Lomax, *Dancing*, Atlas 3-4, MS 39.02.03, Alan Lomax Collection (AFC 2004/004), American Folklife Center, Library of Congress.

79. Lomax, *Dancing*, Atlas 4, MS 39.02.03, Alan Lomax Collection (AFC 2004/004), American Folklife Center, Library of Congress. Emphasis mine.

80. Lomax, "Choreometrics and Ethnographic Filmmaking," 27.

81. McLuhan, *The Gutenberg Galaxy*; Walter Ong, *The Presence of the Word* (Yale University Press, 1967); Ong, *Orality and Literacy*.

82. Igo, *The Averaged American*, 270.

83. Igo, *The Averaged American*, 190.

84. John Bishop, "Alan Lomax (1915–2002): A Remembrance," *Visual Anthropology Review* 17, no. 2 (2001–02): 13–23.

85. "Course Evaluations for 1976 DNB 'Ethnic Dance' Course," 1976, Series 5.6, Pedagogy 3, Irmgard Bartenieff papers, Special Collections in Performing Arts, University of Maryland Libraries.

86. Drid Williams, "Choreometrics Discussion," *CORD News* 6, no. 2 (July 1974): 25–29.

87. William, "Choreometrics Discussion."

88. Joann W. Kealiinohomoku, "Caveat on Causes and Correlations," *CORD News* 6, no. 2 (July 1974): 20–24.

89. "Feedback Week 5," May 10, 1968, 13.01.01, Alan Lomax Collection, Library of Congress.

90. John Martin, "The Dance: A Negro Play," *New York Times*, March 12, 1933; and I.K., "Negro Ballet Has Performance," *New York Sun*, November 22, 1937, quoted in Lynne Fauley Emery, *Black Dance in the United States from 1619 to 1970* (Arno Press, 1980).

91. John Martin, *John Martin's Book of the Dance* (Tudor Publishing Company, 1963).

92. For an account of how Alvin Ailey negotiated these questions, see Thomas DeFrantz, *Dancing Revelations: Alvin Ailey's Embodiment of African American Culture* (Oxford University Press, 2004). See also work on the significance of Africanist elements to American dance more broadly, especially Gottschild, *Digging the Africanist Presence in American Performance*.

93. Emery, *Black Dance in the United States from 1619 to 1970*, 298.

94. Rod Rodgers, "A Black Dancer's Credo: Don't Tell Me Who I Am," *Negro Digest*, July 1968.

95. For more on the material limitations that frequently scuttle universal knowledge projects, see Mary Poovey, "The Limits of the Universal Knowledge Project: British India and the East Indiamen," *Critical Inquiry* 31, no. 1 (2004): 183–202. In 2017, Forrestine Paulay, Anna Lomax Wood, and the Association for Cultural Equity began work on an effort to revise and publish the Choreometrics project and to make at least some of the relevant data available online.

96. Lomax, *Dancing*, 1981, VI/68, MS 39.04.52, Alan Lomax Collection (AFC 2004/004), American Folklife Center, Library of Congress.

97. Scholars in both folklore and sound studies are paying increasing attention to the presence and significance of these kinds of omissions. For one attempt to reconstruct the lives and motivations of musicians who appeared in Library of Congress field recordings, see Stephen Wade, *The Beautiful Music All Around Us: Field Recordings and the American Experience* (University of Illinois Press, 2012).

98. Lomax, *Dancing*, 1981, I/12, MS 39.04.56, Alan Lomax Collection (AFC 2004/004), American Folklife Center, Library of Congress.

99. Lomax, *Dancing*, 1981, "Film Source List," MS 39.04.61, Alan Lomax Collection (AFC 2004/004), American Folklife Center, Library of Congress. For further discussion of Choreometrics and labor, see Whitney Laemmli, "Making Movement Matter," in *Invisible Labor in Modern Science*, ed. Jenny Bangham, Xan Chacko, and Judith Kaplan (Rowman & Littlefield, 2022), 237–46.

100. Lomax, "Choreometrics and Ethnographic Filmmaking."

101. Lomax, "Choreometrics and Ethnographic Filmmaking."

Epilogue

1. Alva Noë, *Strange Tools: Art and Human Nature* (Hill and Wang, 2015), 43.

2. Sealey, "NOTATE," 71. "NOTATE" was a suite of programs designed to create and print scores with a Hewlett-Packard 2648A Graphics Terminal and a 2631 Graphics Printer.

3. Norman I. Badler and Stephen W. Smoliar, "Digital Representations of Human Movement," *ACM Computing Surveys* 11, no. 1 (March 1979): 19.

4. Badler and Smoliar, "Digital Representations of Human Movement," 19.

5. Badler and Smoliar, "Digital Representations of Human Movement," 20.

6. Badler and Smoliar, "Digital Representations of Human Movement," 29.

7. Badler and Smoliar, "Digital Representations of Human Movement," 30.

8. Mubbasir Kapadia et al., "Efficient Motion Retrieval in Large Motion Databases," in *Proceedings of the ACM SIGGRAPH Symposium on Interactive 3D Graphics and Games* (2013), 19–38.

9. Tao Yu et al., "Motion Retrieval Based on Movement Notation Language," *Computer Animation and Virtual Worlds* 16 (2005): 273–82; Yuki Wakayama et al., "IEC-Based Motion Retrieval System Using Laban Movement Analysis," in *KES 2010, Proceedings* (Knowledge-Based and Intelligent Information and Engineering Systems—14th International Conference, Cardiff, United Kingdom, 2010), 251–60; and S. Okajima, Y. Wakayama, and Y. Okada, "Human Motion Retrieval System Based on LMA Features Using Interactive Evolutionary Computation Method," in *Innovations in Intelligent Machines—2*, vol. 376, Studies in Computational Intelligence, 2012, 117–30.

10. Interest in using Labanotation for these purposes continued for some time, particularly in studies of human behavior in space, though the experiments only rarely came to fruition. See, for example, E. C. Wortz et al., "Study of Astronaut Capabilities to Perform Extravehicular Maintenance and Assembly Functions in Weightless Conditions" (Airesearch Manufacturing Company, September 1967), 8–10; "Human Factors in Automated and Robotic Space Systems: Proceedings of a Symposium" (National Aeronautics and Space Administration, January 29, 1987), 332; and T. B. Sheridan, "Human Factors Considerations for Remote Manipulation," in *AGARD Lecture Series 193, Advanced Guidance and Control Aspects in Robotics* (N Advisory Group for Aerospace Research, 1994), 8.1–8.24. See also Norman I. Badler, interview by Joseph Glantz, 2018, http://www.joeglantz.com/Interview_NormBadler.html.

11. N. I. Badler et al., "The Center for Human Modeling and Simulation," *Presence: Teleoperators and Virtual Environments* 4, no. 1 (1995): 81–96.

12. Norman I. Badler, Cary B. Phillips, and Bonnie Lynn Webber, *Simulating Humans: Computer Graphics Animation and Control* (Oxford University Press, 1993), 9.

13. Badler, Phillips, and Webber, *Simulating Humans*, 103, 105, 109, 114.

14. Badler, Phillips, and Webber, *Simulating Humans*, 117.

15. Badler, interview; Strauss, "Computer to Capture Dance before Choreography Is Lost."

16. Indeed, Badler frequently referred to the ongoing influence of Laban-based thinking on his work, noting in 2021, for example, that "most of my (ultimately) important computational insights came from works in Natural Language Processing, dance, ergonomics, and communicative and manipulative behaviors." See J. Z. Wang et al., "Bodily Expressed Emotion Understanding Research: A Multidisciplinary Perspective," 10 (conference panel, 16th European Conference on Computer Vision, January 2021), https://par.nsf.gov/servlets/purl/10295638.

17. Norman I. Badler, Diane M. Chi, and Sonu Chopra-Khullar, "Virtual Human Animation Based on Movement Observation and Cognitive Behavior Models," *Proceedings of Computer Animation*, May 1999, 128–37.

18. Diane Chi et al., "The EMOTE Model for Effort and Shape," *SIGGRAPH '00: Proceedings of the 27th Annual Conference on Computer Graphics and Interactive Techniques* (July 2000), 173.

19. Chi et al., "The EMOTE Model for Effort and Shape," 176.

20. Chi et al., "The EMOTE Model for Effort and Shape," 178. Certified Movement Analyst (CMA) status can be obtained through the completion of an extensive course of study at the Laban/Bartenieff Institute of Movement Studies.

21. Chi et al., "The EMOTE Model for Effort and Shape," 174.

22. Norman I. Badler, "Gesture Expressivity" (conference paper, Computer Animation and Social Agents, 2003); Norman Badler and Dimitris Metaxas, "NSF Award Abstract: Synthesis and Acquisition of Communicative Gestures," September 13, 2002, https://www.nsf.gov/awardsearch/showAward?AWD_ID=0200983.

23. Chi et al., "The EMOTE Model for Effort and Shape."

24. This goal was realized, at least in part, later that same year. See L. Zhao, M. Costa, and N.I. Badler, "Interpreting Movement Manner," *Proceedings of Computer Graphics 2000* (May 2000), 98–103.

25. Chi et al., "The EMOTE Model for Effort and Shape," 181.

26. Funda Durupinar et al., "PERFORM: Perceptual Approach for Adding OCEAN Personality to Human Motion Using Laban Movement Analysis," *ACM Transactions on Graphics* 1 (October 16): Article 6. This work has also been funded by the US Army Research Laboratory under contracts W911NF-07-1-0216 and W911NF-10-2-0016.

27. See, for example, Sinan Sonlu, Uğur Güdükbay, and Funda Durupinar, "A Conversational Agent Framework with Multi-Modal Personality Expression," *ACM Transactions on Graphics* 40, no. 1 (2021): 7.1–7.16.

28. Toru Nakata, Tomomasa Sato, and Taketoshi Mori, "Expression of Emotion and Intention by Robot Body Movement," *5th Conference on Intelligent Autonomous Systems*, 1998.

29. Megha Sharma et al., "Communicating Affect via Flight Path: Exploring Use of the Laban Effort System for Designing Affective Locomotion Paths," *Proc. ACM/IEEE International Conference on Human-Robot Interaction* (2013): 293–300.

30. See, for example, Ed Hooks, *Acting for Animators*, 3rd ed. (Routledge, 2011), 74–77; and Leslie Bishko, "Animation Principles and Laban Movement Analysis: Movement Frameworks for Creating Empathic Character Performances," in *Nonverbal Communication in Virtual Worlds: Understanding and Designing Expressive Characters*, ed. Theresa Jean Tanenbaum, Magy Seif El-Nasr, and Michael Nixon (Carnegie Mellon University ETC Press, 2014), 177–203.

31. Claims about universal, atemporal forms of emotional expression and clear distinctions between thinking and feeling have been challenged in recent years by groups as diverse as scientists, historians, and critical theorists. The literature on this subject is vast, but useful introductions include Lisa Feldman Barrett, Michael Lewis, and Jeannette M. Haviland-Jones, eds., *Handbook of Emotions*, 4th ed. (Guildford Press, 2018); Melissa Gregg and Gregory J. Seigworth, eds., *The Affect Theory Reader* (Duke University Press, 2010); Barbara Rosenwein, "Worrying About Emotions in History," *American Historical Review* 107, no. 3 (2002): 821–45; and Edwin Hutchins, *Cognition in the Wild* (MIT Press, 1996).

32. Nakata, Sato, and Mori, "Expression of Emotion and Intention by Robot Body Movement." The authors specifically cite the work of Paul Ekman and Albert Mehrabian as providing alterative theoretical approaches. As another group of engineers wrote, somewhat ironically: "We use LMA as an intermediary instead of defining a direct mapping between motion parameters and personality to avoid arbitrary parameter selection." See Durupinar et al., "PERFORM," 2.

33. Lamb, *Posture and Gesture*, 185.

34. Lamb, *Posture and Gesture*, 185.

35. Lamb and Watson, *Body Code*, 70. Though efforts to study and record the relationship between electricity and muscular activation have a long history—stretching back to Galvani, du Bois-Reymond, and Marey—the first wave of modern EMG technology was largely developed at the Mayo Clinic in the 1950s, becoming cheaper, more portable, and more widely used by the 1980s.

36. Interview with Warren Lamb quoted in McCaw, *An Eye for Movement*, 179.

37. Lamb was particularly excited by instances in which all fourteen electrodes "bleep[ed]" simultaneously across the page," a phenomenon he saw as electrical proof of a Posture-Gesture Merger. See McCaw, *An Eye for Movement*, 180.

38. Lamb, *Posture and Gesture*, 185.

39. Toru Nakata, Taketoshi Mori, and Tomomasa Sato, "Analysis of Impression of Robot Bodily Expression," *Journal of Robotics and Mechatronics* 14, no. 1 (2002): 35. See also Tino Lourens, Roos van Berkel, and Emilia Barakova, "Communicating Emotions and Mental States to Robots in Real Time Parallel Framework Using Laban Movement Analysis," *Robotics and Autonomous Systems* 58 (2010): 1256–65.

40. Yu Luo et al., "ARBEE: Towards Automated Recognition of Bodily Expression of Emotion in the Wild," *International Journal of Computer Vision* 128 (2020): 1–25.

41. The researchers bemoaned the fact that previous work on bodily expression had been largely limited to laboratory settings and small in scale due to "lengthy human subject regulations." See Luo et al., "ARBEE," 3.

42. Luo et al., "ARBEE," 10.

43. Luo et al., "ARBEE," 2.

44. Luo et al., "ARBEE," 4. For an in-depth analysis of the connections between surveillance technology and the emergence of new forms of capitalism, see Shoshana Zuboff, *The Age of Surveillance Capitalism: The Fight for a Human Future at the New Frontier of Power* (Public Affairs, 2019).

45. Joerg Rett, Luís Santos, and Jorge Dias, "Laban Movement Analysis for Multi-Ocular Systems," in *IEEE/RSJ International Conference on Intelligent Robots and Systems* (Nice, France, 2008), 761–66; and Luís Santos and Jorge Dias, "Laban Movement Analysis towards Behavior Patterns," in *Emerging Trends in Technological Innovation: First IFIP WG 5.5/SOCOLNET Doctoral Conference on Computing, Electrical and Industrial Systems*, ed. Luis M. Camarinha-Matos, Pedro Pereira, and Luis Ribeiro (Springer, 2010), 187–94.

46. Lamb and Watson, *Body Code*, 110.

47. Lamb and Watson, *Body Code*, 112.

48. Rodrique Ngowi, "Behind Those Dancing Robots, Scientists Had to Bust a Move," *The Seattle Times*, January 20, 2021, https://www.seattletimes.com/business/behind-those-dancing-robots-scientists-had-to-bust-a-move.

49. Sydney Skybetter, "Meet the Choreographer Behind Those Dancing Robots," *Dance Magazine*, March 26, 2021, https://www.dancemagazine.com/boston-dynamics-dancing-robots-2651193214.html.

50. "The Twist for Posterity?," *New York Herald Tribune*, February 11, 1962, sec. 4; Mahoney, "How I Got Hooked on Labanotation (Part Two)," 3.

51. "The Twist for Posterity?"; Mahoney, "How I Got Hooked on Labanotation (Part Two)," 3.

52. Jo Floyd, "Some Thoughts Concerning Uses of L/N IBM Element," 1973, Dance Notation Bureau.

53. W. M. Carpenter, "Book Manuscript, Conflict and Harmony Between Man and Woman: A Study in Movement Expression," 1954, L(E) 37:19, Rudolph Laban Archive, University of Surrey.

54. Carpenter, "Book Manuscript, Conflict and Harmony Between Man and Woman."

55. Willson, *In Just Order Move*, 58. In later years, Carpenter's synthesis of Laban movement theory and Jungian analysis (coupled with the acting principles of Konstantin Stanislavski) would become the foundation of an influential school of acting promoted by Yat Malmgren. See, for example, Christopher Fettes, *A Peopled Labyrinth—The Histrionic Sense: An Analysis of the Actor's Craft; The Laban-Carpenter Theory of Movement Psychology Adapted and Brought to Completion by Yat Malmgren* (GCFA Publishing, 2015).

56. Carpenter, "Book Manuscript, Conflict and Harmony Between Man and Woman."

57. Carpenter, "Book Manuscript, Conflict and Harmony Between Man and Woman."

58. Carpenter, "Book Manuscript, Conflict and Harmony Between Man and Woman."

59. Rebecca Milzoff, "Inside 'Single Ladies' Choreographer JaQuel Knight's Quest to Copyright His Dances," *Billboard*, November 7, 2020, https://www.billboard.com/music/music-news/jaquel-knight-beyonce-megan-thee-stallion-billboard-cover-story-interview-2020-9477613.

60. See, for example, Siobhan Burke, "Hear the Dance: Audio Description Comes of Age," *New York Times*, November 11, 2023, https://www.nytimes.com/2023/11/11/arts/dance/dance-and-audio-description.html; Heather Shaw and Krishna Washburn, dirs., *Telephone* (2022), https://telephonefilm.com; Kinetic Light, "Audimance: Imagining the Future of Audio Description," https://kineticlight.org/audimance.

61. See, for example, Esther Geiger, "LMA as a Tool for Developing Audio Description: Making the Arts Accessible to People Who Are Blind," in *Proceedings of the Twenty-Fourth Biennial Conference* (International Council of Kinetography Laban/Labanotation, London, 2005), 2: 119–26; and Wendy Richards, "Music Braille Pedagogy: The Intersection of Blindness, Braille, Music Learning Theory, and Laban" (PhD diss., University of Auckland, 2020).

62. See, among many others, Susan Wendell, *The Rejected Body: Feminist Philosophical Reflections on Disability* (Routledge, 1996); Catherine J. Kudlick, "Disability History: Why We Need Another Other," *The American Historical Review* 108, no. 3 (June 2003): 763–93; Tobin Siebers, *Disability Theory* (University of Michigan Press, 2008); Jonathan Sterne, *Diminished Faculties: A Political Phenomenology of Impairment* (Duke University Press, 2022); Jaipreet Virdi, Mara Mills, and Sarah F. Rose, "Disability and the History of Science," special issue, *Osiris* 39, no. 1 (2024); Serlin, *Replaceable You*; and Linker, *Slouch*.

63. Milzoff, "Inside 'Single Ladies' Choreographer JaQuel Knight's Quest to Copyright His Dances."

64. See, for example, Victoria Watts, "Patterns of Embodiment: A Visual/Cultural Studies Approach to Dance Notation" (PhD diss., George Mason University, 2012); Sarah Burkhalter and Laurence Schmidlin, eds., *Spacescapes, Dance & Drawing since 1962* (JRP Ringier Kunstverlag AG, 2017); and Laban/Bartenieff Institute of Movement Studies, "Diversity, Equity, and Inclusion at LIMS," n.d., https://labaninstitute.org/diversity-equity-and-inclusion-at-lims. As discussed in chapter 5, there is also a sustained engagement with these questions by a burgeoning number of movement therapists. See, for example, Chang, "Cultural Consciousness and the Global Context of Dance/Movement Therapy"; and Caldwell, "Diversity Issues in Movement Observation and Assessment."

65. Amy LaViers and Catherine Maguire, *Making Meaning with Machines: Somatic Strategies, Choreographic Technologies, and Notational Abstractions through a Laban/Bartenieff Lens* (MIT Press, 2023), 17.

66. See, for example, Kelly McGonigal, "The Joy Workout: Six Research-Backed Moves to Improve Your Mood," *New York Times*, May 24, 2022, https://www.nytimes.com/2022/05/24/well/move/joy-workout-exercises-happiness.html. Laban movement theory also retains close, albeit often hidden ties, with these and other practices.

Works Cited

Archives

Ann Hutchinson Guest Papers, Hanya Holm Papers, Kayla Zalk Papers, Helen Priest Rogers dance files, and American Association for Health, Physical Education, and Recreation records, Jerome Robbins Dance Division, the New York Public Library for the Performing Arts, New York City, NY, USA
Bauhaus-Archiv, Berlin, Germany
Dance Notation Bureau papers, New York City, NY, USA
Dartington Hall Trust Records, Devon Heritage Center, Exeter, UK
Irmgard Bartenieff papers and Laban/Bartenieff Institute of Movement Studies (LIMS) records, Special Collections in Performing Arts, University of Maryland Libraries
John Hodgson Collection and Rudolf Laban Collection. Special Collections, University of Leeds, Leeds, UK
Rudolf Laban Papers, Warren Lamb Archive, and Audrey Wethered and Chloë Gardner Collection. Special Collections, Archives and Special Collections, University of Surrey, Guildford, UK
Trinity Laban Archive Collections, London, UK

Secondary Sources

Acton, Ryan M. "The Search for Social Harmony at Harvard Business School, 1919–1942." *Modern Intellectual History* (2022): 1–27.
Adiarte, Sandra. "Movement Analysis in Forensics—An Interdisciplinary Approach." Paper presented at 2018 Academy of Criminalistic and Police Studies, Belgrade, October 2018. https://jakov.kpu.edu.rs/bitstream/id/2232/Major.
Adorno, Theodor, Else Frenkel-Brunswick, Daniel J. Levinson, and Nevitt R. Sanford. *The Authoritarian Personality*. Harper & Row, 1950.
Akhtar, Leena. "From Masochists to Traumatized Victims: Psychiatry, Law, and the Feminist Anti-Rape Movement of the 1970s." PhD diss., Harvard University, 2017.
Alagna, Yvette. *Manuel d'Utilisation de La Sphere d'Impression IBM Labanotation (Cinétographie)*. Centre National d'Ecriture du Mouvement, 1977.

Alder, Ken. *The Lie Detectors: The History of An American Obsession.* Free Press, 2007.

Alexander, Jennifer Karns. "Efficiency and Pathology: Mechanical Discipline and Efficient Worker Seating in Germany, 1929–1932." *Technology and Culture* 47, no. 2 (2006): 286–310.

Alexander, Zeynep Çelik. *Kinaesthetic Knowing: Aesthetics, Epistemology, Modern Design.* University of Chicago Press, 2017.

Alter, Joseph S. *Yoga in Modern India: The Body between Science and Philosophy.* Princeton University Press, 2005.

Anderson, Benedict. *Imagined Communities: Reflections on the Origin and Spread of Nationalism.* Verso, 1983.

Anderson, Warwick. *The Collectors of Lost Souls: Turning Kuru Scientists into Whitemen.* Updated ed. Johns Hopkins University Press, 2019.

Andrews, Lori, and Dorothy Nelkin. "Whose Body Is It Anyway? Disputes over Body Tissues in a Biotechnology Age." *Lancet* 351 (January 3, 1998): 53–60.

Applegate, Celia, and Pamela Potter. "Germans as the 'People of Music': Genealogy of an Identity." In *Music and German National Identity*, edited by Celia Applegate and Pamela Potter, 1–35. University of Chicago Press, 2002.

Aronova, Elena, Christine von Oertzen, and David Sepkoski, eds. "Data Histories." Special Issue, *Osiris* 32, no. 1 (2017).

Averill, Gage. "Ballad Hunting in the Black Republic: Alan Lomax in Haiti, 1936–37." *Caribbean Studies* 36, no. 2 (2008): 3–22.

Badler, Norman I., Diane M. Chi, and Sonu Chopra-Khullar. "Virtual Human Animation Based on Movement Observation and Cognitive Behavior Models." *Proceedings of Computer Animation*, May 1999, 128–37.

Badler, N. I., D. N. Metaxas, B. L. Webber, and M. Steedman. "The Center for Human Modeling and Simulation." *Presence: Teleoperators and Virtual Environments* 4, no. 1 (1995): 81–96.

Badler, Norman I., Cary B. Phillips, and Bonnie Lynn Webber. *Simulating Humans: Computer Graphics Animation and Control.* Oxford University Press, 1993.

Badler, Norman I., and Stephen W. Smoliar. "Digital Representations of Human Movement." *ACM Computing Surveys* 11, no. 1 (March 1979): 19–38.

Baron, Robert. "'All Power to the Periphery': The Public Folklore Thought of Alan Lomax." *Journal of Folklore Research* 49, no. 3 (2012): 275–317.

Barrett, Lisa Feldman, Michael Lewis, and Jeannette M. Haviland-Jones, eds. *Handbook of Emotions.* 4th ed. Guildford Press, 2018.

Bartenieff, Irmgard. "Dance Therapy: A New Profession or a Rediscovery of an Ancient Role of the Dance?" *Dance Scope*, Fall-Winter 1972–73.

Bartenieff, Irmgard. "Renewed Humanity Promised in Choric Dance. Excerpts from Gymnastik Und Tanz." *Effort/Shaper* 1, no. 3 (June 1, 1977): 4.

Bartenieff, Irmgard, with Dori Lewis. *Body Movement: Coping with the Environment.* Gordon and Breach, 1980.

Bartenieff, Irmgard, and Forrestine Paulay. "Choreometric Profiles." In *Folk Song Style and Culture*, edited by Alan Lomax, 248–61. American Association for the Advancement of Science, 1968.

Beebe, Beatrice, Phyllis Cohen, Mark K. Sossin, and Sara Markese, eds. *Mothers, Infants and Young Children Of September 11, 2001: A Primary Prevention Project.* Routledge, 2012.

Beels, C. C. "Israel Zwerling: Lessons from a Career." *New Directions for Mental Health Services* 42 (1989): 3–8.

Bell, Charles. *The Anatomy and Philosophy of Expression as Connected with the Fine Arts*. 3rd ed. London, 1865.
Bell, Charles. *Idea of a New Anatomy of the Brain*. London, 1811.
Bender, Lauretta, and Franziska Boas. "Creative Dance in Therapy." *American Journal of Orthopsychiatry* 11, no. 2 (1941): 235–44.
Benesh, Rudolf, and Joan Benesh. *An Introduction to Benesh Dance Notation*. Adam and Charles Black, 1956.
Bergmann, Martin S., and Milton E. Jucovy, eds. *Generations of the Holocaust*. Basic Books, 1982.
Berkowitz, Carin. *Charles Bell and the Anatomy of Reform*. University of Chicago Press, 2015.
Biagioli, Mario, and Peter Galison, eds. *Scientific Authorship: Credit and Intellectual Property in Science*. Routledge, 2003.
Birdwhistell, Ray L. *Introduction to Kinesics: An Annotation System for Analysis of Body Motion and Gesture*. Department of State, Foreign Service Institute, 1952.
Birdwhistell, Ray L. *Kinesics and Context: Essays on Body Motion Communication*. University of Pennsylvania Press, 1970.
Bishko, Leslie. "Animation Principles and Laban Movement Analysis: Movement Frameworks for Creating Empathic Character Performances." In *Nonverbal Communication in Virtual Worlds: Understanding and Designing Expressive Characters*, edited by Theresa Jean Tanenbaum, Magy Seif El-Nasr, and Michael Nixon, 177–203. Carnegie Mellon University ETC Press, 2014.
Bishop, John. "Alan Lomax (1915–2002): A Remembrance." *Visual Anthropology Review* 17, no. 2 (2001–02): 13–23.
Bittel, Carla, Elaine Leong, and Christine von Oertzen, eds. *Working with Paper: Gendered Practices in the History of Knowledge*. University of Pittsburgh Press, 2019.
Blackie, John. *Inside the Primary School*. Her Majesty's Stationery Office, 1967.
Blatter, Jeremy. "Screening the Psychological Laboratory: Hugo Münsterberg, Psychotechnics, and the Cinema, 1892–1916." *Science in Context* 28, no. 1 (2015): 53–76.
Blayney, Steffan. *Health And Efficiency: Fatigue, the Science of Work, and the Making of the Working-Class Body*. University of Massachusetts Press, 2022.
Boltanski, Luc, and Ève Chiapello. *The New Spirit of Capitalism*. Translated by Gregory Elliott. Verso, 2005.
Bonham-Carter, Victor. *Dartington Hall: The History of an Experiment*. Phoenix House, 1958.
Bouk, Dan. *How Our Days Became Numbered: Risk and the Rise of the Statistical Individual*. University of Chicago Press, 2015.
Bowker, Geoffrey C., and Susan Leigh Star. *Sorting Things Out: Classification and Its Consequences*. MIT Press, 2000.
Boyns, Trevor, and John Richard Edwards. *A History of Management Accounting: The British Experience*. Routledge, 2013.
Bradley, Karen. *Rudolf Laban*. Routledge, 2008.
Brain, Robert. *The Pulse of Modernism: Physiological Aesthetics in Fin-de-Siècle Europe*. University of Washington Press, 2015.
Brain, Robert. "Representation on the Line: Graphic Recording Instruments and Scientific Modernism." In Clarke and Henderson, *From Energy to Information*, 155–77. Stanford University Press, 2002.
Braun, Marta. *Picturing Time: The Work of Étienne-Jules Marey (1830–1904)*. University of Chicago Press, 1995.

Bremmer, Jan. "Walking, Standing, and Sitting in Ancient Greek Culture." In Bremmer and Roodenburg, *A Cultural History of Gesture*, 15–35. Cornell University Press, 1991.

Bremmer, Jan, and Herman Roodenburg, eds. *A Cultural History of Gesture*. Cornell University Press, 1991.

Brick, Howard. "Neo-Evolutionist Anthropology, the Cold War, and the Beginnings of the World Turn in U.S. Scholarship." In *Cold War Social Science: Knowledge Production, Liberal Democracy, and Human Nature*, edited by Mark Solovey and Hamilton Cravens, 155–74. Palgrave Macmillan, 2012.

Brooks, Lynn Matluck. *Women's Work: Making Dance in Modern Europe Before 1800*. University of Wisconsin Press, 2007.

Brown, Maxine D., and Stephen W. Smoliar. "A Graphics Editor for Labanotation." In *SIG-GRAPH*, 60–65. Philadelphia, 1976.

Bryson, Norman. "Cultural Studies and Dance History." In *Meaning in Motion: New Cultural Studies of Dance*, edited by Jane C. Desmond, 55–80. Duke University Press, 1997.

Bücher, Karl. *Arbeit und Rhythmus*. Leipzig, 1899.

Burkhalter, Sarah, and Laurence Schmidlin, eds. *Spacescapes, Dance & Drawing since 1962*. JRP Ringier Kunstverlag AG, 2017.

Butler, Judith. *Gender Trouble: Feminism and the Subversion of Identity*. Routledge, 1990.

Bynum, Caroline. "Why All the Fuss about the Body? A Medievalist's Perspective." *Critical Inquiry* 22, no. 1 (1995): 1–33.

Cage, John, and Alison Knowles. *Notations*. Something Else Press, 1969.

Caldwell, Christine. "Diversity Issues in Movement Observation and Assessment." *American Journal of Dance Therapy* 35 (2013): 183–200.

Callahan, Richard Jr., Kathryn Lofton, and Chad E. Seales. "Allegories of Progress: Industrial Religion in the United States." *Journal of the American Academy of Religion* 78, no. 1 (March 2010): 1–39.

Cambridge University Management Group. *Attitudes to Industry: A Survey of Undergraduate Attitudes to Industry by the Cambridge University Management Group*. BIM Occasional Paper. BKT City Print Ltd., 1969.

Camprubí, Lino. *Engineers and the Making of the Francoist Regime*. MIT Press, 2014.

Carmichael, Leonard. "Sir Charles Bell: A Contribution to the History of Physiological Psychology." *Psychological Review* 33, no. 3 (1926): 188–217.

Cartwright, Lisa. *Screening the Body: Tracing Medicine's Visual Culture*. University of Minnesota Press, 1995.

Chadarevian, Soraya de. "Graphical Method and Discipline: Self-Recording Instruments in Nineteenth-Century Physiology." *Studies in History and Philosophy of Science* 24, no. 2 (1993): 267–91.

Chang, Meg H. "Cultural Consciousness and the Global Context of Dance/Movement Therapy." In *The Art and Science of Dance/Movement Therapy*, edited by Sharon Chaiklin and Hilda Wengrower, 2nd ed., 317–44. Routledge, 2016.

Chang, Meg H. "Dance/Movement Therapists of Color in the ADTA: The First 50 Years." *American Journal of Dance Therapy* 38 (2016): 268–78.

Chaplin, Charles. *My Autobiography*. Simon and Schuster, 1964.

Chi, Diane, Monica Costa, Liwei Zhao, and Norman Badler. "The EMOTE Model for Effort and Shape." *SIGGRAPH '00: Proceedings of the 27th Annual Conference on Computer Graphics and Interactive Techniques*, July 2000, 173–82.

Clarke, Bruce, and Linda Dalrymple Henderson, eds. *From Energy to Information: Representation in Science and Technology, Art, and Literature*. Stanford University Press, 2002.

Cohen, Phyllis, Mark K. Sossin, and Richard Ruth, eds. *Healing After Parent Loss in Childhood and Adolescence: Therapeutic Interventions and Theoretical Considerations*. Rowman & Littlefield, 2014.
Cohen-Cole, Jamie. "The Creative American: Cold War Salons, Social Science, and the Cure for Modern Society." *Isis* 100, no. 2 (2009): 219–62.
Cohen-Cole, Jamie. *The Open Mind: Cold War Politics and the Sciences of Human Nature*. University of Chicago Press, 2014.
Collins, Harry. *Tacit and Explicit Knowledge*. University of Chicago Press, 2010.
Collopy, Peter Sachs. "The Revolution Will Be Videotaped: Making a Technology of Consciousness in the Long 1960s." PhD diss., University of Pennsylvania, 2015.
Colonomos, Ariel, and Andrea Armstrong. "German Reparations to the Jews after World War II: A Turning Point in the History of Reparations." In *The Handbook of Reparations*, edited by Pablo de Grieff, 390–419. Oxford University Press, 2006.
Connors, Brenda. "No Leader Is Ever Off Stage: Behavioral Analysis of Leadership." *Joint Force Quarterly* 43 (2006): 83–87.
Connors, Brenda. "See a Reflection of Saddam's Biography." *Naval War College Review* 57, no. 1 (Winter 2004): 113–16.
Connors, Brenda, Richard Rende, and Timothy Colton. "Decision-Making Style in Leaders: Uncovering Cognitive Motivation Using Signature Movement Patterns." *International Journal of Psychological Studies* 7, no. 2 (2015): 105–12.
Connors, Brenda L., Richard Rende, and Timothy J. Colton. "Predicting Individual Differences in Decision-Making Process from Signature Movement Styles: An Illustrative Study of Leaders." *Frontiers in Psychology* 4 (2013).
Cook, Scott D. N. "Technological Revolutions and the Gutenberg Myth." In *Internet Dreams: Archetypes, Myths, and Metaphors*, edited by Mark Stefik, 67–82. MIT Press, 1997.
Cowan, Michael. *Technology's Pulse: Essays on Rhythm in German Modernism*. IGRS Books, 2011.
Cronin, Blaise. "Collaboration in Art and in Science: Approaches to Attribution, Authorship, and Acknowledgment." *Information and Culture* 47, no. 1 (2012): 18–37.
Cummings, Frederick. "Charles Bell and the Anatomy of Expression." *The Art Bulletin* 46, no. 2 (1964): 191–203.
Curtis, Robin. "An Introduction to Einfühlung." Translated by Richard George Elliott. *Art in Translation* 6, no. 4 (2014): 353–75.
Dagognet, François. *Étienne-Jules Marey: A Passion for the Trace*. Translated by Robert Galeta. Zone Books, 1992.
Dance and Human History. University of California Extension Media Center, 1974.
Das Gupta, Uma. "Tagore's Ideas of Social Action and the Sriniketan Experiment of Rural Reconstruction, 1922–41." *University of Toronto Quarterly* 77, no. 4 (2008): 992–1004.
Daston, Lorraine, ed. *Science in the Archives: Pasts, Presents, Futures*. University of Chicago Press, 2017.
Daston, Lorraine. "Sciences of the Archive." *Osiris* 27, no. 1 (2012): 156–87.
Daston, Lorraine, and Peter Galison. *Objectivity*. Zone Books, 2007.
Davies, Eden. *Beyond Dance: Laban's Legacy of Movement Analysis*. Routledge, 2006.
Davies, Margery. *Woman's Place Is at the Typewriter*. Temple University Press, 1984.
Davis, Crystal U., Selene Carter, and Susan R. Koff. "Troubling the Frame: Laban Movement Analysis as Critical Dialogue." *Journal of Dance Education* (October 18, 2021): 1–9.
Davis, Martha. *Understanding Body Movement: An Annotated Bibliography*. Indiana University Press, 1972.

Davis, Martha, Stan B. Walters, Neal Vorus, and Brenda Connors. "Defensive Demeanor Profiles." *American Journal of Dance Therapy* 22, no. 2 (2000): 103–21.
Deak, Istvan. *Beyond Nationalism: A Social and Political History of the Habsburg Officer Corps, 1848–1918*. Oxford University Press, 1992.
DeFrantz, Thomas. *Dancing Revelations: Alvin Ailey's Embodiment of African American Culture*. Oxford University Press, 2004.
Dinerstein, Joel. *Swinging the Machine: Modernity, Technology, and African American Culture between the World Wars*. University of Massachusetts Press, 2003.
Doerr, Evelyn. *Rudolf Laban: The Dancer of the Crystal*. Scarecrow Press, 2008.
Donaldson, Rachel C. "Broadcasting Diversity: Alan Lomax and Multiculturalism." *The Journal of Popular Culture* 46, no. 1 (2013): 59–78.
Doroshow, Deborah Blythe. *Emotionally Disturbed: A History of Caring for America's Troubled Children*. University of Chicago Press, 2019.
Douard, John W. "E.-J. Marey's Visual Rhetoric and the Graphic Decomposition of the Body." *Studies in History and Philosophy of Science* 26, no. 2 (1995): 175–204.
"Dr. Judith S. Kestenberg Talks to Kristina Stanton." *Free Associations* 2 (1991): 157–74.
Durupinar, Funda, Mubbasir Kapadia, Susan Deutsch, Michael Neff, and Norman I. Badler. "PERFORM: Perceptual Approach for Adding OCEAN Personality to Human Motion Using Laban Movement Analysis." *ACM Transactions on Graphics* 1 (October 16): Article 6.
Dutta, Krishna, and Andrew Robinson. *Rabindranath Tagore: The Myriad-Minded Man*. Bloomsbury, 1995.
Edwards, D. J. A. "Book Reviews: Body Code." *South African Journal of Psychology* 11, no. 4 (1981): 158–59.
Eissler, Kurt R. "Die ermordung von wie vielen seiner Kinder muss ein Mensch symptomfrei ertragen können, um eine normale Konstitution zu haben?" *Psyche* 17, no. 5 (1963): 241–91.
Elcott, Noam. *Artificial Darkness: An Obscure History of Modern Art and Media*. University of Chicago Press, 2016.
Elias, Norbert. *The Civilizing Process*. Blackwell, 1969.
Elmhirst, Leonard K. *Poet and Plowman*. Visva-Bharati, 1975.
Elswit, Kate. *Watching Weimar Dance*. Oxford University Press, 2014.
Emery, Lynne Fauley. *Black Dance in the United States from 1619 to 1970*. Arno Press, 1980.
Emre, Merve. *The Personality Brokers: The Strange History of Myers-Briggs and the Birth of Personality Testing*. Random House, 2018.
Erickson, Paul, Judy L. Klein, Lorraine Daston, Rebecca Lemov, Thomas Sturm, and Michael D. Gordin. *How Reason Almost Lost Its Mind: The Strange Career of Cold War Rationality*. University of Chicago Press, 2015.
Eshkol, Noa, and Avraham Wachman. *Movement Notation*. Weidenfeld & Nicholson, 1958.
Evans, James, and Adrian Johns, eds. "Beyond Craft and Code: Human and Algorithmic Cultures, Past and Present." Special issue, *Osiris* 38, no. 1 (2023).
Falck, Hans. "Review of the Proceedings, Third Annual Conference of the American Dance Therapy Association." *National Association of Social Work Newsletter* IX, no. 3 (1968).
Ferguson, Michael. *The Rise of Management Consulting in Britain*. Ashgate, 2002.
Fettes, Christopher. *A Peopled Labyrinth—The Histrionic Sense: An Analysis of the Actor's Craft; The Laban-Carpenter Theory of Movement Psychology Adapted and Brought to Completion by Yat Malmgren*. GCFA Publishing, 2015.

Feuillet, Raoul-Auger. *Chorégraphie, ou l'art d'écrire la danse*. Paris, 1700.
Flanagan, Steven R., and Leonard Diller. "Dr. George Deaver: The Grandfather of Rehabilitation Medicine." *Physical Medicine and Rehabilitation* 5, no. 5 (May 2013): 355–59.
Fletcher, Sheila. *Women First: The Female Tradition in English Physical Education 1880–1980*. Athlone Press, 1984.
Floyd, Jo. *Manual for Use with the Labanotation-IBM Selectric Typewriter Element*. Dance Notation Bureau, Inc., 1974.
Foster, Susan Leigh. "The Ballerina's Phallic Pointe." In *Corporealities: Dancing Knowledge, Culture and Power*, edited by Susan Leigh Foster, 1–26. New York: Routledge, 1996.
Foster, Susan Leigh. "Choreographing Empathy." *Topoi* 24 (2005): 81–91.
Foster, Susan Leigh. *Choreographing Empathy: Kinesthesia in Performance*. Routledge, 2011.
Foster, Susan Leigh. *Dances That Describe Themselves: The Improvised Choreography of Richard Bull*. Wesleyan University Press, 2002.
Franko, Mark. *The Work of Dance: Labor, Movement, and Identity in the 1930s*. Wesleyan University Press, 2002.
Frazier, Mara Pegeen. "Labanotation Is Creative: How a Systems Perspective Reveals Generativity in Dance Notation and Its Archives." *Journal of Movement Arts Literacy* 7, no. 1 (2021): 105–31.
Fügedi, János. "Dance Notation and Computers." *Yearbook for Traditional Music* 23 (1991): 101–11.
Gaboury, Jacob. *Image Objects: An Archaeology of Computer Graphics*. MIT Press, 2021.
Geiger, Esther. "LMA as a Tool for Developing Audio Description: Making the Arts Accessible to People Who Are Blind." In *Proceedings of the Twenty-Fourth Biennial Conference*, 2:119–26. London, 2005.
Geison, Gerald L. "The Protoplasmic Theory of Life and the Vitalist-Mechanist Debate." *Isis* 60, no. 3 (1969): 272–92.
Geoghegan, Bernard Dionysius. *Code: From Information Theory to French Theory*. Duke University Press, 2023.
George, Alys X. *The Naked Truth: Viennese Modernism and the Body*. University of Chicago Press, 2020.
Gill, Mary Louise, and James G. Lennox, eds. *Self-Motion: From Aristotle to Newton*. Princeton University Press, 1994.
Gillespie, Richard. "Industrial Fatigue and the Discipline of Physiology." In *Physiology in the American Context, 1850–1940*, edited by Gerald L. Geison, 237–62. American Physiological Society, 1987.
Gitelman, Lisa. *Paper Knowledge: Toward a Media History of Documents*. Duke University Press, 2014.
Gleisner, Martin. *Tanz für alle: Von der gymnastik zum gemeinschaftstanz*. Hesse & Becker, 1928.
Goldman, Ellen. *As Others See Us: Body Movement and the Art of Successful Communication*. Routledge, 2003.
Goldschmidt, Victor. *Über harmonie und complication*. Julius Springer, 1901.
González Pendás, María. "Modernity Consecrated: Architectural Discourse and the Catholic Imagination in Franquista Spain." In *Modern Architecture and Religious Communities, 1850–1970*, edited by Kate Jordan and Ayla Lepine, 30–48. Routledge, 2018.
Goody, Jack. *The Domestication of the Savage Mind*. Cambridge University Press, 1977.

Gordin, Michael D. *On the Fringe: Where Science Meets Pseudoscience.* Oxford University Press, 2021.
Gordon, Robert, and Bruce Nemerov, eds. *Lost Delta Found: Rediscovering the Fisk University-Library of Congress Coahoma County Study, 1941–1942.* Vanderbilt University Press, 2005.
Gottschild, Brenda Dixon. *Digging the Africanist Presence in American Performance: Dance and Other Contexts.* Greenwood Press, 1996.
Graff, Ellen. *Stepping Left: Dance and Politics in New York City, 1928–1942.* Duke University Press, 1997.
Green, Martin. *Mountain of Truth: The Counterculture Begins, Ascona, 1900–1920.* University Press of New England, 1986.
Gregg, Melissa, and Gregory J. Seigworth, eds. *The Affect Theory Reader.* Duke University Press, 2010.
Gregory, E. David. "Lomax in London: Alan Lomax, the BBC, and the Folk-Song Revival in London." *Folk Music Journal* 8, no. 2 (2002): 136–69.
Grimshaw, Anna. "Visual Anthropology." In *A New History of Anthropology*, edited by Henrika Kuklick, 293–309. Blackwell, 2008.
Grob, Gerald N. *From Asylum to Community: Mental Health Policy in Modern America.* Princeton University Press, 1991.
Grogan, Jessica. *Encountering America: Humanistic Psychology, Sixties Culture, and the Shaping of the Modern Self.* Harper Perennial, 2013.
Guenther, Katja. *Localization and Its Discontents: A Genealogy of Psychoanalysis and the Neuro-Disciplines.* University of Chicago Press, 2016.
Guilbert, Laure. *Danser avec le IIIe Reich: Les danseurs modernes sous le Nazisme.* Éditions Complexe, 2000.
Haeckel, Ernst. *Kristallseelen: Studien über das anorganische leben.* Alfred Kröner, 1917.
Hall, Fernau. "Review of Posture and Gesture." *Ballet Today*, 1965.
Handler, Richard. "Erving Goffman and the Gestural Dynamics of Modern Selfhood." *Past and Present* 203 (2009): 280–300.
Hanley, Francine. "The Dynamic Body Image and the Moving Body: A Theoretical and Empirical Investigation." PhD diss., Victoria University, School of Social Sciences and Psychology, 2004.
Harrington, Anne. *Mind Fixers: Psychiatry's Troubled Search for the Biology of Mental Illness.* W.W. Norton, 2019.
Harrington, Anne. *Reenchanted Science: Holism in German Culture from Wilhelm II to Hitler.* Princeton University Press, 1999.
Harrison, Tom, ed. *War Factory: A Report by Mass-Observation.* Victor Gollancz Ltd, 1943.
Heims, Steve J. *The Cybernetics Group.* MIT Press, 1991.
Herf, Jeffrey. *Reactionary Modernism: Technology, Culture, and Politics in Weimar and the Third Reich.* Cambridge University Press, 1984.
Herman, Ellen. *The Romance of American Psychology: Political Culture in the Age of Experts, 1940–1970.* University of California Press, 1995.
Herzog, Dagmar. *Cold War Freud: Psychoanalysis in an Age of Catastrophes.* Cambridge University Press, 2017.
Heyck, Hunter. *Age of System: Understanding the Development of Modern Social Science.* Johns Hopkins University Press, 2015.

Hilton, Wendy. *Dance and Music of Court and Theater: Selected Writings of Wendy Hilton.* Pendragon Press, 1997.

Hobsbawm, E. J. "Trends in the British Labor Movement Since 1850." *Science and Society* 13, no. 4 (1949): 289–312.

Hochschild, Arlie Russell. *The Managed Heart: The Commercialization of Human Feeling.* University of California Press, 1979.

Hodgson, John. *Mastering Movement: The Life and Work of Rudolf Laban.* Theatre Arts Books, 2001.

Hoffarth, Matthew. "Building the Hive: Corporate Personality Testing, Self-Development, and Humanistic Management in Postwar America, 1945–2000." PhD diss., University of Pennsylvania, 2018.

Hooks, Ed. *Acting for Animators.* 3rd ed. Routledge, 2011.

Hornsby-Smith, M. P. "Our Industrial Image." *Management Decision* 4, no. 2 (1970): 15–17.

Hornsby-Smith, Michael P. "Student Experiences in Industry." *The Vocational Aspect of Education* 23, no. 55 (1971): 81–89.

Hui, Alexandra, Julia Kursell, and Myles W. Jackson, eds. "Music, Sound, and the Laboratory from 1750–1980." Special issue, *Osiris* 28, no. 1 (2013).

Hulbert, Ann. *Raising America: Experts, Parents, and a Century of Advice about Children.* Vintage, 2004.

Hutchins, Edwin. *Cognition in the Wild.* MIT Press, 1996.

Hutchinson Guest, Ann. "Dance Notation." *Encyclopedia Britannica*, October 2, 2016. https://www.britannica.com/art/dance-notation.

Hutchinson Guest, Ann. "The Jooss-Leeder School at Dartington Hall." *Dance Chronicle* 29 (2006): 161–94.

Hutchinson Guest, Ann. *Labanotation: The System for Recording Movement.* New Directions, 1954.

Huxley, T. H. "On the Physical Basis of Life." *Fortnightly Review* 5 (1869): 129–45.

Hyman, Louis. *Temp: How American Work, American Business, and the American Dream Became Temporary.* Viking, 2018.

Igo, Sarah E. *The Averaged American: Surveys, Citizens, and the Making of a Mass Public.* Harvard University Press, 2007.

Imada, Adria L. *Aloha America: Hula Circuits Through the U.S. Empire.* Duke University Press, 2012.

Imus, Susan D., Aisha Bell Robinson, Valerie Blanc, and Jessica Young. "More Than One Story, More Than One Man: Laban Movement Analysis Re-Examined." *American Journal of Dance Therapy*, no. 44 (2022): 168–85.

Isaac, Joel. "Epistemic Design: Theory and Data in Harvard's Department of Social Relations." In *Cold War Social Science: Knowledge Production, Liberal Democracy, and Human Nature*, edited by Mark Solovey and Hamilton Cravens, 79–95. Palgrave Macmillan, 2012.

Jaques-Dalcroze, Émile. *Méthode Jaques-Dalcroze: Gymnastique rythmique.* Sandoz, Jobin & Cie., 1906.

Jardine, Boris. "Mass-Observation, Surrealist Sociology, and the Bathos of Paperwork." *History of the Human Sciences* 31, no. 5 (2018): 52–79.

Jarvis, Christina. "'If He Comes Home Nervous': U.S. World War II Neuropsychiatric Casualties and Postwar Masculinities." *The Journal of Men's Studies* 17, no. 2 (2010): 97–115.

Jelavich, Peter. *Munich and Theatrical Modernism: Politics, Playwriting, and Performance, 1880–1914.* Harvard University Press, 1985.

Jenner, Mark S. R., and Bertrand O. Taithe. "The Historiographical Body." In *Medicine in the Twentieth Century*, edited by Roger Cooter and John V. Pickstone, 187–200. Routledge, 2003.
Kahn, Robert A. *Holocaust Denial and the Law: A Comparative Study*. Palgrave Macmillan, 2004.
Kaiser, David. *Drawing Theories Apart: The Dispersion of Feynman Diagrams in Postwar Physics*. University of Chicago Press, 2005.
Kant, Marion. "Laban's Secret Religion." *Discourses in Dance* 1, no. 2 (2002): 43–62.
Kapadia, Mubbasir, I-kao Chiang, Tiju Thomas, Norman I. Badler, and Joseph T. Kider Jr. "Efficient Motion Retrieval in Large Motion Databases." In *Proceedings of the ACM SIGGRAPH Symposium on Interactive 3D Graphics and Games* (2013): 19–38.
Karina, Lilian, and Marion Kant. *Hitler's Dancers: German Modern Dance and the Third Reich*. Translated by Jonathan Steinberg. Berghahn, 2003.
Karl-Johnson, Gabriella. "From the Page to the Floor: Baroque Dance Notation and Kellom Tomlinson's The Art of Dancing Explained." *Signs & Society* 5, no. 2 (2017): 269–92.
Kealiinohomoku, Joann W. "Caveat on Causes and Correlations." *CORD News* 6, no. 2 (July 1974): 20–24.
Keilson, Ana Isabel. "Making Dance Modern: Knowledge, Politics, and German Modern Dance, 1890–1927." PhD diss., Columbia University, 2017.
Kern, Stephen. *The Culture of Time and Space, 1880–1918*. Harvard University Press, 1983.
Kestenberg, Judith. *Children and Parents: Psychoanalytic Studies in Development*. Jason Aronson, 1973.
Kestenberg, Judith. "Editorial." *Child Development Research News* 2, no. 1 (Spring/Summer 1979).
Kestenberg, Judith. "Psychoanalytic Contributions to the Problem of Children of Survivors from Nazi Persecution." *Israel Annals of Psychiatry and Related Disciplines* 10 (1972): 311–25.
Kestenberg, Judith. "The Role of Movement Patterns in Development." *Psychoanalytic Quarterly* 34 (1965): 1–36.
Kestenberg, Judith. "The Role of Movement Patterns in Development II: Flow of Tension and Effort." *Psychoanalytic Quarterly* 34 (1965): 517–63.
Kestenberg, Judith, and Ira Brenner. *The Last Witness: The Child Survivor of the Holocaust*. American Psychiatric Press, 1996.
Kestenberg, Judith, and Eva Fogelman, eds. *Children During the Nazi Reign: Psychological Perspectives on the Holocaust*. Praeger, 1994.
Kestenberg, Judith, and Charlotte Kahn, eds. *Children Surviving Persecution: An International Study of Trauma and Healing*. Praeger, 1998.
Kestenberg, Judith, and Mark K. Sossin. *The Role of Movement Patterns in Development*. Dance Notation Bureau, 1977.
Kestenberg, Milton. "Discriminatory Aspects of the German Indemnification Policy: A Continuation of Persecution." In *Generations of the Holocaust*, edited by Martin S. Bergmann and Milton E. Jucovy, 62–79. Columbia University Press, 1982.
Kew, Carole. "From Weimar Movement Choir to Nazi Community Dance: The Rise and Fall of Rudolf Laban's 'Festkultur.'" *Dance Research* 17, no. 2 (1999): 73–96.
Khurana, Rakesh. *From Higher Aims to Hired Hands: The Social Transformation of American Business Schools and the Unfulfilled Promise of Management as a Profession*. Princeton University Press, 2007.
Khurana, Rakesh. *Searching for a Corporate Savior: The Irrational Quest for Charismatic CEOs*. Princeton University Press, 2002.

Kipping, Matthias, and Ove Bjarnar, eds. *The Americanisation of European Business: The Marshall Plan and the Transfer of U.S. Management Models to Europe*. Routledge, 1998.

Kirk, David, and Patricia Vertinsky, eds. *The Female Tradition in Physical Education: Women First Reconsidered (2016)*. Routledge, 2016.

Kirstein, Lincoln. *Ballet Alphabet: A Primer for Laymen*. Kamin Publishers, 1939.

Klein, Ursula. *Experiments, Models, Paper Tools: Cultures of Organic Chemistry in the Nineteenth Century*. Stanford University Press, 2002.

Klingenbeck, Fritz. "Kleiner Rückblick." *Singchor und Tanz* 46, no. 24 (December 15, 1929): 306.

Knapp, B. N. "Review of Effort: Economy of Human Movement (Second Edition)." *Ergonomics* 17, no. 4 (1974): 555–56.

Kowal, Rebekah J., Gerald Siegmund, and Randy Martin, eds. *The Oxford Handbook of Dance and Politics*. Oxford University Press, 2017.

Kracauer, Siegfried. *The Mass Ornament: Weimar Essays*. Translated by Thomas Y. Levin. Harvard University Press, 1995.

Kraut, Anthea. *Choreographing Copyright: Race, Gender, and Intellectual Property Rights in American Dance*. Oxford University Press, 2015.

Kraut, Anthea. *Choreographing the Folk: The Dance Stagings of Zora Neale Hurston*. University of Minnesota Press, 2008.

Kudlick, Catherine J. "Disability History: Why We Need Another Other." *The American Historical Review* 108, no. 3 (2003): 763–93.

Kuhn, Laura, and Nicholas Slonimsky. "History of the Dalcroze Method of Eurythmics." In *Music Since 1900*, 6th ed., edited by Laura Kuhn and Nicholas Slonimsky, 915–16. Schirmer Reference, 2001.

"Labans Wiener Tanzfestzug." *Singchor und Tanz* 46, no. 13 (July 1, 1929): 182–83.

Laban, Rudolf. *Die Welt des Tänzers*. Walter Seifert, 1920.

Laban, Rudolf. *Gymnastik und Tanz*. Gerhard Stalling, 1926.

Laban, Rudolf. *Modern Educational Dance*. Edited by Lisa Ullmann. 2nd ed. Frederick A. Praeger, 1968.

Laban, Rudolf. *Principles of Dance and Movement Notation*. Macdonald & Evans, 1956.

Laban, Rudolf von. *Choreographie*. Eugen Diederichs, 1926.

Laban, Rudolf von. *Ein Leben für den Tanz: Erinnerungen*. Verlag, 1935.

Laban, Rudolf von. *A Life for Dance: Reminiscences*. Translated by Lisa Ullmann. Theatre Arts Books, 1975.

Laban, Rudolf von. *Schrifttanz: Methodik, Orthographie, Erläuterungen*. Universal Edition, 1928.

Laemmli, Whitney. "Making Movement Matter." In *Invisible Labor in Modern Science*, edited by Jenny Bangham, Xan Chacko, and Judith Kaplan, 237–46. Rowman & Littlefield, 2022.

Lamb, Warren. *Posture and Gesture: An Introduction to the Study of Physical Behaviour*. Duckworth, 1965.

Lamb, Warren, and David Turner. *Management Behaviour*. Gerald Duckworth & Co., 1969.

Lamb, Warren, and Elizabeth Watson. *Body Code: The Meaning in Movement*. Routledge, 1979.

Landecker, Hannah. *Culturing Life: How Cells Became Technologies*. Harvard University Press, 2007.

Lange, Carl. *Om sindsbevægelser*. Copenhagen, 1885.

Langley, J. N. *The Autonomic Nervous System, Part I*. W. Heffer & Sons Ltd., 1921.

Latour, Bruno. "Drawing Things Together." *Knowledge and Society* 6 (1986): 1–40.

LaViers, Amy, and Catherine Maguire. *Making Meaning with Machines: Somatic Strategies, Choreographic Technologies, and Notational Abstractions Through a Laban/Bartenieff Lens*. MIT Press, 2023.

Lawrence, Christopher, and Steven Shapin, eds. *Science Incarnate: Historical Embodiments of Natural Knowledge*. University of Chicago Press, 1998.

Leach, Robert. *Theatre Workshop: Joan Littlewood and the Making of Modern British Theatre*. University of Exeter Press, 2006.

Lee, James W. "Dalcroze by Any Other Name: Eurythmics in Early Modern Theatre and Dance." PhD diss., Texas Tech University, 2003.

Leeds-Hurwitz, Wendy. "Notes in the History of Intercultural Communication: The Foreign Service Institute and the Mandate for Intercultural Training." *Quarterly Journal of Speech* 76 (1990): 262–81.

Lemov, Rebecca. *Database of Dreams: The Lost Quest to Catalog Humanity*. Yale University Press, 2015.

Lemov, Rebecca. "Filing the Total Human: Anthropological Archives from 1928 to 1963." In *Social Knowledge in the Making*, edited by Charles Camic, Neil Gross, and Michele Lamont, 119–50. University of Chicago Press, 2011.

Levine, Herbert S. *Hitler's Free City: A History of the Nazi Party in Danzig, 1925–1939*. University of Chicago Press, 1973.

Lévi-Strauss, Claude. *Tristes Tropiques*. Plon, 1955.

Levit, Georgy S., and Uwe Hossfeld. "Natural Selection in Ernst Haeckel's Legacy." In *Natural Selection: Revisiting Its Explanatory Role in Evolutionary Biology*, edited by Richard G. Delisle, 105–36. Springer, 2021.

Levy, Fran J. *Dance/Movement Therapy: A Healing Art*. American Alliance for Health, Physical Education, Recreation and Dance, 1988.

Lewitan, Joseph. "Der Tanz von Morgen." *Der Tanz* 3, no. 11 (November 1930): 2–3.

Light, Jennifer. "When Computers Were Women." *Technology and Culture* 40, no. 3 (July 1999): 455–583.

Lindee, Susan. *Suffering Made Real: American Science and the Survivors at Hiroshima*. University of Chicago Press, 1997.

Link, Adrianna. "Documenting Human Nature: E. Richard Sorenson and the National Anthropological Film Center, 1965–1980." *Journal of the History of the Behavioral Sciences* 52, no. 4 (2016): 371–91.

Link, Adrianna H. "Salvaging a Record for Humankind: Urgent Anthropology at the Smithsonian Institution, 1964–1984." PhD diss., Johns Hopkins University, 2016.

Linker, Beth. *Slouch: Posture Panic in Modern America*. Princeton University Press, 2024.

Lipstadt, Deborah. *Denying the Holocaust: The Growing Assault on Truth and Memory*. Free Press, 1993.

Littler, Craig R. *The Development of the Labour Process in Capitalist Societies*. Heinemann Educational Books, 1982.

Littlewood, Joan. *Joan's Book: Joan Littlewood's Peculiar History As She Tells It*. Methuen, 1994.

Liu, Daniel. "The Cell and Protoplasm as Container, Object, and Substance, 1835–1861." *Journal of the History of Biology* 50, no. 4 (2017): 889–925.

Lofton, Kathryn. "The Spirit in the Cubicle: A Religious History of the American Office." In *Sensational Religion: Sensory Cultures in Material Practice*, edited by Sally M. Promey, 135–58. Yale University Press, 2014.

Lomax, Alan. "Brief Progress Report: Cantometrics-Choreometrics Projects." *Yearbook of the International Folk Music Council* 4, 25th Anniversary Issue (1972): 142–45.
Lomax, Alan. "Choreometrics and Ethnographic Filmmaking." *Filmmakers Newsletter*, February 1971, 22–30.
Lomax, Alan. "Cinema, Science, and Culture Renewal." *Current Anthropology* 13 (October 1973): 474–80.
Lomax, Alan. "Folk Song Style: Musical Style and Social Context." *American Anthropologist* 61, no. 6 (1959): 927–54.
Lomax, Alan. *Mister Jelly Roll: The Fortunes of Jelly Roll Morton, New Orleans Creole and "Inventory of Jazz."* Duell, Sloan and Pearce, 1950.
Lomax, Alan, Irmgard Bartenieff, and Forrestine Paulay. "The Choreometric Coding Book." In *Folk Song Style and Culture*, edited by Alan Lomax, 262–73. American Association for the Advancement of Science, 1968.
Lomax, Alan, Irmgard Bartenieff, and Forrestine Paulay. "Dance Style and Culture." In *Folk Song Style and Culture*, edited by Alan Lomax, 222–47. American Association for the Advancement of Science, 1968.
Lomax, John A., and Alan Lomax. *American Ballads & Folk Songs*. Macmillan, 1934.
Lomax, John A., and Alan Lomax. *Negro Folk Songs as Sung by Lead Belly*. Macmillan, 1936.
Lourens, Tino, Roos van Berkel, and Emilia Barakova. "Communicating Emotions and Mental States to Robots in Real Time Parallel Framework Using Laban Movement Analysis." *Robotics and Autonomous Systems* 58 (2010): 1256–65.
Luo, Yu, Jianbo Ye, Reginald B. Adams Jr., Jia Li, Michelle G. Newman, and James Z. Wang. "ARBEE: Towards Automated Recognition of Bodily Expression of Emotion in the Wild." *International Journal of Computer Vision* 128 (2020): 1–25.
Lussier, Kira. "Temperamental Workers: Psychology, Business, and the Humm-Wadsworth Temperament Scale in Interwar America." *History of Psychology* 21, no. 2 (May 2018): 79–99.
MacKenzie, Donald, and Graham Spinardi. "Tacit Knowledge, Weapons Design, and the Uninvention of Nuclear Weapons." *American Journal of Sociology* 101, no. 1 (1995): 44–99.
Maier, Charles S. "Between Taylorism and Technocracy: European Ideologies and the Vision of Industrial Productivity in the 1920s." *Journal of Contemporary History* 5, no. 2 (1970): 27–61.
Maletic, Vera. *Body, Space, Expression: The Development of Rudolf Laban's Movement and Dance Concepts*. De Gruyter, 1987.
Malnig, Julie. *Dancing Black, Dancing White: Rock 'n' Roll, Race, and Youth Culture of the 1950s and Early 1960s*. Oxford University Press, 2023.
Maloney, Jane. "Further Tributes to Warren Lamb from the Action Profile Community." *Moving On* 13, no. 1 & 2 (2016).
Manning, Susan. *Ecstasy and the Demon: Feminism and Nationalism in the Dances of Mary Wigman*. University of California Press, 1993.
Manning, Susan. *Modern Dance, Negro Dance: Race in Motion*. University of Minnesota Press, 2006.
Manning, Susan. "Modern Dance in the Third Reich, Redux." In Kowal, Siegmund, and Martin, *The Oxford Handbook of Dance and Politics*, 395–416.
Manning, Susan, and Lucia Ruprecht, eds. *New German Dance Studies*. University of Illinois Press, 2012.

Manusov, Valerie. "A History of Research on Nonverbal Communication: Our Divergent Pasts and Their Contemporary Legacies." In *APA Handbook of Nonverbal Communication*, edited by David Matsumoto, Hyisung C. Hwang, and Mark G. Frank, 3–15. American Psychological Association, 2016.

Marion, Sheila. "Recording the Imperial Ballet: Anatomy and Ballet in Stepanov's Notation." In *Dance on Its Own Terms*, edited by Melanie Bates and Karen Eliot, 309–40. Oxford University Press, 2013.

Martin, John. *John Martin's Book of the Dance*. Tudor Publishing Company, 1963.

Martin, John. *The Modern Dance*. A. S. Barnes, 1933.

Matthews, Scott L. "The Alan Lomax Archive." *American Quarterly* 68, no. 2 (June 2016): 429–37.

Mauss, Marcel. "Les Techniques Du Corps." *Journal de Psychologie Normal et Pathologique* 32 (1935): 271–93.

Mauss, Marcel. "Techniques of the Body." *Economy and Society* 2 (1973): 70–88.

May, Elaine Tyler. *Homeward Bound: American Families in the Cold War Era*. Basic Books, 1998.

Mayer, Andreas. *The Science of Walking: Investigations into Locomotion in the Long Nineteenth Century*. University of Chicago Press, 2020.

McCarren, Felicia. *Dancing Machines: Choreographies of the Age of Mechanical Reproduction*. Stanford University Press, 2003.

McCaw, Dick. *An Eye for Movement: Warren Lamb's Career in Movement Analysis*. Brechin, 2006.

McCaw, Dick. *The Laban Sourcebook*. Routledge, 2011.

McCaw, Dick, and Pat Lehner, eds. *The Art of Movement: Rudolf Laban's Unpublished Writings*. Routledge, 2024.

McCray, W. Patrick. *Making Art Work: How Cold War Engineers and Artists Forged a New Creative Culture*. MIT Press, 2020.

McElvenny, James, and Andrea Ploder, eds. *Holisms of Communication: The Early History of Audio-Visual Sequence Analysis*. Language Science Press, 2021.

McIvor, Arthur J. *A History of Work in Britain, 1880–1950*. Palgrave, 2001.

McLuhan, Marshall. *The Gutenberg Galaxy: The Making of Typographic Man*. University of Toronto Press, 1962.

McNeill, William H. *Keeping Together in Time: Dance and Drill in Human History*. Harvard University Press, 1995.

Menosky, Joseph. "Video Graphics & Grand Jetés." *Science* 82, May 1982.

Merton, Robert K. "The Normative Structure of Science." In *The Sociology of Science: Theoretical and Empirical Investigations*, 267–78. University of Chicago Press, 1973.

Metzl, Jonathan M. *The Protest Psychosis: How Schizophrenia Became a Black Disease*. Beacon Press, 2009.

Milam, Erika. *Creatures of Cain: The Hunt for Human Nature in Cold War America*. Princeton University Press, 2019.

Mills, C. Wright. *White Collar: The American Middle Classes*. 50th anniversary ed. Oxford University Press, 2002 [1956].

Mitman, Gregg, and Kelley Wilder. *Documenting the World: Film, Photography, and the Scientific Record*. University of Chicago Press, 2016.

Modern Times. United Artists, 1936.

Moore, Carol-Lynne. *Executives in Action: A Guide to Balanced Decision-Making in Management*. Macdonald and Evans, 1982.

Moore, Carol-Lynne. *Movement and Making Decisions: The Body-Mind Connection in the Workplace*. Dance & Movement Press, 1995.
Moreton, Bethany. *To Serve God and Walmart: The Making of Christian Free Enterprise*. Harvard University Press, 2010.
Morgan, Mary. "Economics." In *The Cambridge History of Science: Modern Social Sciences*, edited by Theodore M. Porter and Dorothy Ross, 7:275–305. Cambridge University Press, 2003.
Morrison, Mark. *Modern Alchemy: Occultism and the Emergence of Atomic Theory*. Oxford University Press, 2017.
Mosse, George L. *The Nationalization of the Masses: Political Symbolism and Mass Movements in Germany from the Napoleonic Wars Through the Third Reich*. H. Fertig, 1975.
Mullaney, Thomas S. *The Chinese Typewriter: A History*. MIT Press, 2017.
Müller, Hedwig, and Patricia Stöckemann. *". . . Jeder Mensch ist ein Tänzer": Ausdruckstanz in Deutschland zwischen 1900 und 1945*. Anabas-Verlag, 1993.
Murrow, Edward R. *In Search of Light: The Broadcasts of Edward R. Murrow, 1938–1961*. Edited by Edward Bliss. Da Capo Press, 1997.
Nakata, Toru, Taketoshi Mori, and Tomomasa Sato. "Analysis of Impression of Robot Bodily Expression." *Journal of Robotics and Mechatronics* 14, no. 1 (2002): 27–36.
Neima, Anna. "Dartington Hall and the Quest for 'Life in Its Completeness,' 1925–45." *History Workshop Journal* 88 (2019): 111–33.
Nelson, Daniel. *Frederick W. Taylor and the Rise of Scientific Management*. University of Wisconsin Press, 1980.
Noë, Alva. *Strange Tools: Art and Human Nature*. Hill and Wang, 2015.
Nye, David. *American Technological Sublime*. MIT Press, 1994.
Nye, Robert A. *The Origins of Crowd Psychology: Gustave Le Bon and the Crisis of Mass Democracy in the Third Republic*. Sage, 1975.
Oforlea, Aaron N. "[Un]Veiling the White Gaze: Revealing Self and Other in the Land Where the Blues Began." *Western Journal of Black Studies* 36, no. 4 (2012): 289–300.
Okajima, S., Y. Wakayama, and Y. Okada. "Human Motion Retrieval System Based on LMA Features Using Interactive Evolutionary Computation Method." In *Innovations in Intelligent Machines—2* 376: 117–30. Studies in Computational Intelligence, 2012.
Oldenziel, Ruth. *Making Technology Masculine: Men, Women, and Modern Machines in America, 1870–1945*. Amsterdam University Press, 2004.
Olenina, Ana Hedberg. *Psychomotor Aesthetics: Movement and Affect in Modern Literature and Film*. Oxford University Press, 2020.
Ong, Walter. *Orality and Literacy: The Technologizing of the Word*. 2nd ed. Routledge, 2002.
Ong, Walter. *The Presence of the Word*. Yale University Press, 1967.
Otis, Laura. "The Metaphoric Circuit: Organic and Technological Communication in the Nineteenth Century." *Journal of the History of Ideas*, 63, no. 1 (2002): 105–28.
Partsch-Bergsohn, Isa, and Harold Bergsohn. *The Makers of Modern Dance in Germany: Rudolf Laban, Mary Wigman, Kurt Jooss*. Princeton Book Company, 2002.
Paterson, Mark. *How We Became Sensorimotor: Movement, Measurement, Sensation*. University of Minnesota Press, 2021.
Pearl, Sharrona. *About Faces: Physiognomy in Nineteenth Century Britain*. Harvard University Press, 2010.
People in Production: An Enquiry into British War Production; A Report Prepared by Mass Observation for the Advertising Services Guild. Penguin, 1942.

Peter, Frank-Manuel. "The German Dance Archive, Cologne." *Dance Chronicle* 32, no. 3 (2009): 476–89.

Pettit, Michael. *The Science of Deception: Psychology and Commerce in America*. University of Chicago Press, 2012.

Picart, Caroline Joan S. *Critical Race Theory and Copyright in American Dance: Whiteness as Status Property*. Palgrave Macmillan, 2013.

Pick, Daniel. "Freud's Group Psychology and the History of the Crowd." *History Workshop Journal* 40 (Autumn 1995): 39–61.

Pickering, Andrew. *The Cybernetic Brain: Sketches of Another Future*. University of Chicago Press, 2010.

Pierce, Ken. "Dance Notation Systems in Late 17th-Century France." *Early Music* 26, no. 2 (1998): 286–99.

Polanyi, Michael. *The Tacit Dimension*. University of Chicago Press, 2009.

Poovey, Mary. "The Limits of the Universal Knowledge Project: British India and the East Indiamen." *Critical Inquiry* 31, no. 1 (2004): 183–202.

Popoff, Alexandre. "John Cage's Number Pieces: The Meta-Structure of Time-Brackets and the Notion of Time." *Perspectives of New Music* 48, no. 1 (2010): 65–82.

Porter, Roy. "History of the Body Reconsidered." In *New Perspectives on Historical Writing*, edited by Peter Burke, 2nd ed., 233–60. Pennsylvania State University Press, 2001.

Porter, Theodore M. "Forward: Positioning Social Science in Cold War America." In *Cold War Social Science: Knowledge Production, Liberal Democracy, and Human Nature*, edited by Mark Solovey and Hamilton Cravens, ix–xvi. Palgrave Macmillan, 2012.

Potter, Pamela M. *Art of Suppression: Confronting the Nazi Past in Histories of the Visual and Performing Arts*. University of California Press, 2016.

Preston-Dunlop, Valerie, and Susanne Lahusen, eds. *Schrifttanz: A View of German Dance in the Weimar Republic*. Dance Books Ltd., 1990.

Price, Brian. "Frank and Lillian Gilbreth and the Motion Study Controversy, 1907–1930." In *A Mental Revolution: Scientific Management Since Taylor*, edited by Daniel Nelson, 58–76. Ohio State University Press, 1992.

Pritchett, James. *The Music of John Cage*. Cambridge University Press, 1996.

Pross, Christian. *Paying for the Past: The Struggle Over Reparations for Surviving Victims of the Nazi Terror*. Translated by Belinda Cooper. Johns Hopkins University Press, 1998.

Quinton, Laura. "Britain's Royal Ballet in Apartheid South Africa." *The Historical Journal* 64, no. 3 (2021): 750–73.

Rabinbach, Anson. "The Aesthetics of Production in the Third Reich." *Journal of Contemporary History* 11, no. 4 (1976): 43–74.

Rabinbach, Anson. *The Crisis of Austrian Socialism: From Red Vienna to Civil War, 1927–1934*. University of Chicago Press, 1983.

Rabinbach, Anson. *The Human Motor: Energy, Fatigue, and the Origins of Modernity*. Basic Books, 1990.

Radin, Joanna. *Life on Ice: A History of New Uses for Cold Blood*. University of Chicago Press, 2017.

Ramsden, Pamela. *Top Team Planning: A Study of the Power of Individual Motivation in Management*. Wiley, 1973.

Ramsden, Pamela, and Jody Zacharias. *Action Profiling: Generating Competitive Edge through Realizing Management Potential*. Gower Press, 1993.

Randall, Tresa. "Cultural Modernity, the Wigman School, and the Modern Girl." *Feminist Modernist Studies* 4, no. 3 (2021): 360–74.
Reardon, Jenny. *Race to the Finish: Identity and Governance in an Age of Genomics.* Princeton University Press, 2004.
Redman, Samuel J. *Prophets and Ghosts: The Story of Salvage Anthropology.* Harvard University Press, 2021.
Rett, Joerg, Luís Santos, and Jorge Dias. "Laban Movement Analysis for Multi-Ocular Systems." In *IEEE/RSJ International Conference on Intelligent Robots and Systems*, 761–66. Nice, France, 2008.
Rhythms of Earth. Media Generation, Association for Cultural Equity, 1974.
Richards, Wendy. "Music Braille Pedagogy: The Intersection of Blindness, Braille, Music Learning Theory, and Laban." PhD diss., University of Auckland, 2020.
Rickford, John R. "Labov's Contributions to the Study of African American Vernacular English: Pursuing Linguistic and Social Equity." *Journal of Sociolinguistics* 20, no. 4 (2016): 561–80.
Rieck, Miriam, and Gali Eshet. "Die Bürden der Experten: Gespräche mit Deutschen und Israelischen Psychiatern über ihre Rolle as Gutachter in Entschädigungsverfahren." In *Die Praxis der Wiedergutmachung*, edited by Norbert Frei, José Brunner, and Constantin Goschler, 452–69. Wallstein Verlag, 2009.
Robinson, Danielle. "The Ugly Duckling: The Refinement of Ragtime Dancing and the Mass Production and Marketing of Modern Social Dance." *Dance Research* 28, no. 2 (2010): 179–99.
Robinson, Danielle. *Modern Moves: Dancing Race During the Ragtime and Jazz Eras.* Oxford University Press, 2015.
Rockman, Seth. "Forum: The Paper Technologies of Capitalism." *Technology and Culture* 58, no. 2 (2017): 487–505.
Rogers, Naomi. *Polio Wars: Sister Kenny and the Golden Age of American Medicine.* Oxford University Press, 2013.
Rolfe, Sidney E. "Manpower Allocation in Great Britain During World War II." *ILR Review* 5, no. 2 (January 1952): 173–94.
Rose, Hilary. "Hand, Brain, and Heart: A Feminist Epistemology for the Natural Sciences." *Signs* 9, no. 1 (1983): 73–90.
Rose, Nikolas. *Governing the Soul: The Shaping of the Private Self.* 2nd ed. Free Association Books, 1999.
Rosen, Elizabeth. *Dance in Psychotherapy.* Teachers College, Columbia University, 1957.
Rosenwein, Barbara. "Worrying About Emotions in History." *American Historical Review* 107, no. 3 (2002): 821–45.
Ruprecht, Lucia. *Gestural Imaginaries: Dance and Cultural Theory in the Early Twentieth Century.* Oxford University Press, 2019.
Sandelowski, Margarete. *Devices and Desires: Gender, Technology, and American Nursing.* University of North Carolina Press, 2000.
Santos, Luís, and Jorge Dias. "Laban Movement Analysis towards Behavior Patterns." In *Emerging Trends in Technological Innovation: First IFIP WG 5.5/SOCOLNET Doctoral Conference on Computing, Electrical and Industrial Systems*, edited by Luis M. Camarinha-Matos, Pedro Pereira, and Luis Ribeiro, 187–94. Springer, 2010.
Schilder, Paul. *The Image and Appearance of the Human Body.* Kegan Paul, 1935.
Schmais, Claire, and Elissa Q. White. "An Interview with Irmgard Bartenieff." *American Journal of Dance Therapy* 4, no. 1 (1981): 5–24.

Schmitt, Jean-Claude. *La raison des gestes dans l'occident Médiéval*. Gallimard, 1990.

Schmitt, Jean-Claude. "The Rationale of Gestures in the West: Third to Thirteenth Centuries." In Bremmer and Roodenburg, *A Cultural History of Gesture*, 59–70.

Schur, Richard L. *Parodies of Ownership: Hip-Hop Aesthetics and Intellectual Property Law*. University of Michigan Press, 2009.

Schwall, Elizabeth B. *Dancing with the Revolution: Power, Politics, and Privilege in Cuba*. University of North Carolina Press, 2021.

Schwartz, Hillel. "Torque: The New Kinaesthetic of the Twentieth Century." In *Incorporations*, edited by Jonathan Crary and Sanford Kwinter. Zone Books, 1992, 71–126.

Sealey, D. "NOTATE: Computerized Programs for Labanotation." *Journal for the Anthropological Study of Human Movement* 1, no. 2 (Autumn 1980): 70–74.

Sekora, Terese. "Dance Notation: A History of the Dance Notation Bureau, 1940–1952." Master's thesis, Texas Woman's University, 1979.

Sepkoski, David. *Catastrophic Thinking: Extinction and the Value of Diversity from Darwin to the Anthropocene*. University of Chicago Press, 2020.

Serlin, David. *Replaceable You: Engineering the Body in Postwar America*. University of Chicago Press, 2004.

Shannon, Claude. "Communication in the Presence of Noise." *Proceedings of the IRE* 37, no. 1 (1949): 10–21.

Shapira, Michal. *The War Inside: Psychoanalysis, Total War, and the Making of the Democratic Self in Postwar Britain*. Cambridge University Press, 2013.

Sharma, Megha, Dale Hildebrandt, Gem Newman, James E. Young, and Rasit Eskicioglu. "Communicating Affect via Flight Path: Exploring Use of the Laban Effort System for Designing Affective Locomotion Paths." *Proc. ACM/IEEE International Conference on Human-Robot Interaction* (2013): 293–300.

Shaw, Heather, and Krishna Washburn, dirs. *Telephone*. 2022. https://telephonefilm.com.

Shephard, Ben. *A War of Nerves: Soldiers and Psychiatrists in the Twentieth Century*. Harvard University Press, 2003.

Siebers, Tobin. *Disability Theory*. University of Michigan Press, 2008.

Singleton, Mark. *Yoga Body: The Origins of Modern Posture Practice*. Oxford University Press, 2010.

Sirotkina, Irina, and Roger Smith. *The Sixth Sense of the Avant-Garde: Dance, Kinaesthesia and the Arts in Revolutionary Russia*. Bloomsbury, 2017.

Smith, Pamela H. *The Body of the Artisan: Art and Experience in the Scientific Revolution*. University of Chicago Press, 2004.

Smith, Pamela H. *From Lived Experience to the Written Word: Reconstructing Practical Knowledge in the Early Modern World*. University of Chicago Press, 2022.

Smith, Roger. *The Sense of Movement: An Intellectual History*. Process Press, 2019.

Sofer, Cyril. *Students and Industry: Essays by Members of the Cambridge Balance Group*. W. Heffer & Sons Limited, 1966.

Solnit, Rebecca. *River of Shadows: Eadweard Muybridge and the Technological Wild West*. Penguin Books, 2003.

Sonlu, Sinan, Uğur Güdükbay, and Funda Durupinar. "A Conversational Agent Framework with Multi-Modal Personality Expression." *ACM Transactions on Graphics* 40, no. 1 (2021): 7.1–7.16.

Sossin, K. Mark. "Reliability of the Kestenberg Movement Profile." *Movement Studies*, no. 2 (January 1987): 23–28.

Sossin, Mark. "Development of Aggressive Patterns in Childhood." *Child Development Research News* 1, no. 1 (1978): 1–3.

Spaulding, John, and Philip Balch. "A Brief History of Primary Prevention in the Twentieth Century: 1908 to 1980." *American Journal of Community Psychology* 11, no. 1 (1983): 59–80.

Spencer, Brett. "Rise of the Shadow Libraries: America's Quest to Save Its Information and Culture from Nuclear Destruction during the Cold War." *Information and Culture* 49, no. 2 (2014): 145–76.

Stanger, Arabella. "Dancing Nature, Dancing Artifice: Laban, Schlemmer, and Reactionary Living Diagrams." In *Dancing on Violent Ground: Utopia as Dispossession in Euro-American Theater Dance*, 89–124. Northwestern University Press, 2021.

Sterne, Jonathan. *The Audible Past: Cultural Origins of Sound Reproduction*. Duke University Press, 2003.

Sterne, Jonathan. *Diminished Faculties: A Political Phenomenology of Impairment*. Duke University Press, 2022.

Stocking, George W. *The Ethnographer's Magic and Other Essays in the History of Anthropology*. University of Wisconsin Press, 1992.

Stocking, George W., Jr. *Race, Culture, and Evolution: Essays in the History of Anthropology*. New ed. University of Chicago Press, 1982.

Streeter, Carrie. "Breathing Power and Poise: Black Women's Movements for Self-Expression and Health, 1880s–1900s." *Australasian Journal of American Studies* 39, no. 1 (2020): 5–46.

Svec, Henry Adam. "Folk Media: Alan Lomax's Deep Digitality." *Canadian Journal of Communication* 38, no. 2 (2013): 227–44.

Swanson, Gillian. "Collectivity, Human Fulfilment and the 'Force of Life': Wilfred Trotter's Concept of the Herd Instinct in Early 20th-Century Britain." *History of the Human Sciences* 27, no. 1 (2014): 21–50.

"Symbols in the Orient." *Dance Notation Record* IX, no. 3/4 (Fall & Winter 1958): 29.

Szwed, John. *Alan Lomax: The Man Who Recorded the World*. Penguin Books, 2011.

Tagore, Rabindranath, and L. K. Elmhirst. *Rabindranath Tagore, Pioneer in Education: Essays and Exchanges between Rabindranath Tagore and L. K. Elmhirst*. John Murray, 1961.

"Tanzformen—Tanzprache—Tanznotation: Eine Rundfrage." *Singchor und Tanz* 45, no. 12 (1928): 170–74.

Thompson, Emily. *The Soundscape of Modernity: Architectural Acoustics and the Culture of Listening in America, 1900–1933*. The MIT Press, 2004.

Tiratsoo, Nick. "The 'Americanization' of Management Education in Britain." *Journal of Management Inquiry* 13, no. 2 (June 2004): 118–26.

Tiratsoo, Nick, and Jim Tomlinson. "Exporting the 'Gospel of Productivity': United States Technical Assistance and British Industry 1945–1960." *The Business History Review* 71, no. 1 (1997): 41–81.

Toepfer, Karl. *Empire of Ecstasy: Nudity and Movement in German Body Culture, 1910–1935*. University of California Press, 1997.

Tomko, Linda. "Dance Notation and Cultural Agency: A Meditation Spurred by 'Choreo-Graphics.'" *Dance Research Journal* 31, no. 1 (1999): 1–3.

Tresch, John. "Leroi-Gourhan's Hall of Gestures." In *Energies in the Arts*, edited by Douglas Kahn, 193–238. MIT Press, 2019.

Turner, Fred. *The Democratic Surround: Multimedia and American Liberalism from World War II to the Psychedelic Sixties*. University of Chicago Press, 2013.

Turner, Fred. "Millenarian Tinkering." *Technology and Culture* 59, no. 4 supplement (2018): S160–82.

Valent, Paul. "A Child Survivor's Appraisal of His Own Interview." In Kestenberg and Fogelman, *Children During the Nazi Reign*, 121–35.

Varmer, Borge. "Copyright Law Revision, Study No. 28: Copyright in Choreographic Works." Studies Prepared for the Subcommittee on Patents, Trademarks, and Copyrights. Washington, DC: Committee on the Judiciary, United States Senate, 1961.

Veder, Robin. *The Living Line: Modern Art and the Economy of Energy*. Dartmouth College Press, 2015.

Venable, Lucy. "Labanwriter: There Had to Be a Better Way." *Dance Research* 9, no. 2 (Autumn 1991): 76–88.

Venable, Lucy, and George Karl. *Manual for the LabanWriter Program*. Department of Dance, College of the Arts, The Ohio State University, 1987.

Vertinsky, Patricia. "Movement Practices and Fascist Infections: From Dance Under the Swastika to Movement Education in the British Primary School." In *Physical Culture, Power and the Body*, edited by Jennifer Hargreaves and Patricia Vertinsky, 25–51. Routledge, 2007.

Vertinsky, Patricia, and Sherry McCay, eds. *Disciplining Bodies in the Gymnasium: Memory, Monument and Modernism*. Routledge, 2004.

Vidal, Fernando, and Nélia Dias. *Endangerment, Biodiversity and Culture*. Routledge, 2016.

Virdi, Jaipreet, Mara Mills, and Sarah F. Rose. "Disability and the History of Science." Special issue, *Osiris* 39, no. 1 (2024).

Wade, Stephen. *The Beautiful Music All Around Us: Field Recordings and the American Experience*. University of Illinois Press, 2012.

Wakayama, Yuki, Seiji Okajima, Shigeru Takano, and Yoshihiro Okada. "IEC-Based Motion Retrieval System Using Laban Movement Analysis." In *KES 2010, Proceedings*, 251–60. Cardiff, United Kingdom, 2010.

Walker, Julia A. "Eurhythmics and Bohemian Models of Affiliation." In *Performance and Modernity: Enacting Change on the Globalizing Stage*, 106–59. Cambridge University Press, 2021.

Wang, J. Z., N. Badler, N. Berthouze, R. O. Gilmore, K. L. Johnson, A. Lapedriza, X. Lu, and N. Troje. "Bodily Expressed Emotion Understanding Research: A Multidisciplinary Perspective." Conference panel, 16th European Conference on Computer Vision, January 2021. https://par.nsf.gov/servlets/purl/10295638.

Warren, Louis S. *God's Red Son: The Ghost Dance Religion and the Making of Modern America*. Basic Books, 2017.

Wassmann, Claudia. "Reflections on the 'Body Loop': Carl Georg Lange's Theory of Emotion." *Cognition and Emotion* 24, no. 6 (2010): 974–90.

Watson, W.F. *The Work and Wage Incentives: The Bedaux and Other Systems*. Day to Day Pamphlets. Hogarth Press, 1934.

Watter, Seth Barry. "Scrutinizing: Film and the Microanalysis of Behavior." *Grey Room* 66 (2017): 32–69.

Watts, Victoria. "Patterns of Embodiment: A Visual/Cultural Studies Approach to Dance Notation." PhD diss., George Mason University, 2012.

Watts, Victoria. "The Perpetual 'Present' of Dance Notation." *Ekphrasis* 2 (2014): 180–99.

Weber, Lynne. "Ann Hutchinson Guest, Dancer, Notator, Founder, Educator, Innovator (1918–2022)." *Library News From the Dance Notation Bureau* 16, no. 3 (2022): 6–14.

Weber, Max. *The Protestant Ethic and the Spirit of Capitalism*. Translated by Stephen Kalberg. Oxford University Press, 2010.

Weidman, Nadine. "Between the Counterculture and the Corporation: Abraham Maslow and Humanistic Psychology in the 1960s." In *Groovy Science: Knowledge, Innovation, and American Counterculture*, edited by David Kaiser and W. Patrick McCray, 109–41. University of Chicago Press, 2016.

Weinstein, Deborah. *The Pathological Family: Postwar America and the Rise of Family Therapy*. Cornell University Press, 2013.

Welshman, John. "Physical Culture and Sport in Schools in England and Wales, 1900–1940." *The International Journal of the History of Sport* 15, no. 1 (1998): 54–75.

Wendell, Susan. *The Rejected Body: Feminist Philosophical Reflections on Disability*. Routledge, 1996.

White, Elissa Queyquep, "Effort-Shape: Its Importance to Dance Therapy and Movement Research." In *Dance Therapy: Focus on Dance VII*, edited by Kathleen Criddle Mason, 33–38. American Association for Health, Physical Education, and Recreation, 1974.

Whyte, William H., Jr. *The Organization Man*. Simon and Schuster, 1956.

Widman, Rosa. "The Modern 'Absolute' Dance." *Journal of Physical Education and School Hygiene* 16, no. 48 (1924): 138–41.

Wigman, Mary. "Rudolf von Laban." *Singchor und Tanz* 46, no. 24 (December 15, 1929): 295.

Williams, Drid. "Choreometrics Discussion." *CORD News* 6, no. 2 (July 1974): 25–29.

Williams, Drid. "A Note on Human Action and the Language Machine." *Dance Research Journal* 7, no. 1 (Autumn 1974): 8–9.

Williams, Valarie. "Writing Dance: Reflexive Processes-at-Work Notating New Choreography." *Journal of Movement Arts Literacy* 4, no. 1 (2018): Article 7.

Willson, F. M. G. *In Just Order Move: The Progress of the Laban Centre for Movement and Dance, 1946–1996*. Athlone Press, 1997.

Wilson, Sloan. *The Man in the Gray Flannel Suit*. Simon and Schuster, 1955.

Wir Tanzen. Reichsbund für Gemeinschaftstanz in der Reichstheaterkammer, 1936.

Wisnioski, Matthew. "Why MIT Institutionalized the Avant-Garde: Negotiating Aesthetic Virtue in the Postwar Defense Institute." *Configurations* 21, no. 1 (2013): 85–116.

Woodmansee, Martha. "The Genius and the Copyright: Economic and Legal Conditions of the Emergence of the 'Author.'" *Eighteenth-Century Studies* 17, no. 4 (1984): 425–48.

Wylie, Alison. "Doing Science as a Feminist: The Engendering of Archeology." In *Feminism in Twentieth-Century Science, Technology, and Medicine*, edited by Angela N. H. Creager, Elizabeth Lunbeck, and Londa L. Schiebinger, 23–45. University of Chicago Press, 2001.

Yates, JoAnne. *Control Through Communication: The Rise of System in American Management*. Johns Hopkins University Press, 1989.

Yates, JoAnne, and Craig N. Murphy. *Engineering Rules: Global Standard Setting Since 1880*. Johns Hopkins University Press, 2021.

Young, Allan. *The Harmony of Illusions: Inventing Post-Traumatic Stress Disorder*. Princeton University Press, 1995.

Young, Michael. *The Elmhirsts of Dartington*. Routledge & Kegan Paul, 1982.

Yu, Tao, Xiaojie Shen, Qilei Li, and Weidong Geng. "Motion Retrieval Based on Movement Notation Language." *Computer Animation and Virtual Worlds* 16 (2005): 273–82.

Zegart, Dorothy. "Dance Groups for Psychotic Patients: A Survey Study." Smith College School of Social Work, 1956.

Zhao, L., M. Costa, and N. I. Badler. "Interpreting Movement Manner." *Proceedings of Computer Graphics 2000* (May 2000): 98–103.

Zuboff, Shoshana. *The Age of Surveillance Capitalism: The Fight for a Human Future at the New Frontier of Power.* Public Affairs, 2019.

Zwerling, Israel. "The Creative Arts Therapies as 'Real Therapies.'" *American Journal of Dance Therapy* 11, no. 1 (1989): 19–26. Reprinted from *Hospital and Community Psychiatry*, 1979.

Zwerling, I., and J. F. Wilder. "An Evaluation of the Applicability of the Day Hospital in Treatment of Acutely Disturbed Patients." *The Israel Annals of Psychiatry and Related Disciplines* 2 (1964): 162–85.

Index

Page numbers followed by "f" indicate figures.

Action Profilers International, 156, 159
Action Profiling. *See* Aptitude Assessment
Adams, Diana, 222
Adamson, Andy, 124
ADTA. *See* American Dance Therapy Association (ADTA)
Agon (1957), 221–22
Aiken Computation Laboratory, 197
Air Freight, 144
ALEADMOVE (program), 162, 235
Alger, Ian, 175
Amazon Mechanical Turk, 234
American Ballads and Folk Songs (Lomax and Lomax), 202
American Council of Learned Societies, 203
American Dance Therapy Association (ADTA), 175–76, 190, 196–97
Anderson, Benedict, 49
animation, 15, 229–33
Aptitude Assessment, 2, 13–14, 130–31, 133–63
ARBEE (Automated Recognition of Bodily Expression of Emotion) system, 235
Arbeit und Rythmus (Bücher), 45
"Are These Two Pictures the Same to You?" (promotional material), 110f
Art of Movement Studio, 93, 132, 142, 239
Arthur Murray diagrams, 27
Association for the Promotion of Laban Movement Theory, 50f
Ausdruckstanz, 6, 25, 27, 57, 59
Authoritarian Personality, The (Adorno et al.), 185–86
AutoCAD, 124

Bach, Rudolf, 60
Badler, Norman, 226–32
Balanchine, George, 13, 104–5, 110–11, 221–22
ballet, 21, 25, 45, 60, 115
Bartenieff, Irmgard, 99f, 166–72, 174–78, 180–81, 185–87, 189, 192–96, 193f, 198, 204, 207–8, 207f, 219–20
Bartenieff, Michal, 167
Bauhaus, 26, 42
BBC Computer Literacy Project, 124
Beauchamp-Feuillet notation, 27–28, 28f
Bedaux, Charles, 63, 69, 74
Beels, C. C., 197
Bell, Charles, 38–40
Bereska, Dussia, 30, 35, 66, 105
Berlin Olympics (1936), 54
Bevin, Ernest, 76
Bewegungschor. *See* movement choirs
Beyoncé, 241
Billion Dollar Baby (1945), 100
biology, 8, 39–43, 52–53, 135
Birdwhistell, Ray, 14, 132, 142, 203, 207
Bishop, John, 219, 220f
Boas, Franziska, 187
Bode, Rudolf, 40
Body Code (Lamb and Watson), 154–55, 157, 158f
"Body Leads" (project), 161–62
Böhme, Fritz, 51–53
Boltanski, Luc, 160
Borde, Percival, 222
Boston Dynamics, 235–36, 236f
Brandenburg, Hans, 19, 34–35, 53
Bronx State Hospital, 175

Cage, John, 126–27, 217
Calaban (Computer Aided Labanotation) (program), 124
Campbell, Joseph, 175
Carpenter, William, 238–40
Center for Human Modeling and Simulation, 229
Center for Parents and Children, 179–80, 186
Central Advisory Council for Science and Technology, 145
Champernowne, Irene, 238–39
Chaplin, Charlie, 62
Checker, Chubby, 3, 236–37
Chiapello, Ève, 160
child development, 177–79, 184
choirs, movement. *See* movement choirs
Choreographie (Laban), 27, 29f, 43f
Choreometrics, 14–15, 200–224, 206f, 207f, 209f, 213f, 220f, 233, 240. *See also* Lomax, Alan
chronophotography, 4, 25–26
Cohen, Fritz, 64
Cohen, Selma Jeanne, 102
Cold War, 108, 185, 194, 217
computerization, 15, 197–98, 122–26, 225–43. *See also specific programs and systems*
Congress of Berlin, 21–22
Connors, Brenda, 161–62
copyright, 98, 108–9, 110, 113–15, 125–26, 241
Copyright Office (US), 13, 98, 113–14
cost and works accountants, 68–69
Csuri, Charles, 124
Cunz, Rudolf, 57–58

Dalcroze eurythmics, 21
Dance and Human History (film), 214–15
Dance Notation Bureau (DNB), 12–13, 17, 98–128, 99f, 101f, 104f, 142, 170, 175, 178, 219–20, 236, 238, 240–41
Dance Notation Record (journal), 103
Dance Observer (Balanchine), 104
Dance Theatre of Harlem, 222
Dance Therapy (journal), 176f
Dancers' Congress, 19, 25, 30, 34, 52, 170
Dancing (Lomax, unpublished), 215, 218, 222
Danzig, Free City of, 54
Dartington Hall, 64–67, 73, 91–92, 170
Das Korsett (Moholy-Nagy), 44f
Davis, Martha, 172, 196
Decision Dynamics, 155
Dee, Joey, 237
Delsartian theater, 9
democratic personality, 185, 217
Der Tanz (journal), 46
Die Neue Tanzbühne (company), 64
Die Welt des Tänzers (Laban). *See World of the Dancer, The* (*Die Welt des Tänzers*) (Laban)
digital age. *See* computerization

Dinerstein, Joel, 75
disability arts movement, 241
DNB. *See* Dance Notation Bureau (DNB)
Duchamp, Marcel, 25–26
Duncan, Isadora, 34, 92
Duval, Mathias-Marie, 37

Effort (Lawrence and Laban), 63, 94
Effort Graph, 71–73, 136–37
Effort/Shape (E/S) notation, 137, 166–67, 174–75, 194, 197, 207, 229–31
Effort/Shaper (newsletter), 193
Ein Leben für den Tanz (Laban), 22–23
Eissler, Kurt, 182
electromyography (EMG), 233–34
electronic curtain, 199–200
Elias, Norbert, 148
Elmhirst, Leonard, 64–67, 91–92, 239
Elswit, Kate, 21
EMOTE (program), 231–32
employee evaluation. *See* Aptitude Assessment; Industrial Rhythm
Eshkol-Wachman notation, 7
Essays on the Anatomy of Expression in Painting (Bell), 38
Executive Search (company), 140

Faith and Works at Dartington (Elmhirst), 91–92
Fechner, Gustav, 36
femininity, 2, 5, 34–35, 60, 105, 108, 156
Festzug der Gewerbe (Pageant of the trades), 45–47, 47f
film, 27, 70, 108–11, 114–15, 199–201, 204–5, 207, 213–15, 217–20, 223, 226, 228, 234
FitBit, 8
Floyd, Jo, 238
Foster, Susan Leigh, 108
Freikörperkultur, 9, 21

Gemeinschaftstanz (community dance), 53
gender, 5, 11, 34–35, 60, 78, 83–84, 105–6, 108, 126–27, 156–58, 161–62, 195–96, 212, 242
Gentry, Eve, 97, 100
German Dance Community, 58
German Dance Theater, 52
German Dancers' Congress, 34
German Expressionism, 6, 19
German National Dancers' Congresses, 25
Gestapo, 183
gestural theater, 21
Gilbreth, Frank, 4, 63, 69
Gilbreth, Lillian, 4, 63
Glaxo Laboratories, 81, 133
Gleisner, Martin, 49–50, 53
globalization, 216–19
Goebbels, Joseph, 2, 51, 55–58, 60

INDEX 327

Goodman, Erica, 101f
Graham, Martha, 110
graphic method in physiology, 25
Green Table, The (Jooss), 64, 66
Gregory, Winifred, 112
Guest, Ivor, 128
Gymnastik und Tanz (Gymnastics and dance) (Laban), 38, 87

Haeckel, Ernst, 40–41
Hagel, Chuck, 162
Hain, Helga, 49
Hamilton, Howard, 108
Hamilton, P. G., 83–84
Harless, C. F., 37
Helmholtz, Hermann von, 36
Hemming, Jeremy, 150
Henry, Charles, 37
Hitler, Adolf, 51, 54, 57–59
Hitler Youth, 54, 58
Hitler's Dancers (Karina and Kant), 6
Holm, Hanya, 100, 113–15
Holocaust, the, 59, 165–68, 177–85
Honda, Charlotte, 159
Human Relations Area Files (Yale University), 204
Humphrey, Doris, 107
Hurston, Zora Neale, 202
Hutchinson Guest, Ann, 16f, 17, 99–105, 107, 109–1, 113–15, 122, 127–28, 135
Huxley, T. H., 40

IBM, 117–22, 238. *See also* Selectric
IBM News, 117f
icosahedron, 42, 43f, 148
Igo, Sarah, 219
Image and Appearance of the Human Body, The (Schilder), 177
Industrial Notation, 2, 10, 70, 73, 88, 92, 135–36
Industrial Rhythm, 12, 63–96, 70f, 80f, 82f, 95f, 132–34, 137, 240
Institut Jaques-Dalcroze, 24
International Study of Organized Persecution of Children, 184

J. Lyons (company), 63, 76, 82, 90–91
"Jack" movement simulator, 229
James Thornhill Consultancy, 155
Jarrett, Clare, 155, 158f
Joint Commission Report on Mental Illness (1963), 172–73
Jooss, Kurt, 6, 25, 62, 64, 66–67, 93
Journal of School Hygiene and Physical Education (journal), 93

Kahn, Gustave, 37
Kant, Marion, 6, 57, 59

Kealiinohomoku, Joann W., 220–21
Kestenberg, Judith, 14, 166, 177–86, 193, 196, 198
Kestenberg, Milton, 180–84
Kestenberg Movement Profile (KMP), 178–79, 182
Keudell, Otto von, 52, 55, 57
Kew, Carole, 6
Kiernan, Kip, 112
kinaesthetic knowing, 9
kinesics, 200, 203
kinesphere, 42, 75
kinesthesia, 8, 11
Kinetographie, 1–2, 7–8, 11–12, 18–35, 30–34f, 42, 45–61, 50f, 63, 70–71, 98–102, 105, 136, 194
KINOTATE (program), 124
Kirstein, Lincoln, 105, 110
Kiss Me, Kate, 113–15
Klingenbeck, Fritz, 45
Knight, JaQuel, 241
Knust, Albrecht, 45, 54–55, 100–101
Kollman, Julius, 37
Körperkreuz (bodycross) method, 27–29
Kuklick (Takiff), Riki, 101f
Kummel, Herbert, 118, 128
Kurras, Hans-Joachim, 49

LABA (Labanotation printing system), 122–23
Laban, Rudolf, 1–2, 6–8, 11–12, 16, 18–31, 29f, 33–61, 43f, 63–68, 70–76, 78–96, 98–105, 107, 132–35, 139f, 147, 169–70, 193–94, 201, 239–40
Laban Institute of Movement Studies (LIMS), 175–76, 178, 194
Laban Movement Analysis (LMA), 2, 7, 10, 228, 231–33
Laban-Lawrence Test for Selection and Placement, 134–35
Labanotation (Hutchinson), 103
Labanotation, 1–2, 7–10, 12–13, 14–17, 16f, 80f, 98–128, 112f, 168, 178, 194, 205–6, 226–29, 236, 238, 241–43
LabanWriter, 124–25, 226
Lamb, Warren, 13, 130–58, 139f, 143f, 153f, 158f, 160–65, 170, 178, 186, 196, 233–35, 240
Lange, Carl, 38–40
Langley, John Newport, 38–40
Last Witness, The (Kestenberg and Brenner), 184
Latour, Bruno, 128, 141
LaViers, Amy, 242
Lawrence, F. C., 12, 63 64, 67 74, 75 76, 78–79, 81–95, 133–35. *See also* Industrial Rhythm
Les Noces, 106
LIMS. *See* Laban Institute of Movement Studies (LIMS)
Ling, Per Henrik, 93
LMA. *See* Laban Movement Analysis (LMA)
L/N Ball (IBM), 118, 120, 122

Lomax, Alan, 5, 14–16, 199–224, 233. *See also* Choreometrics
Longest Trail, The (1984), 214
Loring, Eugene, 114
Lösser, Gert Ruth, 22f

MacCartney, Richard S., 114
Macintosh (Apple), 124–25
Maguire, Catherine, 242
Mainly for Women (television series), 140
Maloney, Jane, 159
"Man and the Commonwealth" (Laban and Lawrence), 85, 87
Man in the Gray Flannel Suit, The (Wilson), 129, 165
management, scientific. *See* scientific management
Management Behavior (Lamb), 142, 146, 149, 152
managers, 91, 133, 144–45, 149, 152, 160, 165
Manning, Susan, 59, 107–8
Marey, Étienne-Jules, 4, 25
Mars (company), 63, 78–79, 83, 85–86
Martin, C. C., 67–68, 100
Martin, John, 42, 97–98, 104–5, 109, 113, 115, 221
Mauss, Marcel, 3–4
McIvor, Arthur, 78
McLuhan, Marshall, 218
Mead, Margaret, 175, 202–3, 214, 217
military, 21–24, 59, 108, 131, 161, 163, 166, 186, 202, 205, 229, 238
Mills, C. Wright, 160
Mister Jelly Roll (Lomax), 202
Mitchell, Arthur, 222
modern dance, 6, 44, 58, 75, 92–93, 108, 110–12, 115, 174
Modern Dance, The (Bode), 40
Modern Educational Dance (Laban), 94
Modern Times (film), 62, 95–96
modernism, 6, 9, 37, 107
Moholy-Nagy, László, 42
Moore, Carol-Lynne, 162
Morris, Desmond, 155
motography, 7
movement choirs, 48–51, 48f, 50f, 53–54, 60, 67, 90
Movement Pattern Analysis (MPA), 13, 130, 161, 233. *See also* Aptitude Assessment
"Movement Script is Easily Learned by Everyone, The" (Knust), 45
movement therapy, 14, 165–98, 176f, 188f, 191f, 193f, 239–40
MPA. *See* Movement Pattern Analysis (MPA)
Murdock, George, 204
Murrow, Edward R., 108
Musée de l'Homme, 204
Muybridge, Eadweard, 25

Nathanson, Lucile, 115
National Institute of Mental Health, 191f
natural language processing, 226–27, 231
Nazi Germany, 2, 51–61, 66, 98, 167, 181–83, 185, 240. *See also* Holocaust, the; World War II
Negro Folk Songs as Sung by Lead Belly (Lomax and Lomax), 202
neo-evolutionism, 210–13
Neoplatonism, 42
Newlove, Jean, 132
Nietzsche, Friedrich, 54, 57
Noë, Alva, 225
nonverbal behavior, 134–35, 142, 155, 172, 203, 214, 226
Norris, Lynne, 215
NOTATE (program), 226

objectivity, 8, 12, 16, 36, 103, 105, 127–28, 163
occupational settings. *See* Aptitude Assessment; Industrial Rhythm
Of the Warm Wind (Laban), 55, 60
Olympia (film), 57
"On Emotions" (Lange), 38
One Touch of Venus (1943), 100
Ong, Walter, 218
Opus Dei, 92
Organization Man, The (Whyte), 130, 146

Palm Play (film), 214
Paton Lawrence & Co., 69, 82–83, 133–35, 140
Paulay, Forrestine, 15, 200, 203–8, 207f, 214–15, 248
Pavlova, Anna, 21
Peck, Gregory, 129
Peggy (World War II conscript), 77
PERFORM (program), 232
personality styles, 134–39, 146–48, 151, 160–61
Pforsich, Janis, 231
PGM. *See* Posture-Gesture Mergers (PGM)
physical education, 6, 93–94, 170
physiological aesthetics, 9, 37
physiology, 25, 36–40
Porter, Cole, 113
Posture and Gesture (Lamb), 140, 152, 154, 157, 233
Posture-Gesture Mergers (PGM), 138–40, 152, 154, 231
Price, Janey, 100
Professional People Development, 155
protoplasm, 40–41
Pyramid Climbers, The (Packard), 146

race, 11–12, 51–53, 59, 75, 98, 108, 126–27, 158–59, 191, 195–96, 199, 202, 211–12, 217, 221–22, 233, 235, 240, 242
Ramsden, Pamela, 142, 156, 160
Reich Culture Chamber, 51, 57

INDEX 329

Reich League for Community Dance, 53, 57–58
Reich Theater Chamber, 58
Reid, John, 140
religion and spirituality, 3, 12, 33, 35–36, 42, 63–65, 67, 87–93, 255
Rhythm, Industrial. *See* Industrial Rhythm
Riefenstahl, Leni, 57
robotics, 233, 235–36, 236f, 240, 242
Rodgers, Rod, 222
Rogers, Helen Priest, 99–100

Sachs, Felicia, 59
Scheflen, Albert, 175
Schlee, Alfred, 20
Schlemer, Oskar, 26
Schmikl, Erik, 159
Schrifttanz (Written dance) (journal), 8, 19–20, 26, 26f, 30
Schwartz, Hillel, 9
scientific management, 2–5, 8, 35–44, 50–51, 69, 74, 103–5, 118, 127, 142, 225, 240
Selectric (IBM), 117–22, 238
SIGGRAPH International Conference, 229
Singchor und Tanz (journal), 19, 46, 52
Smith, Roger, 11
Smoliar, Stephen, 226–29
sound studies, 10–11
sphygmograph, 25
spirituality. *See* religion and spirituality
St. Olave's, 76, 81
Stanford, Leland, 25
Stanger, Arabella, 6
statistical subject, 219
Step Play (film), 214
Sterne, Jonathan, 205
Steward, Julian, 211
Swann Report (1985), 145
swing dance, 52

Tagore, Rabindranath, 65
Taketomo, Yaushiko, 173–74
Taper, Bernard, 107
Taylor, Frederick Winslow, 63, 69, 92
Taylor, Joyhope, 190–91
"Techniques of the Body" (Mauss), 3–4, 9
television, 135, 140, 164, 199, 214, 217–18, 222
Theatre Workshop, 131
Thomas, Monica, 236
Thus Spoke Zarathustra (Nietzsche), 54
Tiller Girls, 42
Titan (1927), 51

Tomorrow's World (television show), 141
Topaz, Muriel, 106, 107, 111, 125
torque, 9
trauma, 85, 165–67, 180–84, 198, 240
Treaty of Versailles (1919), 54
Triadic Ballet (Schlemmer), 26
Tudor, Antony, 110f
twist, the (dance), 3, 236, 237f
Tyresoles (company), 76, 83–87

Ubell, Earl, 116–17, 125, 128
Ullmann, Lisa, 67, 93, 132f
unconscious, the, 130, 143–45, 161

Varmer, Borge, 115
Veder, Robin, 9
Venable, Lucy, 106, 110, 123, 125, 128
Vertinsky, Patricia, 6
Vom Tauwind und der neuen Freude (Of the warm wind and the new joy) (Laban), 54–56, 56f

W. Heffer & Sons, 145–46
Wagner festival, 54
Warren Lamb Associates, 130, 135, 141
Watts Rebellion (1965), 191
Weber, Lynn, 241
Welles, Orson, 205
West Germany, 182
Wethered, Audrey, 189
White, Leslie, 211
Whitney, Dorothy Payne, 64–67, 92
Wiener Neustadt academy, 23
Wigman, Mary, 6, 25, 100
Williams, Drid, 220–21
Wir Tanzen (pamphlet), 55
Withymead Center, 238–39
Wood, Michael, 141
World of the Dancer, The (Die Welt des Tänzers), 35, 37–38, 41, 43–44, 87–88
World War I, 24, 37, 49, 65, 67–68
World War II, 1–2, 12, 76–93, 108, 179. *See also* Holocaust, the; Nazi Germany
Writing on Body Movement and Communication (ADTA), 188f
Wundt, Wilhelm, 36–40

Yamano Hakudai, 103
yoga, 9
Young, Allan, 181

Zwerling, Israel, 172

www.ingramcontent.com/pod-product-compliance
Lightning Source LLC
Chambersburg PA
CBHW022031290426
44109CB00014B/819